Aero-Optical Effects

WILEY SERIES IN PURE AND APPLIED OPTICS

Founded by Stanley S. Ballard, University of Florida
EDITOR: Glenn Boreman, University of North Carolina at Charlotte

ASSANTO · *Nematicons: Spatial Optical Solitons in Nematic Liquid Crystlas*
BARRETT AND MYERS · *Foundations of Image Science*
BEISER · *Holographic Scanning*
BERGER-SCHUNN · *Practical Color Measurement*
BOYD · *Radiometry and The Detection of Optical Radiation*
BUCK · *Fundamentals of Optical Fibers*, Second Edition
CATHEY · *Optical Information Processing and Holography*
CHUANG · *Physics of Optoelectronic Devices*
DELONE AND KRAINOV · *Fundamentals of Nonlinear Optics of Atomic Gases*
DERENIAK AND BOREMAN · *Infrared Detectors and Systems*
DERENIAK AND CROWE · *Optical Radiation Detectors*
DE VANY · *Master Optical Techniques*
ERSOY · *Diffraction, Fourier Optics and Imaging*
GASKILL · *Linear Systems, Fourier Transform, and Optics*
GOODMAN · *Statistical Optics*
GROBE AND EISELT · *Wavelength Division Multiplexing: A Practical Engineering Guide*
HOBBS · *Building Electro-Optical Systems: Making It All Work*, Second Edition
HUDSON · *Infrared System Engineering*
IIZUKA · *Elements of Photonics, Volume I: In Free Space and Special Media*
IIZUKA · *Elements of Photonics, Volume II: For Fiber and Integrated Optics*
JUDD AND WYSZECKI · *Color in Business, Science, and Industry*,Third Edition
KAFRI AND GLATT · *The Physics of Moire Metrology*
KAROW · *Fabrication Methods for Precision Optics*
KASUNIC · *Optomechanical Systems Engineering*
KLEIN AND FURTAK · *Optics*, Second Edition
MA AND ARCE · *Computational Lithography*
MALACARA · *Optical Shop Testing*, Third Edition
MILONNI AND EBERLY · *Lasers*
NASSAU · *The Physics and Chemistry of Color: The Fifteen Causes of Color*, Second Edition
NIETO-VESPERINAS · *Scattering and Diffraction in Physical Optics*
OSCHE · *Optical Detection Theory for Laser Applications*
O'SHEA · *Elements of Modern Optical Design*
OZAKTAS · *The Fractional Fourier Transform*
PRATHER, SHI, SHARKAWY, MURAKOWSKI, AND SCHNEIDER · *Photonic Crystals: Theory, Applications, and Fabrication*
SALEH AND TEICH · *Fundamentals of Photonics*,Second Edition
SCHUBERT AND WILHELMI · *Nonlinear Optics and Quantum Electronics*
SHEN · *The Principles of Nonlinear Optics*
STEGEMAN AND STEGEMAN · *Nonlinear Optics: Phenomena, Materials, and Devices*
UDD · *Fiber Optic Sensors: An Introduction for Engineers and Scientists*
UDD · *Fiber Optic Smart Structures*
VANDERLUGT · *Optical Signal Processing*
VEST · *Holographic Interferometry*
VINCENT · *Fundamentals of Infrared Detector Operation and Testing*
VINCENT, HODGES, VAMPOLA, STEGALL, AND PIERCE · *Fundamentals of Infrared and Visible Detector Operation and Testing*, Second Edition
WEINER · *Ultrafast Optics*
WILLIAMS AND BECKLUND · *Introduction to the Optical Transfer Function*
WU AND REN · *Introduction to Adaptive Lenses*
WYSZECKI AND STILES · *Color Science: Concepts and Methods, Quantitative Data and Formulae*, Second Edition
XU AND STROUD · *Acousto-Optic Devices*
YAMAMOTO · *Coherence, Amplification, and Quantum Effects in Semiconductor Lasers*
YARIV AND YEH · *Optical Waves in Crystals*
YEH · *Optical Waves in Layered Media*
YEH · *Introduction to Photorefractive Nonlinear Optics*
YEH AND GU · *Optics of Liquid Crystal Displays*, Second Edition
GORDEYEV, JUMPER, AND WHITELEY · *Aero-Optical Effects: Physics, Analysis and Mitigation*

Aero-Optical Effects

Physics, Analysis and Mitigation

Stanislav Gordeyev
Department of Aerospace and Mechanical Engineering
University of Notre Dame
Notre Dame, IN, USA

Eric J. Jumper
Department of Aerospace and Mechanical Engineering
University of Notre Dame
Notre Dame, IN, USA

Matthew R. Whiteley
MZA Associates Corporation
Dayton, OH, USA

Published by John Wiley & Sons, Inc., Hoboken, New Jersey.

Published simultaneously in Canada.

For general information on our other products and services or for technical support, please contact our Customer Care Department within the United States at (800) 762-2974, outside the United States at (317) 572-3993 or fax (317) 572-4002.

Wiley also publishes its books in a variety of electronic formats. Some content that appears in print may not be available in electronic formats. For more information about Wiley products, visit our web site at www.wiley.com.

A catalogue record for this book is available from the Library of Congress

Hardback ISBN: 9781119037170; ePub ISBN: 9781119037217; ePDF ISBN: 9781119037156; oBook ISBN: 9781119037064

Cover image: Courtesy of Aero-Optics group
Cover design: Wiley

Set in 9.5/12.5pt STIXTwoText by Integra Software Services Pvt. Ltd, Pondicherry, India

Contents

Acknowledgements

No research effort can proceed without financial support. There are many program managers in the Air Force Research Laboratory (AFRL), the Air Force Office of Naval Research (AFOSR), the Office of Naval Research (ONR), the Naval Air Warfare Center Weapons Division (NAWCWD), the Defense Advanced Research Projects Agency (DARPA), the High Energy Laser Joint Technology Office (HEL JTO) and the Directed Energy Joint Transition Office (DE JTO) to which we are thankful; however, we are particularly thankful to a few key individuals for their financial and moral support at critical points in the long research history from which the contents of this book came. First among these is Dr. James McMichael at the Air Force Office of Scientific Research, who supported our early efforts at Notre Dame. This support was continued under Dr. Thomas Beutner with follow-on support later in our research with his move to the Office of Naval Research. The biggest growth in our efforts came with a large grant through Mark Neice, then director of the High-Energy Laser Joint Technology Office. In recent years, that funding has continued due to the ardent support from Dr. Lawrence Grimes, heading the office under the name change to Directed Energy Joint Transition Office. Under the Office of Naval Research, we owe a special thanks to Dr. Lewis DeSandre in Washington and Dr. C Denton Marrs at the Naval Warfare Center at China Lake. We would also like to send our special thanks to Dr. Bill Bower at Boeing Corporation for supporting many turret-related studies. Finally, it is impossible to not acknowledge the special working relationship we have had with the Dayton office of MZA Associates Corporation.

1

Introduction

We considered the title Modern Aero-Optics, but because we believe that this book will become a quintessential reference far into the future, the term "modern" would quickly lose its significance. What we have done here is gathered what is now known about the field of Aero-Optics, including how to measure it, how to estimate its deleterious effects on laser beams, and a discussion of how to mitigate these effects. Any attempt to do this must begin by defining what is meant and what is not meant by aero-optics. In the strictest sense, the term aero-optics refers to the optical effects that a variable-index-of-refraction flow has on the wavefront figure of a laser beam. In aero-optics we are concerned with laser beam propagated through variable-index flows with the understanding that the extent of the propagation through the flow is on the order of the beam's aperture. This assumes that the extent of the variable-index flow is from one to ten times the beam's diameter. This last supposition is what distinguishes aero-optics from atmospheric propagation. For atmospheric propagation, the propagation path through the atmosphere ranges from kilometers to hundreds of kilometers, and the optical-turbulence structures as large as hundreds of meters in size (Tatarski 1961; etc.). In fact, because of this it is important to note that it is not possible to cast aero-optical environments into the theoretical constructs of atmospheric optical turbulence. The variability in index-of-refraction of the flow can be caused by any number of conditions that range from the mixing of streams of fluid with different indexes of refraction, to compressibility effects due to the acceleration and deceleration of the flow. In general, aero-optics deals with what is known in atmospheric turbulence as the "outer scale," that is to say the turbulence has not cascaded to the "inertial range" of the turbulence; thus, C_n^2 cannot be used to quantify aero-optical turbulence. Of course, the fundamental understanding of aero-optics is as applicable to imaging as it is to laser propagation, but from an analytic point of view it is more straightforward to confine our discussion in this book to laser propagation. Although in its strict definition aero-optics refers to propagation through the

Aero-Optical Effects: Physics, Analysis and Mitigation, First Edition. Stanislav Gordeyev, Eric J. Jumper, and Matthew R. Whiteley.

variable-index flow, we will also include another aero-induced effect referred to as aero buffet, which causes considerable forcing on a beam director. The unsteady force imposed on the beam director results in beam jitter, which directly causes a time-averaged reduction in intensity in the far-field. The most common applications where aero-optics/aero effects play a critical performance role are airborne laser systems and airborne laser free-space communication; thus, some of the material in this book is also applicable to communication.

This is not the first book to deal with aero-optics. The American Institute of Aeronautics and Astronautics published a Progress in Astronautics and Aeronautics Series Volume in 1982, *Aero-Optical Phenomena*, edited by K.G. Gilbert and L.J. Otten (Gilbert and Otten 1982). That book was made up of a collection of individual chapters written by various authors active in the field in the 1970s. To this day, there is useful information in that book; however, most of the information and data presented in it has been proven to be incorrect, primarily due to the instrumentation limitations and methods of reducing data available at the time. We hope that this book will replace the ubiquitous use of the information in the previous book, much of which is not only incorrect but also misleading.

It is interesting to note that essentially all funding in aero-optics had ended by the end of the 1980s because the work performed in the area was considered mature and could not be further extended due to limited measurement methods. In the early 1990s a new initiative at the Air Force Office of Scientific Research (AFOSR) began to revisit aero-optics. What would become the Aero-Optics Group at Notre Dame was the first to receive funding in AFOSR's decision to again fund the topic. In this chapter, we will discuss the rationale for revisiting the topic of aero-optics.

1.1 Motivation for Revisiting Aero-Optics

The original motivation for emphasizing aero-optics in the 1970s was to support the first airborne directed-energy system, The Airborne Laser Laboratory or ALL (Duffner 1997), shown in a top left picture in Figure 1.1. A rotating turret, mounted on top of the aircraft, was used to point the laser beam in a desired direction. Lasers had only recently been demonstrated, and the first truly high-energy laser was the Carbon Dioxide Gas Dynamic Laser, GDL, developed by AVCO, which lased at a wavelength of 10.6 μm. The aero-optics book edited by Gilbert and Otten was based solely on work funded in support of the ALL program. What we now refer to as aero-optics was first addressed by Liepmann in 1952 to assess the effect of high-speed boundary layers in supersonic tunnels to investigate the sharpness of schlieren images (Leipmann 1952). In the mid-1950s, Stine and Winovich performed experiments attempting to validate Liepmann's theory (Stine and Winovich 1956).

Figure 1.1 Airborne Laser Laboratory (ALL) (USAF Museum / Wikimedia Commons), Airborne Laser Testbed (formally AirBorne Laser or ABL) (Bobby Jones / Wikimedia Commons Public domain) and Advanced Tactical Laser (ATL) (U.S. Air Force / Wikimedia Commons Public domain).

Early papers on aero-optics in support of the ALL made use of methods introduced by Tatarski (Tatarski 1961; Sutton 1969). It should be noted that virtually all the theoretical constructs for estimating aero-optical effects were for attached turbulent boundary layers, but were applied to shear layers by default. Near the end of the 1980s, a memo was circulated inside the Air Force Weapons Laboratory that categorically stated that aero-optic aberrations for a short-wavelength (i.e., lasing at a wavelengths much smaller than 10 microns) laser-beam directed from an aircraft flying at Mach 0.8 and 33,000 ft would suffer a reduction in intensity due to aero-optic effects of about 5 percent at the most. At the same time, several prominent people in the field of aero-optics had suggested that the field of aero-optics was essentially mature, and that the opportunities for further research were to polish up the techniques already in existence and maybe apply them to a few more flows (Sutton 1985). Thus, aero-optics was considered simply irrelevant. Others were willing to assume that aero-optical effects may have degraded the performance of the ALL at certain pointing directions, but results showed only between 1 and 10 percent reduction in intensity at the target. However, aero-optical effects strongly depend on the laser wavelength. While all these estimates were correct for the long laser wavelength of 10.6 μm, for shorter wavelengths, the aero-effects result in much more severe intensity reduction; however, misguided by the circulating memo, a system hit of let's only about 5% was adapted for system performance predictions.

While the ALL was shown to be effective at shooting down surface-skimming cruise missiles and missiles shot directly at it (i.e., self-defense), range was a concern. The damage mechanism for a laser weapon is producing and maintaining an average intensity, I, at the aim point on a target above a minimum threshold and maintaining it for sufficient time, t, to exceed a minimum accumulated fluence $I \cdot t$ (Bloembergen et al. 1987). The diffraction-limited intensity is the highest physically possible intensity, achievable only when no optical distortions are present along the propagation of the laser beam. For a diffraction-limited laser beam, the maximum intensity on the target is

$$I_{\max} \sim \frac{P D^2}{F^2 \lambda^2}, \tag{1.1}$$

where P is the power, D is the diameter of the beam aperture at the exit pupil of the beam director, F is the range (effective focal length) and λ is the laser wavelength. Equation (1.1) is fully derived in Chapter 2, Equation (2.25). The effect of wavelength has clear implications. First, for the same power and beam diameter decreasing the wavelength directly increases the range to obtain the same maximum intensity. But, it also implies that for the same range, the required power is dramatically reduced. So, under the misimpression that aero-optics would reduce the intensity on target by 5%, at most, the emphasis was placed on developing a high-power laser at shorter wavelength. In fact, the Chemical Oxygen-Iodine Laser, COIL, which lased at 1.315 μm, developed in the late 1980s and early 1990s, met that goal.

In the same memo circulating in the late 1980s that categorically stated that aero-optical effects would reduce performance above 33,000 ft by 5% at most, based on "measurements," also stressed the importance of reducing imperfections of the optical components in the beam-control system to a very small number (Gilbert 2013). To address this concern, it is necessary to make use of some terms and equations that will be addressed in more detail in Chapter 2. The first of these is the definition of Strehl Ratio, SR, which is defined as the instantaneous intensity on the target in the far-field, $I(t)$, or its time average, $\overline{I(t)}$ divided by the maximum value of so-called diffraction-limited intensity, $I_{\max}$. A simple estimate based on OPD_{rms} is given by the Large-Aperture Approximation,

$$SR = \frac{\overline{I(t)}}{I_{\max}} = \exp\left[-\left(\frac{2\pi OPD_{rms}}{\lambda}\right)^2\right], \tag{1.2}$$

where OPD_{rms} is a time average of the spatial rms of the optical path difference, OPD, over the aperture. The optical path difference will be properly defined in Chapter 2; however, here consider it a measure of the magnitude of the aberrations present. More discussion about the assumptions behind Equation (1.2) will also be provided in Chapter 2.

New funding made available by AFOSR in the early 1990s addressed the possibility that the aero-optic effects, assumed to reduce the intensity by 5% at most, might be wrong. Without any of the new information we now have, it is instructive to plot these two equations and examine their implications. To do this, let us suppose that the maximum intensity of the 10.6 μm for CO_2 Gas Dynamic Laser (GDL), was reduced by 5% due to aero-optical effects from the flow over the beam director. Figure 1.2 plots first the effect of reducing the wavelength for a diffraction-limited beam, given by Equation (1.1); the second plot is the effect the same aberrations that reduced the 10.6 μm performance by 5% has on the shorter wavelengths, given by Equation (1.2). The wavelength of various types of lasers are also marked on the figures, CO_2, Hydrogen-Fluoride (HF), Deuterium-Fluoride (DF) and the Chemical Oxygen-Iodine Laser (COIL). It should be noted that COIL laser with its wavelength of 1.315 μm, was the High Energy Laser (HEL) used in 2000s in both the AirBorne Laser (ABL), shown in Figure 1.1, and Advanced Tactical Laser (ATL), also shown in Figure 1.1. The ABL used a nose-mounted turret and the ATL used a side-mounted turret underneath the aircraft. Solid-state lasers at this writing have wavelengths of approximately 1 μm. Figure 1.2, clearly shows that a laser wavelength near 1 μm would drive the Strehl ratio to near zero.

Until the early 1990s, the Weapons Laboratory memo from the 1980s made consideration of aero-optics for new airborne laser systems a simple 5% Strehl ratio budget hit. In 1992, Notre Dame received a small grant from AFOSR to try to measure the amplitude, temporal and spatial frequency content of aero-optical flows of various types. This led to new types of wavefront sensors based on small-aperture laser beams, which will be discussed in Chapter 3. A serendipitous opportunity to make wavefront measurements in a Mach 0.8 shear layer at the

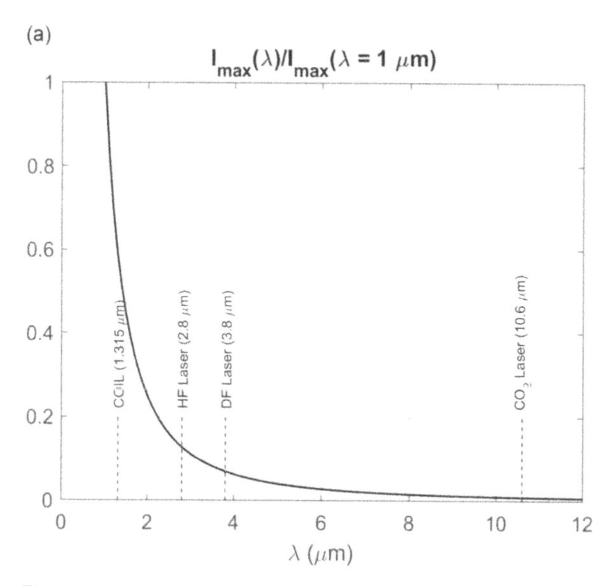

Figure 1.2 (a) Ratio of maximum diffraction limited intensity on target to that for a laser with the wavelength of 1 μm based on Equation (1.1). *Source:* Jumper et al. (2013) / With permission of SPIE.

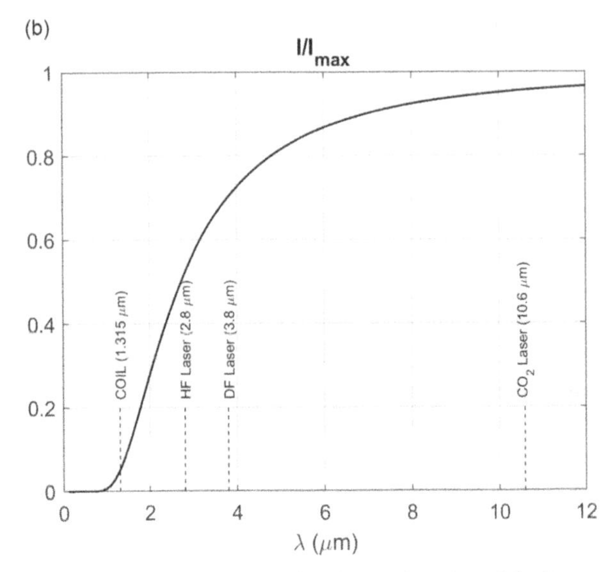

Figure 1.2 (Cont'd) (b) Strehl ratio as a function of the laser wavelength, Equation (1.2), based on optical aberrations at the exit pupil that degrade the intensity on target by 5% for a 10.6 μm laser. *Source:* Jumper et al. (2013) / With permission of SPIE.

Arnold Engineering Development Center (AEDC) found that rather than being relatively benign, the shear layer, which will always be present for flows separating off a turret, produced devastating aero-optical effects that had high spatial and temporal frequencies (Jumper and Fitzgerald 2001; Wang et al. 2012; Jumper and Gordeyev 2017). The AEDC experiment turned out to be the single most important aero-optical experiment to date. The data showed that the OPD_{rms} for an aperture of at least 20 cm was 0.43 μm. For a 1 μm laser wavelength this would result in a Strehl ratio of only 0.0007! Thus, began a now 25-year program of research that has moved from basic laboratory research to flight testing.

This book attempts to collect these 25 years of progress in aero-optics into a useful collection of material that begins with a minimum introduction into optics needed to understand the chapters that follow. This book will include chapters on the physical cause of aero-optics in air, and several methods of measuring aero-optical flows to estimate the amplitude of the disturbances, as well as spatial and temporal frequencies. We will include a chapter on the quantification and scaling of aero-optics for several flow types encountered by airborne systems, including aero-optical effects induced by turbulent flows around turrets. These chapters will be followed by a chapter on aero-optical jitter and a transitional chapter on some aspects of application to adaptive optics systems. We then proceed to a full treatment of adaptive optics analysis highlighting spatial and temporal limits to system performance.

2

Fundamentals

This chapter gives a minimal background in optics to understand the aero-optics material to follow in later chapters. We will introduce wavefront distortions and their components, and quantitative ways to characterize them and several definitions. We will discuss far-field intensity patterns on a target, and how different wavefront components affect it. We will also derive several useful relationships between the density field, wavefronts and the resulting far-field intensity.

2.1 Wavefronts and Index of Refraction

There are many excellent textbooks on the development of propagation rules for electromagnetic waves; see Klein (1970), for instance. Analysis shows that light propagates in a medium at some velocity, v, which is a property of the medium. Light always travels slower in a medium, where the photons are constantly absorbed and re-emitted an instant later by atoms or molecules. In a vacuum, the speed of light is constant, c (approximately, $c = 3.0 \cdot 10^8$ m/s). Since the speed in vacuum is constant and known, index-of-refraction, n, is defined as the ratio of the speed in vacuum divided by the speed of light, v, in the medium, $n = c / v$. Consequently, the index-of-refraction is always greater than or equal to unity, $n \geq 1$.

Consider light propagating through a medium of variable index-of-refraction. For simplicity, consider a monochromatic point light source emitting electromagnetic waves at a wavelength, λ, and a frequency, $f = v / \lambda$, as schematically shown in Figure 2.1. By definition, a wavefront is a simply connected surface of the constant phase at some fixed moment of time, $\Phi(x,y,z,t = fixed) = const$. For example, the solid and dashed lines in Figure 2.1 represent the connected points where the light amplitude reaches either a maximum or a minimum. When light propagates outward, points on the wavefront propagate with the local speed of light. If there is vacuum between points A and A', it would take time $T = |AA'| / c$ to get from point

Aero-Optical Effects: Physics, Analysis and Mitigation, First Edition. Stanislav Gordeyev, Eric J. Jumper, and Matthew R. Whiteley.

A to point A'. If a different point B at the same wavefront surface travels through a region of some media, with $u < c$, for the same traveled time, T, the wavefront will travel less distance, $|BB'| = T \cdot v = |AA'| v / c = |AA'| / n < |AA'|$, compared to the would-have-been undistorted wavefront, indicated as a dotted-dashed green line in Figure 2.1. Thus, the otherwise spherical wavefront becomes distorted in space. The distortions are always negative (lagging), compared to the undistorted wavefront in vacuum.

For compressible flows the index-of-refraction depends on the media density, ρ, via the Gladstone-Dale relation (Gladstone and Dale 1863),

$$n = 1 + n' = 1 + K_{GD}\rho, \tag{2.1}$$

where K_{GD} is the Gladstone-Dale constant. This constant depends on the gas mixture and the laser wavelength (Gardiner et al. 1980); for dry air over the infrared, visible and into the infrared range K_{GD} for room temperatures can be approximated by the following equation (Barrell and Sears 1939),

$$K_{GD} = 2.223 \times 10^{-4} \left\{1 + 5.669 \times 10^{-3} / \lambda[\mu m]^2\right\} [m^3 kg^{-1}]. \tag{2.2}$$

For the double-frequency Yag:Nd laser, $\lambda = 532\text{nm}$, $K_{GD} = 2.267 \times 10^{-4} m^3 kg^{-1}$, and for a laser with $\lambda = 1\mu\text{m}$, $K_{GD} = 2.236 \times 10^{-4} m^3 kg^{-1}$.

2.2 Huygens' Principle

For distorted wavefronts, light travels along the original undistorted straight lines but changes its local direction. This phenomenon is known as refraction. For instance, a lens can change a divergent beam into a beam with a plane wavefront, often called a collimated beam, as different parts of the wavefront travel through the parts of the lens with different thicknesses. The refraction is a consequence of

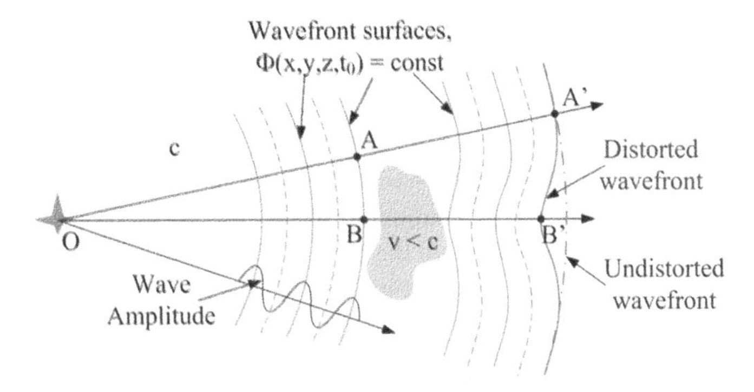

Figure 2.1 Distortions of monochromatic light wavefronts.

the Huygens Principle, which states that a wavefront propagating through a variable index-of-refraction media can be broken up into an infinite number of self-emitting point sources along the wavefront surface, as shown in Figure 2.2. After a small increment time, Δt, the sources of the radiating spheres, according to wave optics, sum into a surface of constant phase exactly parallel to the previous surface aberrated only by any speeding up or slowing down of the local spherical waves emanating from each point source. A shorthand version of Huygens' principle is that a wavefront always propagates locally normal to itself.

As mentioned before, a wavefront is defined as a surface of the constant phase, $\Phi(x,y,z)=const$. It is more convenient, however, to define the wavefront as a distance from a known surface, usually a (x,y)-plane with a fixed $z=L=const$. An example for a collimated beam is shown in Figure 2.3. Let us say it takes an amount of time, $T=L/c$, for the light to travel to the (x,y)-surface in a vacuum. If the medium with $v<c$ is present along the beam, during the same time the light will travel the smaller distance, $|AA'|=\int_{t=0}^{T} v(x,y,z)dt < L = \int_{t=0}^{T} cdt$. Thus, the distance from the (x,y)-plane to the wavefront surface will be

$$W(x,y)=|AA'|-L=\int_{t=0}^{T} v(x,y,z)dt-\int_{t=0}^{T} cdt=\int_{t=0}^{T}\left[v(x,y,z)-c\right]dt. \tag{2.3}$$

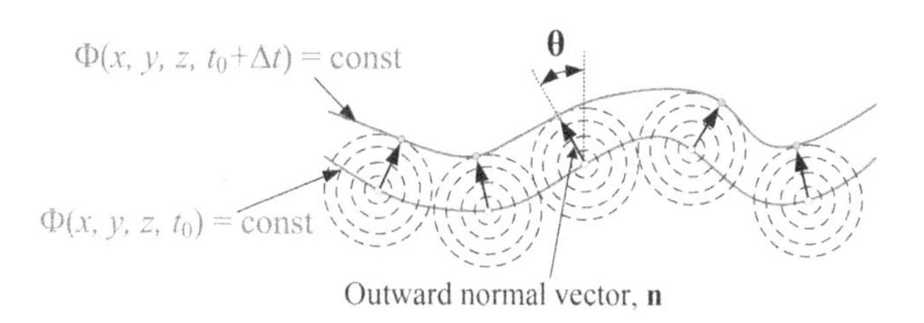

Figure 2.2 Hyugens' principle.

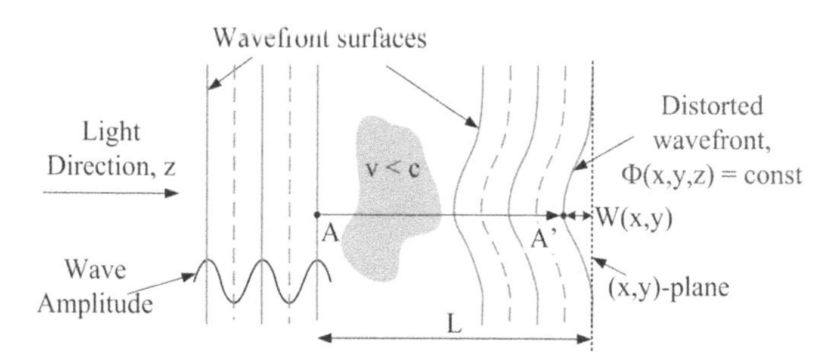

Figure 2.3 Distortions imposed on the collimated beam.

Using the definition of the index-of-refraction and recalling that $n' = n - 1 \ll 1$, this equation can be re-written as

$$W(x,y) = \int_{t=0}^{T} [1 - n(x,y,z)] v(x,y,z) dt \approx \int_{t=0}^{T} [1 - n(x,y,z)] c dt = \int_{0}^{L} [1 - n(x,y,z)] dz.$$

Finally, using the Gladstone-Dale relation, Equation (2.2), it can be written as

$$W(x,y,t) = -\int_{0}^{L} n'(x,y,z,t) dz = -K_{GD} \int_{0}^{L} \rho(x,y,z,t) dz. \quad (2.4)$$

Here, we allowed the density field to depend on time, as the turbulent field can be safely treated as a frozen medium during beam propagation.

As mentioned previously, Huygens' principle states that, locally, the wavefront always travels normal to itself. Another consequence is that the local direction of the propagation of light, relative to the undistorted direction, z, is given by an outward normal vector, indicated in Figure 2.2, which is equal to a local gradient of the wavefront,

$$\nabla \Phi = \mathbf{n} \quad (2.5)$$

In the case of a collimated beam, shown in Figure 2.3, it is easy to see that the phase function $\Phi(x,y,z)$ is related to the wavefront, $W(x,y)$, as

$$\Phi(x,y,z) = z - W(x,y) = L = const$$

Plugging this equation into Equation (2.5) gives the components of the normal vector as

$$\begin{aligned} n_x &= \frac{\partial \Phi(x,y,z)}{\partial x} = -\frac{\partial W(x,y)}{\partial x} \\ n_y &= \frac{\partial \Phi(x,y,z)}{\partial y} = -\frac{\partial W(x,y)}{\partial y}. \\ n_z &= 1 \end{aligned}$$

It is convenient to define the local direction of the light as the angles between the z-direction and the normal vector; these angles are known as the local deflection angles, θ_x and θ_y,

$$\begin{aligned} \theta_x &= \tan \frac{n_x}{n_z} = -\tan \frac{\partial W(x,y)}{\partial x} \\ \theta_y &= \tan \frac{n_y}{n_z} = -\tan \frac{\partial W(x,y)}{\partial y} \end{aligned}.$$

In most cases, the distorted wavefronts are on the order of a micron or less, while all typical scales in the (x, y)-plane are on the order of a millimeter or more. Thus, the partial derivatives in the above equation are small and the tangent function can be dropped to get the final set of equations, relating the wavefront and the local deflection angles,

$$\begin{cases} \dfrac{\partial W(x,y,t)}{\partial x} = -\theta_x(x,y,t) \\ \dfrac{\partial W(x,y,t)}{\partial y} = -\theta_y(x,y,t) \end{cases}. \tag{2.6}$$

Equation (2.6) forms a mathematical background for several types of the wavefront sensors, such as a Shack-Hartmann wavefront sensor and a Malley probe. See Chapter 3 for a detailed discussion of their principle of operation.

2.3 Basic Equations and Optical Path Difference

Aero-optics involves the refracting effect on a laser's wavefront as it propagates through a variable density flow field. The theoretical foundation for electromagnetic wave propagation in a turbulent medium can be found in Monin and Yaglom (1975). Here we briefly outline a basic derivation, assumptions and definitions in the context of aero-optics. In the most general sense, the propagation of electromagnetic waves is governed by the Maxwell equations. For aero-optical problems, the timescale for optical propagation is negligibly short relative to flow timescales, and hence optical propagation can be solved under the assumption of frozen flow at each time instant. If the optical wavelength is much shorter than the smallest flow scale (Kolmogorov scale), which is generally the case, and the effect of depolarization is negligible, the Maxwell equations are reduced to a vector wave equation in which all three components of the electromagnetic field, $U(x, y, z, t)$, are decoupled. In particular, a scalar component of the electric field at a fixed frequency, ω, becomes $U(x, y, z, t) = E(x, y, z)\exp(i\omega t)$ where the spatial distribution of the electric field, $E(x, y, z)$, is governed by the following equation,

$$\nabla^2 E(x,y,z) + \left(\frac{\omega n}{c}\right)^2 E(x,y,z) = 0. \tag{2.7}$$

Here again, n is the index-of-refraction and c is the speed of light in vacuum. Unless stated otherwise, in this book we will denote the direction of the beam propagation as the z-direction and the (x, y)-plane as the plane normal to the z-direction, as shown in Figure 2.4. If the beam amplitude does not change significantly over the wavelength, which is almost always the case in practical applications, the electric field can be approximated as a fast-changing component in the z-direction, multiplied by a slowly changing envelope function, $A(x,y,z)$, (the so-called paraxial approximation),

$$E(x,y,z) = A(x,y,z)\exp(-ikz), \tag{2.8}$$

where $\left|\partial^2 A(x,y,z)/\partial z^2\right| << \left|k\partial A(x,y,z)/\partial z\right|$, and $k = \omega / c = 2\pi / \lambda$ is the wavenumber. Substituting this approximation into Equation (2.7) gives the following equation for the A-function,

$$-2ik\frac{\partial A(x,y,z)}{\partial z} + \frac{\partial^2 A(x,y,z)}{\partial x^2} + \frac{\partial^2 A(x,y,z)}{\partial y^2} + k^2(n^2-1)A(x,y,z) = 0. \tag{2.9}$$

As the density variations generally occur near the aircraft inside the turbulent region, the related aero-optical distortions are also confined to this region. Thus, it can be assumed that the laser beam gets distorted along a finite distance, called the aero-optical region, as shown in Figure 2.4. Assuming that typical turbulent scales inside the aero-optical region are much larger than the laser wavelength, then $\left|\frac{\partial^2 A(x,y,z)}{\partial x^2} + \frac{\partial^2 A(x,y,z)}{\partial y^2}\right| << \frac{2\pi}{\lambda}\left|\frac{\partial A(x,y,z)}{\partial z}\right|$. With this approximation, Equation (2.9) becomes

$$-2ik\frac{\partial A(x,y,z)}{\partial z} + k^2(n^2-1)A(x,y,z) = 0. \tag{2.10}$$

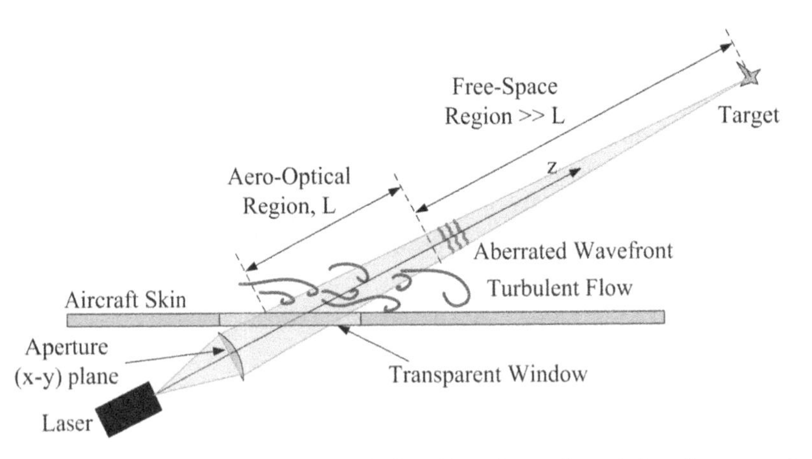

Figure 2.4 Schematic of the aero-optical problem due to the turbulent flow around an aircraft.

Using Equation (2.1), and recalling that $n' \ll 1$, the coefficient in front of the second term in Equation (2.10) can be re-written as $k^2(n^2-1) = k^2(n+1)(n-1) = 2k^2 n'$. Integrating Equation (2.10) along the z-direction inside the aero-optical region gives the following solution for the A-function,

$$A(x,y,z) = A(x,y,0)\left[-ik\int_0^z n'(x,y,z)dz\right].$$

Substituting the above equation into Equation (2.8) gives the final solution for the laser beam transmitted through the aero-optical region, $z < L$.

$$E(x,y,z) = A(x,y,0)\left[-ik\int_0^z n'(x,y,z)dz\right]\exp(-ikz) =$$
$$A(x,y,0)\left[-ik\int_0^z \left(n'(x,y,z)+1\right)dz\right] = E(x,y,0)\left[-ik\int_0^z n(x,y,z)dz\right]. \quad (2.11)$$

The intensity of the electric field can be found by computing the square of the electric field amplitude,

$$I(x,y,z) = \left|E(x,y,z)\right|^2.$$

If the initial intensity of the beam, $I_0(x,y) = I(x,y,z=0)$, and its initial wavefront, $W_0(x,y) = W(x,y,z=0)$, over the aperture are given, the initial electric field over that aperture becomes, $E(x,y,0) = \sqrt{I_0(x,y)}\exp(ikW_0(x,y))$. Knowing the initial electric field, the resulting electric field after passing through the aero-optical region is given by Equation (2.11). Note that the term in the square brackets is pure imaginary, and it can be treated as an additional phase distortion added to the initial wavefront, $W(x,y,z=L) = W_0(x,y) - \int_0^z n(x,y,z)dz$. As discussed before, the aero-optical distortions due to the aero-optical effects are always negative (lagging), compared to the undistorted laser beam wavefront. Another conclusion is that the aero-optical distortions do not change the initial intensity distribution. Thus, the aero-optical effect after transmission through the turbulent region is simply an additional negative distortion of the optical wavefront with the amplitude being approximately constant.

Without loss of generality, we will assume that the initial beam over the aperture has a plane wavefront; thus, the beam is collimated. In this case, the initial wavefront, $W_0(x,y)$, is zero, and the electric field due to the aero-optical distortions is given by

$$E(x,y,z=L)=\sqrt{I_0(x,y)}\exp\left[ikW(x,y)\right]. \quad (2.12)$$

Once the solution past the aero-optical region, $E(x,y,z=L)$, is determined, it can be used as the initial condition to propagate the optical beam into the far-field region using the free-space wave equation, as it will be shown later in this chapter.

The integral in Equation (2.11) is known is Optical Path Length or OPL,

$$\begin{aligned} OPL(x,y,t) &= \int n(x,y,z,t)dz = \int [1+K_{GD}\rho(x,y,z,t)]dz \\ &= L + K_{GD}\int \rho(x,y,z,t)dz. \end{aligned} \quad (2.13)$$

Here we recall the density field can also be a function of time, as the turbulent field is treated as a frozen medium during the beam propagation, since the speed of light is much faster than the flow speed inside the turbulent region. Unlike the distorted wavefront, *OPL* is always positive over the aero-optical region, as the density is always positive.

As will be discussed later in this chapter, we are always interested in the wavefront deviation (error) about a plane wavefront rather than just the absolute optical distance. So, we can remove the spatially averaged (a spatial mean) wavefront location or, similarly, the spatially averaged optical path. The averaging is performed over the beam aperture. This mean-removed *OPL* is used often in optical problems, and it is called Optical Path Difference or more commonly, *OPD*,

$$\begin{aligned} OPD(x,y,t) &= OPL(x,y,t) - \langle OPL(x,y,t)\rangle_{\text{over aperture}} \\ &= K_{GD}\int_A^B \rho'(x,y,z,t)ds, \end{aligned} \quad (2.14)$$

where the angular brackets denote the spatial average over the aperture and ρ' are mean-removed density fluctuations. Thus, by definition, the spatially averaged value of *OPD* is zero, $\langle OPD(x,y,t)\rangle_{\text{over aperture}}=0$. Once obtained, $OPD(x,y,t)$ and the beam intensity profile over the aperture can be used to compute the intensity distribution at the target due only to the aero-optic field using free-space wave optics. A measurement of $OPD(x,y,t)$ is clearly imperative for understanding how various types of flow fields found around beam directors affect the airborne optical system's performance.

Comparing Equations (2.4) and (2.13) leads us to the conclusion that *OPL* is related to a negative distorted wavefront, $OPL(x,y)=L-W(x,y)$. If we remove the spatial mean components from both parts, we will get that *OPD* is the conjugate (negative) of the mean-removed wavefront aberrations. For most of the data analysis, the sign difference between the wavefront and *OPD* is not important, so in this book we will be using the word "wavefront" to imply both *W* and *OPD*.

In addition to the instantaneous OPD, the spatial root-mean-square of OPD at each instant in time, $OPD_{rms}(t)$,

$$OPD_{rms}(t) = \left(\overline{\langle OPD(x,y,t)^2 \rangle}_{\text{aperture}} \right)^{1/2}, \tag{2.15}$$

and the time-averaged spatial root-mean-square of OPD, $OPD_{rms} \equiv \overline{OPD_{rms}(t)}$, are typically computed. Here and elsewhere in this book the overbar denotes a time averaging. Later these properties will be shown to be directly related to the far-field intensity of the laser beam.

2.4 Linking Equation

Several theoretical efforts resulted in formulas for estimating the level of aero-optical distortions from other measures of the flow. The first theoretical study of optical distortions caused by compressible boundary layers was done by Liepmann (1952) and made use of the local deflection angle of a thin beam of light as it traveled through the compressible boundary layer on the sides of high-speed wind tunnels to quantify the crispness of Schlieren photographs. After propagating through a turbulent region of thickness δ, in the propagation direction, y, Liepmann's analysis gave

$$\langle \theta^2 \rangle = \int_0^{\delta} \int_0^{\delta} \left\langle \left(\frac{\partial n'}{\partial y} \right)^2 \right\rangle R_{n'}\left(\left| y - \varsigma \right| \right) dy \, d\varsigma, \tag{2.16}$$

where $n' = n - 1$ is the spatially varying fluctuating index of refraction, and $R\left(\left|y-\varsigma\right|\right)$ is the correlation function for the index variation. When n' is replaced by the density fluctuation, ρ', using the Gladstone-Dale relation, $n' = K_{GD}\rho'$, Equation (2.16) becomes

$$\langle \theta^2 \rangle = K_{GD}{}^2 \int_0^{\delta} \int_0^{\delta} \left\langle \left(\frac{\partial \rho'}{\partial y} \right)^2 \right\rangle R_{\rho}\left(\left| y - \varsigma \right| \right) dy \, d\varsigma. \tag{2.17}$$

As it was suggested by Liepmann, Equation (2.16) has always held the promise of being able to use the deflection angle data as a nonintrusive flow diagnostic tool (Hugo and Jumper 2000).

Another major theoretical work was done by George Sutton. Based heavily on the approach taken by Tatarski (1961) for electromagnetic waves propagated through the atmosphere, Sutton (1969) produced the most-widely referenced theoretical formulation on the aberrating effects of turbulent boundary layers. Using statistical measures of the turbulence, he developed a "linking equation" between the turbulence quantities and the optical distortions. In the general form the equation is given as,

$$OPD_{rms}^2 = K_{GD}^2 \int_0^L \int_0^L \text{cov}_{\rho'}(y_1, y_2) dy_1 dy_2, \tag{2.18}$$

where $\text{cov}_{\rho'}(y_1, y_2) = \overline{(\rho(y_1,t) - \bar{\rho}(y_1,t))(\rho(y_2,t) - \bar{\rho}(y_2,t))}$ is a two-point density covariance function and the overbar denotes time averaging. Because the covariance function requires a two-point measurement, the density covariance function is usually modeled by either an exponential or Gaussian functional form based on a local characteristic length scale, $\Lambda(y)$, and the square of the fluctuating density, $\rho_{rms}^2(y)$, to arrive at a relationship between the OPD_{rms} and quantities presumed to be extracted from measurements. It leads to a simplified version of the linking equation, which is commonly used,

$$OPD_{rms}^2 = 2K_{GD}^2 \int_0^L \rho_{rms}^2(y) \Lambda_y(y) dy. \tag{2.19}$$

Equation (2.19) assumes that the density covariance can be approximated by the exponential form. If the covariance is approximated by the Gaussian function, the pre-multiplier in Equation (2.19) should be replaced with $\sqrt{\pi}$ (Hugo and Jumper 2000). Typically, the value of 2 is used for the pre-multiplier.

Malley et al. (1992) realized that aberrations produced by a laser propagated through convecting flow structures convect themselves. It can be shown that both Liepmann's Equation (2.16), and Sutton's Equation (2.19), formulations are identical (Jumper and Fitzgerald 2001).

Although the notion that aero-optical aberrations convect with the flow appears trivial, it was a breakthrough for the field. It is so commonly used that this supposition is now referred to as the *Malley Principle*. This principle is at the heart of several wavefront measurement techniques. Malley et al. (1992) invented an OPD_{rms} measurement device based on this supposition that made use of a single small-aperture laser beam projected through a turbulent flow and measured the time-varying displacement of the beam on a position sensing device. Based on the Hartmann sensor ideas, Malley et al. (1992) projected a single small aperture laser beam to be sent through a shear layer and then applied the Reynolds transport theory to the resulting deflection angles, imposed on the beam, to trade time and position. The Malley principle was further matured by Jumper and Hug (1995) for large apertures using the Small-Aperture Beam Technique (SABT) sensor. Later, this principle became a base for another wavefront sensor, fittingly called a Malley Probe (Gordeyev et al. 2014). Both sensors made use of multiple small-aperture laser beams so that both wavefront slope and aberration convection speed could be collected at very-high sampling rates up to 1 MHz. We will further discuss SABT and the Malley Probe in Chapter 3.

2.5 Image at a Focal Plane (Far-field Propagation)

Based on the Huygen-Fresnel principle, if the electromagnetic field, E, is known over a finite surface, $S(x,y)$, it is uniquely defined at any image point (x',y') via Equation (2.20),

$$E(x',y') = \frac{1}{i\lambda}\int_S E(x,y)\frac{\exp(ikR(x,y;x',y'))}{R(x,y;x',y')}ds, \tag{2.20}$$

where $R(x,y;x',y')$is the distance between points (x,y) and (x',y') and $k = 2\pi/\lambda$ is a wavenumber.

Consider a typical situation when the field properties, $E(x,y) = \sqrt{I_0(x,y)}$ $\exp(ikW(x,y))$, both the initial intensity, $I_0(x,y)$, and phase, $kW(x,y)$, are known over a finite source aperture and a single lens focusing system is used to focus the light on a distant target. It is shown schematically in Figure 2.5, where the source plane is described by a coordinate (x,y), the image plane is described by a coordinate (x',y'), the focal distance is denoted by F and the image plane is offset from the focal point by distance L. Also, we will assume that $A/F \ll 1$ and $L/F \ll 1$.

The distance between two arbitrary points (x,y) and (x',y') is

$$R^2 = (x-x')^2 + (F-L-x^2/(2F))^2 + (y-y')^2 + (F-L-y^2/(2F))^2 + H.O.T.$$

Expanding this expression over (x,y) and (x',y') gives,

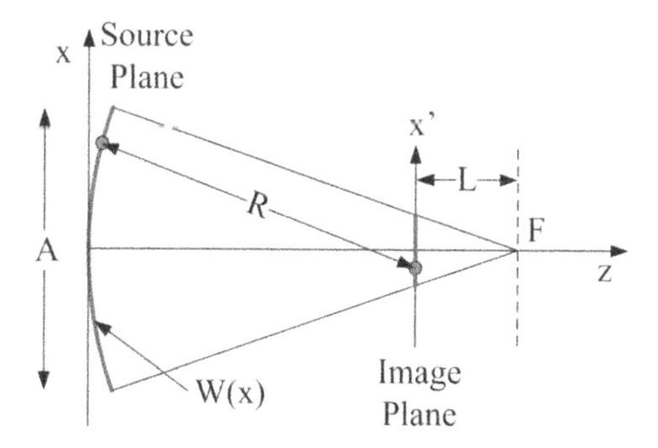

Figure 2.5 Schematic of the source and the image planes with a focusing system.

$$R(x,x') = F + \frac{1}{2}\frac{L}{F^2}x^2 - \frac{xx'}{F} + \frac{1}{2}\frac{x'^2}{F} + \frac{1}{2}\frac{L}{F^2}y^2 - \frac{yy'}{F} + \frac{1}{2}\frac{y'^2}{F} + H.O.T. \quad (2.21)$$

Substituting Equations (2.21) into (2.20) gives the following expression for the field at the image plane,

$$E(x',y') \approx \frac{1}{i\lambda F}\int_S E(x,y)\exp\left[ik\left(F + \frac{1}{2}\frac{L}{F^2}x^2 - \frac{xx'}{F} + \frac{1}{2}\frac{x'^2}{F} + \frac{1}{2}\frac{L}{F^2}y^2 - \frac{yy'}{F} + \frac{1}{2}\frac{y'^2}{F}\right)\right]ds =$$

$$\frac{1}{i\lambda F}\exp\left[ik\left(F + \frac{1}{2}\frac{(x'^2+y'^2)}{F}\right)\right]\int_S E(x,y)\exp\left[ik\left(\frac{1}{2}\frac{L}{F^2}(x^2+y^2)\right)\right]\exp\left[-ik\frac{(xx'+yy')}{F}\right]ds =$$

$$\frac{1}{i\lambda F}\exp\left[ik\left(F + \frac{1}{2}\frac{(x'^2+y'^2)}{F}\right)\right]\int_S \tilde{E}(x,y)\exp\left[-ik\frac{(xx'+yy')}{F}\right]ds$$

where

$$\widetilde{E}(x,y) \equiv E(x,y)\exp\left[ik\left(\frac{1}{2}\frac{L}{F^2}\left(x^2+y^2\right)\right)\right]. \quad (2.22)$$

Thus, the field at the image plane is simply a Fourier transformation of the modified source field, $\widetilde{E}$, a well-known result from Fourier Optics (Goodman 1996; Born and Wolf 1999).

In many practical cases, the beam aperture is round. Let D be the aperture diameter. Consider undistorted case with a planar wavefront, $W = 0$. Finally, we will assume that the initial intensity I_0 is constant inside the aperture and zero outside. If $\frac{k}{2}\frac{L}{F^2}(D/2)^2 << 1$, the so-called far-field or a Fraunhofer regime, the exponent in the definition of $\widetilde{E}(x)$ in Equation (2.22) can be ignored and the intensity at the image plane near the target can be calculated in the polar frame of reference with a help of integral identities for the Bessel functions,

$$J_0(t) = \frac{1}{2\pi}\int_0^{2\pi}\exp[it\cos(\theta)]d\theta \text{ and } \int_0^u tJ_0(t)dt = uJ_1(u),$$

$$I(x',y') \equiv |E(x',y')|^2 = \frac{1}{(\lambda F)^2}\left|\int_{Aperture}\sqrt{I_0}\exp\left[-ik\frac{(xx'+yy')}{F}\right]ds\right|^2$$

$$= \frac{I_0}{(\lambda F)^2}\left|\int_0^{D/2}\int_0^{2\pi}\exp\left[-ik\frac{rr'\cos\theta}{F}\right]rdrd\theta\right|^2 = \frac{I_0}{(\lambda F)^2}\left|2\pi\int_0^{D/2}J_0\left(k\frac{rr'}{F}\right)rdr\right|^2 \quad (2.23)$$

$$= I_0\frac{(2\pi)^2}{(\lambda F)^2}\left|\left(\frac{F}{kr'}\right)^2\left(k\frac{Dr'}{2F}\right)J_1\left(k\frac{Dr'}{2F}\right)\right|^2 = I_0\left(\frac{\pi D^2}{4\lambda F}\right)^2\left(\frac{2J_1(z)}{z}\right)^2.$$

Here $z = \pi r' D / (\lambda F)$ and J_0 and J_1 are Bessel functions of the zeroth and the first order, respectively. This resultant image at the target for the round undistorted laser beam is axisymmetric and known as an Airy pattern. It represents the best-focused diffraction-limited pattern near the focal plane. The radial intensity distribution of the Airy pattern and the two-dimensional image are shown in Figure 2.6.

The Airy disk is defined as a circle with the radius at the location of the first intensity minimum. From Figure 2.6, the first minimum is located at $z_{min} = \pi R_{Airy} D / (\lambda F) \approx 3.83$, giving the radius of the Airy disk to be

$$R_{Airy} \approx 1.22(\lambda F) / D. \tag{2.24}$$

The size of the Airy disk is proportional to the distance to the target and the wavelength is inversely proportional to the aperture diameter. As an example, for $F = 10$ km, $\lambda = 1$ micron and $D = 0.2$ m, $R_{Airy} = 6.1$ cm. Equivalently, the angular radius of the Airy disk is $\theta_{Airy} = R_{Airy} / F \approx 1.22\lambda / D$.

The maximum intensity inside the Airy disk occurs at its center, $r' = 0$,

$$I_{max} = I(r' = 0) = I_0 \left(\frac{\pi D^2}{4\lambda F} \right)^2 \approx I_0 (0.957 D / R_{Airy})^2.$$

The initial intensity depends on the total transmitted power, P_{total}, and the size of the aperture. Thus, it is more convenient to describe the maximum intensity

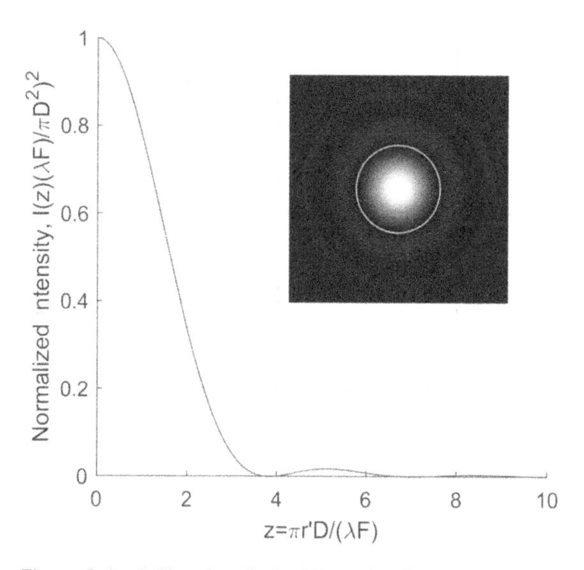

Figure 2.6 Diffraction-limited intensity distribution at the focal plane (Airy pattern). The radius of the Airy disk is indicated by a circle in the insert.

in terms of the transmitted power. Recall that the total transmitted power is the aperture area, multiplied by the initial intensity, $P_{total} = I_0(\pi D^2/4)$. Expressing the initial intensity in the above equation in terms of the total power gives the following expression of the maximum intensity at the center of the Airy disk,

$$I_{\max} = P\frac{\pi}{4}\left(\frac{D}{\lambda F}\right)^2. \tag{2.25}$$

The maximum intensity is proportional to the square of the initial beam diameter and inversely proportional to the square of the laser wavelength or the square of distance to the target. Equation (2.25) was already used in the form of Equation (1.1) in Chapter 1 to estimate the maximum diffraction-limited intensity on the target at different laser wavelengths.

The intensity distribution inside the Airy disk can be integrated to compute the power contained inside the disk,

$$P_{Airy} = \int_0^{R_{Airy}} I(r)2\pi r dr = \int_0^{R_{Airy}} I_{\max}\left(\frac{2J_1(z)}{z}\right)^2 2\pi r dr = \frac{P}{2}\int_0^{z_{\min}}\left(\frac{2J_1(z)}{z}\right)^2 z dz \approx 0.84P.$$

In other words, the Airy disk contains approximately 84% of the transmitted power.

As a final comment, the Airy pattern occurs near the focal plane, where, as stated before, $\frac{k}{2}\frac{L}{F^2}(D/2)^2 << 1$, or, for $L << \frac{4\lambda F^2}{\pi D^2}$. For the same representative values of $F = 10$ km, $\lambda = 1$micron and $D = 0.2$ m, $L \ll 3$km. For larger L or shorter F values, the exponential term of the modified definition of the source field $\tilde{E}$, the so-called Fresnel regime, in Equation (2.22) should be included.

2.6 Far-field Intensity in the Presence of Near-field Distortions

In the presence of optical distortions in the near-field region, the intensity distribution near the focal plane will be,

$$I(x', y') = \frac{1}{(\lambda F)^2}\left|\int_S \sqrt{I_0(x,y)}\exp\left[ikW(x,y)\right]\exp\left[-ik\frac{(xx' + yy')}{F}\right]ds\right|^2. \tag{2.26}$$

To understand how these distortions affect the far-field intensity, we will consider on-axis intensity, or, rather, the ratio of the on-axis far-field intensity to the diffraction-limited intensity, called on-axis Strehl Ratio, $SR = I(0,0)/I_0(0,0)$. Using Equation (2.26), the equation for SR becomes,

$$SR = \frac{I}{I_0} = \left| \frac{\int_S \sqrt{I_0(x,y)} \exp\left[ikW(x,y)\right] ds}{\int_S \sqrt{I_0(x,y)} ds} \right|^2. \tag{2.27}$$

Let us consider an arbitrary time-changing wavefront, $W(x,y,t)$, with a uniform initial intensity, $I_0 = const$, across the aperture with area A. The on-axis Strehl ratio in the far-field becomes,

$$SR(t) = \left| \frac{\exp(2\pi i \bar{W})/\lambda) \int_A \exp(2\pi i(W(x.y,t) - \bar{W})/\lambda) ds}{A} \right|^2$$
$$= \left| \frac{\int_A \exp(2\pi i W(x,y,t)/\lambda) ds}{A} \right|^2, \tag{2.28}$$

where $W(x,y,t) = W(x,y,t) - \bar{W}(t)$ is a mean-removed wavefront, and $\bar{W}(t) = \int_A W(x,y,t) ds / A$ is the instantaneous spatial mean. Strehl Ratio depends on the *relative* variance of the root-mean-squared of aero-optical phase, $\sigma = 2\pi OPD_{rms}/\lambda$ and can be different for the same values of OPD_{rms}, but different laser wavelengths.

First, let us consider the case of small optical distortions, $2\pi W/\lambda \ll 1$ and expand the exponent in the integral in Taylor series,

$$SR(t) = \left| \frac{\int_A \left(1 + 2\pi i W(x,y,t)/\lambda - (2\pi W(x,y,t)/\lambda)^2 + \ldots\right) ds}{A} \right|^2 =$$
$$\left| 1 - \frac{1}{2}\left(\frac{2\pi}{\lambda}\right)^2 \frac{1}{A} \int_A W^2(x,y,t) ds + \ldots \right|^2 \approx \left(1 - \frac{1}{2}\left(\frac{2\pi W_{rms}(t)}{\lambda}\right)^2\right)^2, \tag{2.29}$$

where $W_{rms}(t) \equiv OPD_{rms}(t) = \left(\frac{1}{A} \int_A W^2(x,y,t) ds\right)^{1/2}$ is the instantaneous spatial root-mean-squared of the wavefront. The Strehl ratio is *always* less than unity in the presence of optical distortions. Equation (2.29) is known as Maréchal formula. Mahajan (1982, 1983) proposed another expression for the Strehl Ratio,

$$SR(t) = \exp\left[-\sigma^2(t)\right] = \exp\left[-(2\pi OPD_{rms}(t)/\lambda)^2\right], \tag{2.30}$$

and found that it better describes experimental results than the Maréchal formula for small aberrations $OPD_{rms} < 0.1\lambda$. Note that Equation (2.30) was originally simply postulated by Mahajan to empirically describe experimentally observed data.

For arbitrary large wavefronts, the integral in Equation (2.28) is hard to evaluate in general due to a nonlinear nature of the complex exponent. One way to get around this problem is to re-write it in terms of a probability function of the wavefront W. Let us fix a value W_0 for the wavefront and consider the small range of values dW around it, $W = [W_0, W_0 + dW]$. The contribution to the integral 2.28 from this range is the relative area occupied by W from this range, see Figure 2.7, multiplied by a phase shift, $\exp(2\pi i W_0/\lambda)$. But the relative area is nothing else than the probability density function, $PDF(W)$, of $W(x,y)$ to be W multiplied by dW. Thus, the Equation (2.28) can be rewritten as,

$$SR(t) = \left|\int_{-\infty}^{\infty} \exp(2\pi i W(\vec{x},t)/\lambda) PDF(W) dW\right|^2.$$

This integral is a Fourier transform of the probability distribution $PDF(W)$ and is known in probability theory as a characteristic function, $\varphi(t) = \int \exp(itx) PDF(x) dx$ (Abramowitz and Stegun 1972); thus, the above equation can be written as $SR = |\varphi(2\pi/\lambda)|^2$. For one important case of spatially Gaussian processes, the spatial probability density function is $PDF(W) = \exp\left[-W^2/(2W_{rms}^2)\right]$ and the instantaneous Strehl ratio becomes

$$SR(t) = \exp\left[-(2\pi W_{rms}(t)/\lambda)^2\right] = \exp\left[-(2\pi OPD_{rms}(t)/\lambda)^2\right], \tag{2.31}$$

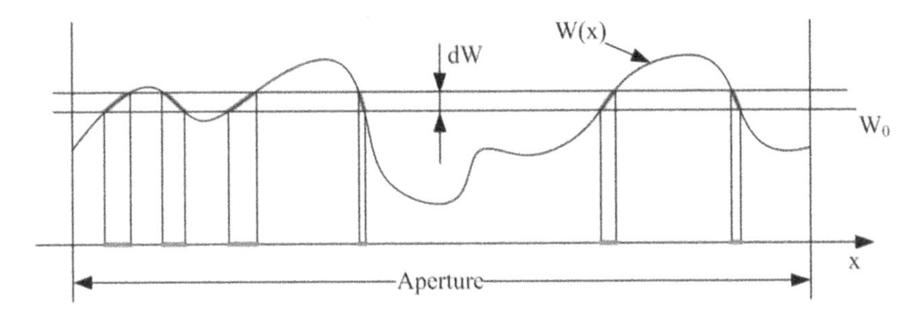

Figure 2.7 Spatial occurrences of $W(x) = W_0$.

the same expression as in Equation (2.30). The same result was derived, although somewhat differently, by Ross (2009). It should be emphasized that as long as the *spatial* statistics of wavefronts are Gaussian, Equation (2.30) remains *accurate, regardless* of the amplitude of OPD_{rms}.

Steinmetz (1982) showed that in the limit when the aperture size is much larger than the characteristic turbulence structure size, Equation (2.30) is approximately valid for any statistical processes, if the time-averaged OPD_{rms} is used,

$$\overline{SR} \approx \exp\left[-(2\pi\overline{OPD_{rms}} / \lambda)^2\right]. \tag{2.32}$$

This approximation is called the Large-Aperture Approximation (LAA) and is widely used to estimate far-field effects on a target.

So, on one hand, decreasing the laser wavelength decreases the size of the Airy disk and increases the intensity on the target. On the other hand, aero-optical effects become significant for shorter wavelengths for the same level of OPD_{rms}, since the ratio OPD_{rms} / λ increases and the Strehl Ratio decreases. This was demonstrated in Figure 1.2 (b), for a specific aberration level at $\lambda = 10.6$ microns. Taken together for a number of aberrations is demonstrated in Figure 2.8, where the on-axis intensity, normalized by the on-axis intensity for $\lambda = 10.6$ microns, is plotted for different wavelengths and levels of aero-optical distortions. As mentioned in Chapter 1, at the large wavelength near 10 microns, aero-optical effects are negligible, but the diffraction-limited intensity is also low. Using shorter wavelengths would result in much larger intensities on the target, but the aero-optical effects will significantly reduce this intensity at smaller wavelengths. For example, for a laser wavelength of 1 micron, the diffraction-limited intensity is 100 times larger than for the wavelength of 10 microns, but this number drops to 70 for $OPD_{rms} = 0.1$ micron. At the wavelength of 1 micron, in the presence of aero-optical distortions with the level of $OPD_{rms} = 0.2$ micron, the diffraction-limited intensity is increased only by a factor of 20 and is only 3 times larger for $OPD_{rms} = 0.3$ micron. Therefore, mitigating aero-optical effects is critical to increasing the intensity on the target.

The presented analysis considers primarily the time-averaged intensity losses on the target. However, for communication systems it is more important to reduce instantaneous variations of the intensity for a stable communication link. As it will be discussed in Chapter 6, even simple flows like turbulent boundary layers might create significant intensity variations or drop-outs on the target.

As a final comment, although the LAA is commonly used to estimate intensity reductions, it is important to remember that it provides only an approximation at large OPD_{rms}, while the instantaneous version of it, Equation (2.30), is valid for all OPD_{rms}, as long as the spatial distribution of the wavefront is Gaussian. One way to test validity of Equation (2.30) is to recognize the non-linear nature of the relation between the instantaneous OPD_{rms} and the Strehl ratio, Equation (2.30), from which it follows that

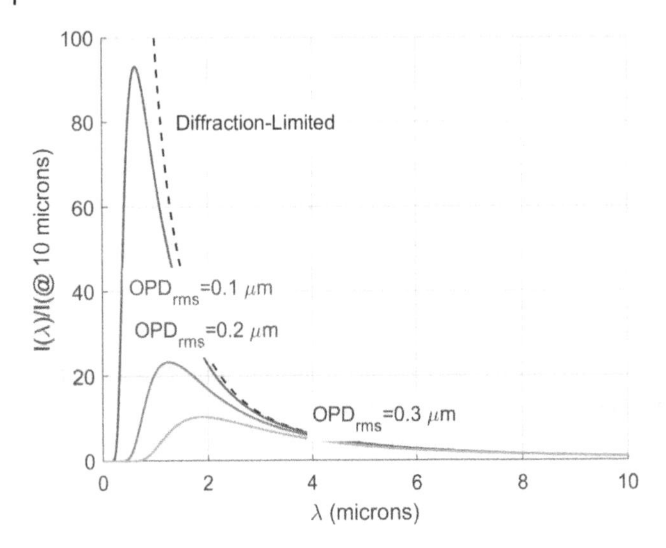

Figure 2.8 Far-field intensity for different levels of aero-optical distortions, using the Large Aperture Approximation.

$$\overline{SR(t)} = \overline{\exp\left[-(2\pi OPD_{rms}(t)/\lambda)^2\right]} \neq \exp\left[-(2\pi \overline{OPD_{rms}(t)}/\lambda)^2\right].$$

Porter et al. (2013a), had proposed a correction to the LAA,

$$\overline{SR} = \int \exp\left[-(2\pi OPD_{rms}/\lambda)^2\right] PDF(OPD_{rms}) d(OPD_{rms}), \tag{2.33}$$

where $PDF(OPD_{rms})$ is the *temporal* probability density function of the time series of $OPD_{rms}(t)$. They used experimental data collected in-flight on Airborne Aero-Optics Laboratory (described in more detail in Chapter 6) and had showed that this approximation provides a better estimate of the Strehl Ratio, within 10% up to $OPD_{rms}/\lambda = 0.2$, while the LAA is only within 20% at these levels of aero-optical distortions. For larger distortions, exact equations, Equations (2.27) or (2.28), should be used.

While the LAA has some limitations, its functional form has important system-performance implications. As long as the contribution to the overall level of optical distortions from various sub-systems are independent, then the total "optical energy" $OPD^2_{rms,total}$ is a sum of individual "optical energies" from sub-systems, see the additional discussion in Chapter 3,

$$OPD^2_{rms,total} = OPD^2_{rms,1} + OPD^2_{rms,2} + OPD^2_{rms,3} + \dots$$

This nonlinear combination of wavefront errors becomes linear through the Large-Aperture Approximation, as follows

$$\begin{aligned} SR_{total} &= \exp\left[-\left(\frac{2\pi OPD_{rms,total}}{\lambda}\right)^2\right] \\ &= \exp\left[-\left(\frac{2\pi}{\lambda}\right)^2 \left(OPD_{rms,1} + OPD_{rms,2} + OPD_{rms,3} + \ldots\right)\right] \\ &= \exp\left[-\left(\frac{2\pi OPD_{rms,1}}{\lambda}\right)^2\right] \exp\left[-\left(\frac{2\pi OPD_{rms,2}}{\lambda}\right)^2\right] \exp\left[-\left(\frac{2\pi OPD_{rms,3}}{\lambda}\right)^2\right] \ldots \\ SR_{total} &= SR_1 \cdot SR_2 \cdot SR_3 \ldots \end{aligned} \tag{2.34}$$

In the form of Equation (2.34), it is easy to see how Strehl ratios can be computed separately for each of the various contributors to the overall Strehl ratio. Other "hits" on the Strehl ratio could include the atmosphere, for example. Often specific Strehl losses can be assigned as an error budget.

2.6.1 Temporal Intensity Variation

For directed energy applications, the system performance is primarily defined by the time-averaged Strehl ratio on the target, as engagement times are typically few seconds. For the reliable free-space communication system, it is more critical to ensure that the signal at the target does not significantly fluctuate in time, that is, it does not have significant intensity drop-outs. The temporal intensity variation is typically described by the temporal variance of the logarithm of the intensity, $\sigma^2_{\ln(I)} = \overline{\{\ln(I(t))\}^2}$. Again, considering the case when the wavefront spatial statistics are Gaussian, the instantaneous Strehl Ratio is given by Equation (2.30) and $\sigma^2_{\ln(I)}$ becomes

$$\begin{aligned} \sigma^2_{\ln(I)} &= \overline{\left\{-\left(2\pi OPD_{rms}(t)/\lambda\right)^2\right\}^2} = \left(\frac{2\pi}{\lambda}\right)^4 \overline{OPD^4_{rms}(t)} \\ &= \left(\frac{2\pi}{\lambda}\right)^4 \int OPD^4_{rms} PDF(OPD_{rms}) d(OPD_{rms}), \end{aligned}$$

where the 4th-moment of the temporal dependence of OPD_{rms} is given in terms of $PDF(OPD_{rms})$. In a later section it will be shown that the statistics of $OPD(x, y, t)$ significantly depends on the beam aperture, or, more formally, on the ratio of the beam aperture and the characteristic spatial size of the optically aberrating structure.

2.7 Wavefront Components

Before identifying different wavefront components, it is useful to talk about different types of aero-optical distortions and how these distortions can be compensated for or mitigated. Consider a simple optical system, which is intended to send an unaberrated laser beam with a spherical wavefront that should be focused on a distant target, as indicated in Figure 2.9(a). In case of a distant target, the imposed spherical component is very small. For simplicity, we will ignore this spherical component, and treat the wavefront as planar. Without any distortions, the beam will reach the target with the diffraction-limited intensity. In the presence of the optical distortions, both due to a turbulent compressible flow around the transmitting platform (in our case, a turret) and temperature-induced density variations in the atmosphere, the laser beam will be aberrated. One way to distort the beam is to add small-scale distortions to the otherwise planar wavefront. As we had discussed in the previous section, it will result in a reduction of the maximum intensity on the target. But even in the absence of these small-scale aberrations, the beam can still be distorted when the optical aberrations have a linear gradient component, called *tip/tilt*. From Equation (2.6) it follows that, in this case, the entire beam will change the intended direction, causing the beam to deflect away from the desired aim-point. If the optical distortions change in time, relative to the transmitting platform, the amount of tip/tilt due to aero-optical distortions will also change in time. This time changing tip/tilt imposed on the outgoing beam is called *jitter*.

If these optical distortions are known, it is possible to compensate for them using a deformable mirror (DM) and a fast steering mirror (FSM), as shown in Figure 2.9(b). The DM is typically a thin metal reflective membrane, which can be spatially deformed in the normal direction by multiple translation actuators mounted on the back of the membrane. Recall that a wavefront is defined as a distance that light travels from the source. When the laser beam is reflected off the spatially deformed mirror, additional optical distortions are added to the beam. If the overall optical distortions outside the turret are known, the deformable mirror can be programmed to add these distortions with an opposite sign (conjugate distortions). These "anti"-distortions will effectively cancel the optical distortions present outside the turret, restoring a planar wavefront and the maximum intensity on the target. To compensate for the tip/tilt component, a fast-steering mirror (FSM), capable of quickly changing its angular position, is added to the beam train to recover the correct direction to the target.

The optically compensating system with the DM is called an adaptive optics system. Originally it was introduced in the 1950s to improve the optical performance of large ground-based telescopes in the presence of slowly changing atmospheric optical distortions. Later adaptive optics systems, with an added FSM to stabilize the beam direction, were implemented to enable various free-space laser-based systems. Detailed discussion of adaptive optics systems with various applications in aerospace

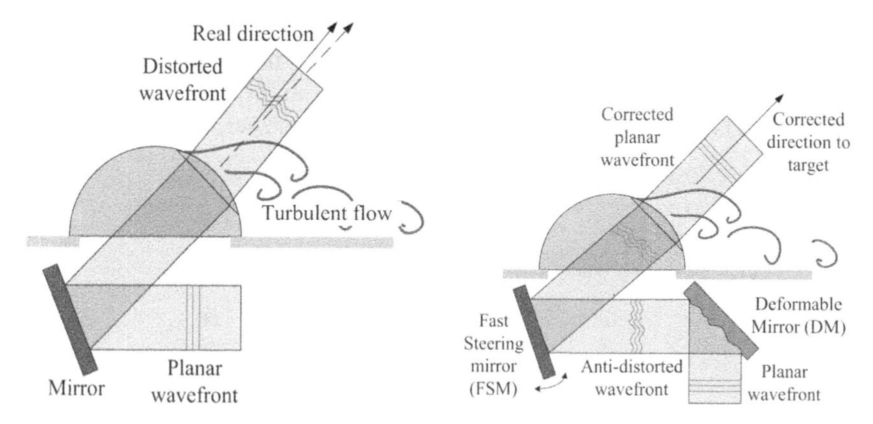

Figure 2.9 Left: Nominal optical system. Right: An adaptive optics system with a Fast Steering Mirror (FSM) to correct the beam direction and a Deformable Mirror (DM) to introduce anti-distortions into the beam.

can be found in several textbooks, such as Tyson (1997) or Merritt (2012). We will return to the discussion of some aspects of adaptive optics systems in Chapter 8.

Since the density field in general has both a steady, or time-independent component, $\rho(x,y,z)$, and a fluctuating part, $\rho'(x,y,z)$, the resulting wavefronts will also have a steady component, usually called a *steady lensing* term and a time-dependent or unsteady component as

$$W(x,y,t) = W_{SL}(x,y) + W_{unsteady}(x,y,t).$$

The steady lensing, by definition, can be computed by averaging all resulting wavefronts in time $W_{\mathrm{SL}}(x,y) = \overline{W(x,y,t)}$. The steady lensing term will also include any steady optical distortions from optical components present in the optical beam train. Examples of such optical distortions are residual distortions from improperly aligned lenses, non-flat mirrors, and glass pieces with non-uniform thicknesses. The time-dependent component can also be decomposed into several terms as,

$$W_{unsteady}(x,y,t) = A(t) + B_1(t)x + B_2(t)y + W_{H.O.T.}(x,y,t). \tag{2.35}$$

The unsteady term, $A(t)$, is an instantaneous spatially averaged value of the wavefront and is referred to as a piston component. Terms $B_1(t)$ and $B_2(t)$ are related to the time-changing wavefront gradient and are called instantaneous tip/tilt components or unsteady beam jitter. We will discuss the explicit equations for all these terms later in Chapter 5, when addressing the finite aperture effects. Finally, the rest of the aero-optical distortions are included in the *higher-order term*, $W_{H.O.T.}(x,y,t)$.

To discuss the contribution of all these terms to the far-field intensity pattern, we can substitute the wavefront decomposition (with the steady lensing term) into Equation (2.26). After some manipulations, we can get the following expression for the far-field pattern, with F denoting a focal distance,

$$I(x',y') \sim \underbrace{\left|\exp\left[ikA(t)\right]\right|^2}_{(1)} \left| \int \sqrt{I(x,y)} \underbrace{\exp\left[ikW_{SL}(x,y)\right]}_{(2)} \underbrace{\exp\left[ikW_{HOT}(x,y,t)\right]}_{(3)} \underbrace{\exp\left\{-ik\frac{\left(\left[x'-F\cdot B_1(t)\right]x+\left[y'-F\cdot B_2(t)\right]y\right)}{F}\right\}}_{(4)} ds \right|^2 \tag{2.36}$$

The piston component, term (1) in Equation (2.36), does not affect the far-field pattern, as $\left|\exp\left[ikA(t)\right]\right|^2 = 1$. For this reason, the instantaneous piston mode is usually removed from the instantaneous wavefronts in the post-process analysis. The unsteady tip/tilt component, term (4) in Equation (2.36), simply shifts the whole far-field pattern to a new location, $(x'',y'') = (x' - F\cdot B_1(t), y' - F\cdot B_2(t))$. The only terms affecting the actual shape of the far-field pattern and decreasing the peak intensity on the target are the steady lensing component, term (2), and the higher-order term, term (3). The higher-order term (3) in Equation (2.36) depends on the laser wavenumber, $k = 2\pi/\lambda$, and consequently, the far-field intensity pattern depends to the laser wavelength. At the same time, the tip/tilt related shift is independent of the wavelength. So, it is important to not only accurately measure or predict the higher-order distortions, but also to know the amount of aero-optical tip/tilt present in the wavefront. Even a relatively small amount of tip/tilt might result in significant displacement of the beam on the target. For instance, if B is equal to only 10 microrad, at the distance of $F = 10$ km it will shift the beam off the target by 10 cm. More importantly, for the unsteady tip/tilt or jitter, the focal spot will constantly move over the target, significantly reducing the time-averaged intensity. Thus, instantaneous tip/tilt statistics, like amplitude and the temporal spectra, are important to know to properly design adaptive optics systems. These statistics are also useful for extracting some information about the dynamics of the aero-optical effects and underlying density fields. We will discuss the dependence of the aero-optical jitter on the aperture size and statistics of the jitter for different fundamental flows in Chapter 7.

As mentioned before, the tip/tilt-related shift in the far-field intensity pattern on the target technically does not depend on the laser wavelength. Nevertheless, it has a more significant effect on reducing the time-averaged intensity, when shorter wavelength lasers are used, compared to longer wavelength lasers. This is because far-field intensity patterns, either diffraction-limited or distorted by the higher-order wavefront term, are smaller for shorter wavelengths, as described by

Equations (2.24) and (2.36). To demonstrate this effect, the same amount of time-dependent tilt, indicated as horizontal bars in Figure 2.10, was imposed on the diffraction-limited (that is, no higher-order distortions) far-field intensity, indicated by dashed lines in Figure 2.10, and the resultant time-averaged intensity

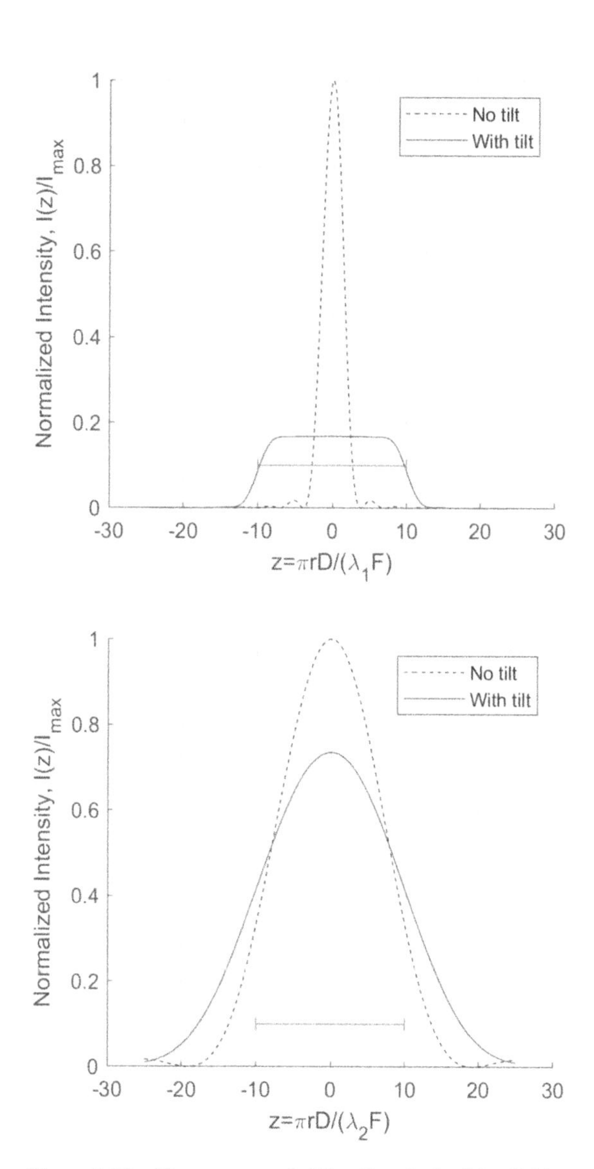

Figure 2.10 Time-averaged diffraction-limited far-field pattern without and with time-changing tilt for (a) short and (b) long laser wavelength. Range of tilt-related beam shift is indicated by the horizontal bar.

patterns were computed for short and long laser wavelengths. The results are presented in Figure 2.10. For a small wavelength, the same time-changing tip/tilt-related beam displacement will significantly "smear out" the small far-field pattern, resulting in smaller time-averaged intensity on the target, as shown in Figure 2.10(a). For a long wavelength, as seen in Figure 2.10(b), the far-field pattern is larger, and the same "smearing" will produce less intensity reduction on the target. Clearly, the time-averaged intensity on the target is significantly reduced in the case of a short wavelength, compared to the case of a long wavelength.

Any mirrors, both flat and curved, which are present in the optical beam train, might and often do introduce an additional tip/tilt component into the wavefronts. If the mirrors are not perfectly stationary, but move or vibrate, they will add the vibration-related tip/tilt into the wavefronts. More importantly, this vibration-induced jitter is indistinguishable from the turbulence-induced jitter, thus corrupting the measurements of the aero-optical tip/tilt. Technically, if the mirror motion is known or measured somehow, it is possible to remove the mechanical component from the overall jitter. In practice, however, if several mirrors are present in the optical beam train, it becomes very difficult to decouple mechanical jitter resulting from vibrations of tunnel/model/optical-table from the turbulence induced jitter. Also, as discussed before, most adaptive optics systems have a fast-steering mirror, designed to eliminate the overall tip/tilt on the outgoing beam. Thus, instantaneous tip/tilt is typically presumed to be removed from the measured wavefronts.

Similarly, to accurately measure the steady lensing due to only the aero-optical flow, all other contributions to the steady lensing from the optical set-up (lenses, mirrors, etc.) should be negligible. In practice, it requires high quality, expensive optical components and a controlled optical environment. For most typical set-ups, though, the steady lensing term is always corrupted by non-ideal optical components. In addition, it is often presumed that the steady-lensing component will be removed by the deformable mirror, if the adaptive optics system is implemented. For this reason, the steady lensing term is usually removed from the wavefronts in post-processing, to analyze the effects of the higher-order term only.

The final wavefronts with removed steady lensing, piston and tip/tilt components, for simplicity denoted again as $W(x,y,t)$, contain only the unsteady tilt-removed aberrations. They reveal the character of the convective, optically active structures passing over the aperture.

3

Measuring Wavefronts

There are many experimental techniques for measuring the distortions imposed on a wavefront of a collimated beam. They fall into three major categories.

3.1 Interferometry Methods

Interferometers are one of the most accurate techniques for measuring the wavefronts. Briefly, interferometry is a method of measuring Optical Path Distances based on the principle of interference or the interaction of two or more coherent light waves. In the case of the monochromatic light with the same frequency, the intensity of the superposition of monochromatic waves, $A_1 \exp(i(\omega t + \phi_1)$ and $A_2 \exp(i(\omega t + \phi_2)$, at a given point in space is

$$I = \left|A_1 \exp(i(\omega t + \phi_1) + A_2 \exp(i(\omega t + \phi_2)\right|^2 = A_1^2 + A_2^2 + 2A_1 A_2 \cos(\phi_1 - \phi_2), \quad (3.1)$$

where phase difference is related to OPD as $\phi_1 - \phi_2 = 2\pi OPD / \lambda$ and λ is the laser wavelength. If amplitudes A_1 and A_2 are constant, then the intensity reaches its maximum at points where the two waves are in phase (the phase difference is an integer multiple of 2π) and its minimum when the waves are out of phase (the phase difference is a half-integer multiple of 2π). By measuring the resultant intensity in different spatial points, one can compute the phase and the wavefront at these spatial points, if the phase of the second beam, the reference beam, is known.

One of the most common interferometers is a Michelson interferometer, shown in Figure 3.1(a). In the Michelson interferometer, a single incoming beam of coherent light is split into two identical beams by a beam splitter. One beam, called the reference beam, stays unaberrated, and another beam travels through a medium of variable index-of-refraction. After being reflected by return mirrors back to the beam splitter, both beams interact on a beam splitter before being

Aero-Optical Effects: Physics, Analysis and Mitigation, First Edition. Stanislav Gordeyev, Eric J. Jumper, and Matthew R. Whiteley.

(a)

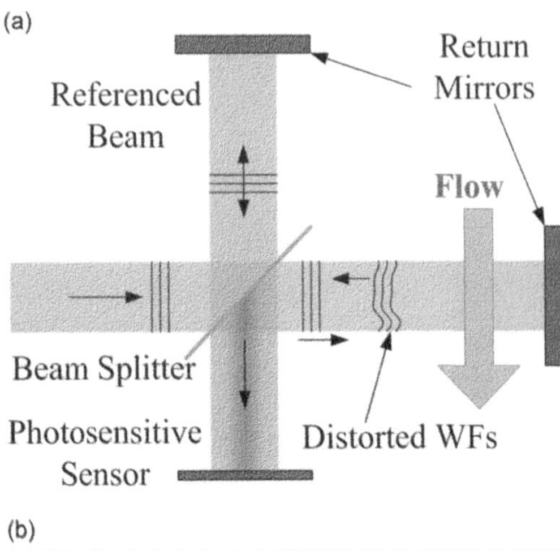

(b)

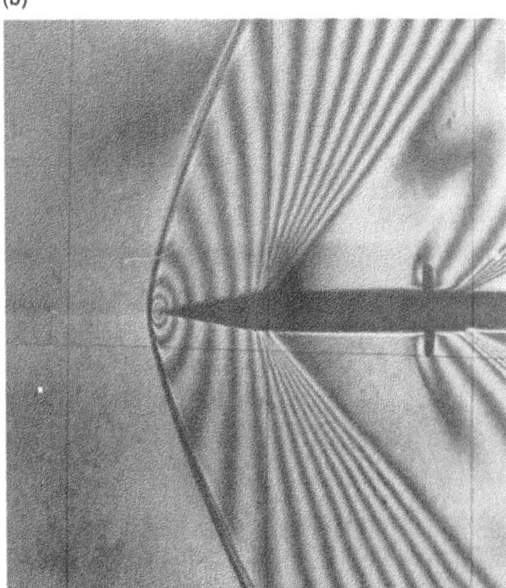

Figure 3.1 (a) Michelson-type interferometer. (b) A fringe pattern resulted from interference between two laser beams. *Source:* Ashkenas et. al., (1951).

forwarded to a photosensitive detector. The Optical Path Difference, introduced by the flow, creates a phase difference and, as a result, the variable intensity with bright and dark regions, called a fringe pattern, shown in Figure 3.1(b), on the sensor. Using Equation (3.1), the fringe pattern can be processed to extract the measured wavefronts. More information about different types of the interferometers, and various algorithms to reconstruct the wavefronts, can be found in many textbooks on the subject (Malacara 1978; Geary 1995; Poon and Liu 2014; Schnars et al. 2015 etc.).

Interferometry is advantageous due to its sensitivity and accuracy. One of the most sensitive interferometry measurements is in the Laser Interferometer Gravitational-Wave Observatory (LIGO). With a complex set-up and a lot of data processing, LIGO can detect relative changes in the space metric, caused by a collapse of distant black holes, with astonishing accuracy of 10^{-18} meters. However, with extreme sensitivity comes sensitivity to interfering effects, such as nonideal optical elements or mechanical vibrations. The interferometry set-ups usually require components with high optical quality, stable lasers (with large coherence length), and a vibration-free environment.

3.2 Wavefront Curvature Methods

A second class of wavefront sensors are sensitive to the local curvature of the wavefront, $\nabla^2 W(x,y)$. The shadowgraph is the simplest form of an optical system capable of observing a variable index of refraction flow. In principle, this system does not need any optical component except a light source, a collimating lens, and

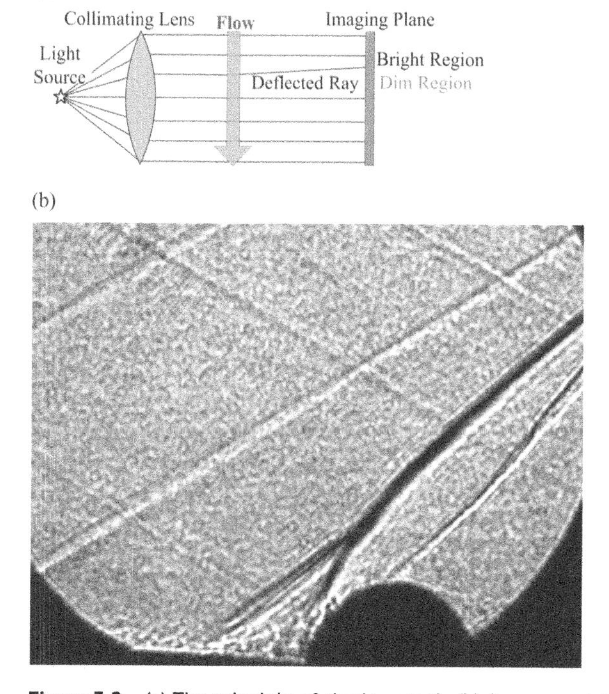

Figure 3.2 (a) The principle of shadowgraph. (b) A representative shadowgraph image of a supersonic flow around an obstacle. *Source:* Courtesy of Stanislav Gordeyev.

a screen on which to project the shadow of the varying density field, see Figure 3.2(a). A shadow effect is produced when a light ray is refractively deflected by the non-uniform density medium, so the position on the recording plane where the undeflected ray would arrive now remains dimmer. At the same time, the locations where the deflected ray arrives appear brighter than the undisturbed image. A visible pattern of variations of light intensity is produced on the screen; see Figure 3.2(b). More information on the shadowgraph technique and various applications can be found in Settles (2001).

Another important class of the curvature-based sensors are the Distorted Grating Wavefront Sensors (DGWFS). The principle of operation of the DGWFS is described in detail in Blanchard et al. (2000) and Woods and Greenaway (2003), so only essential information will be given here. The DGWFS relies on the Intensity Transport Equation, which describes how intensities at different distances, z, along the beam propagation, $I_z(\mathbf{r})$, are related to the wavefront, $\phi_z(\mathbf{r})$, for a given wavenumber, $k = 2\pi / \lambda$.

$$-k\frac{\partial I_z(\mathbf{r})}{\partial z} = \nabla \cdot [I_z(\mathbf{r})\nabla\varphi_z(\mathbf{r})] = I_z(\mathbf{r})\nabla^2\varphi_z(\mathbf{r}) + \nabla I_z(\mathbf{r}) \cdot \nabla\varphi_z(\mathbf{r}) \tag{3.2}$$

Consider a uniform-intensity beam, $I = I_0 W_A$, where W_A is unity inside the beam and zero outside the beam; the intensity gradient becomes $\nabla I = -I_0\delta_C\mathbf{n}$, where δ_C is the Delta-function along the beam boundary and $\mathbf{n}$ is the outward normal vector. Equation (3.2) can be simplified to,

$$-k\frac{\partial I_z(\mathbf{r})}{\partial z} = I_0 W_A \nabla^2\varphi_z(\mathbf{r}) - I_0\delta_C\mathbf{n} \cdot \nabla\varphi_z(\mathbf{r}). \tag{3.3}$$

This equation shows that the change in the intensity along the beam is proportional to the local wavefront curvature (the first term in right-hand-side of Equation (3.3)) and the change in the beam shape depends on the local wavefront slope on the beam boundary (the second term in right-hand-side of Equation (3.3)).

One way to simultaneously record intensities at two different z-planes is to use a quadratic grading combined with a lens; see Figure 3.3. It creates a pair of images corresponding to beam intensities at known distances z_1 and z_2 which are defined by the lens-grating geometry. In addition, the image has a central focused spot, which is irrelevant to the measurements. Subtracting one side image from another and knowing the distance between image planes, $\Delta z = z_1 - z_2$, the left-hand-side term in Equations (3.2) or (3.3) can be approximated as $\frac{\partial I_z(\mathbf{r})}{\partial z} \approx \frac{I_{z1}(\mathbf{r}) - I_{z2}(\mathbf{r})}{\Delta z}$. Equation (3.3) can be solved iteratively (Gonsalves 1982; Baba et al. 1994), by expanding the solution into Zernike polynomials (Gureyev et al. 1995) or by applying a Green's function approach (Woods and Greenaway 2003).

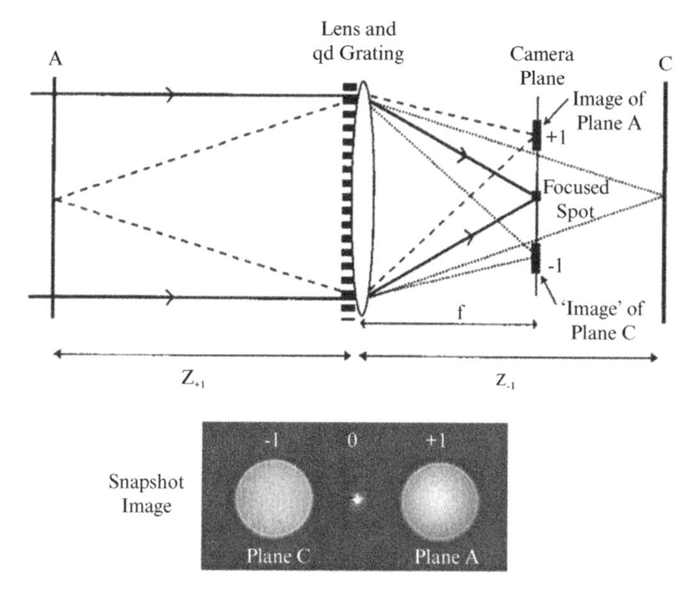

Figure 3.3 Schematic of the optical system used for distorted grading wavefront sensing and an example recorded image. *Source:* Blanchard et. al. (2000), reproduced with permission from The Optical Society.

DGWFS has a good spatial resolution since the wavefront is reconstructed at every pixel in the image. The main drawback of the DGWFS is that from Equation (3.3) it follows that the DGWFS relies on the information inside the beam and along the beam boundary, specifically, the changes in the boundary shape in time. The beam boundary can be easily corrupted by real-life issues, like beam jitter, or distortions from nonideal optics. These changes usually lead to the presence of nonphysical, low-order modes in wavefronts. Thus, the DGWFS generally does not perform well under the presence of mechanical corruption and, as a rule, requires a vibration-free environment and a set-up with a good optical quality collimated beam. Another issue is that the Green's function algorithm is useful only for a few simple beam shapes (circular, annulus, or rectangular), where the Green's function is known analytically. It becomes very difficult to use the Green's function approach for arbitrary beam shapes. Finally, Zernike-based algorithm can be used only for circular or annulus beam shapes.

3.3 Gradient-based Wavefront Sensors

A third class of wavefront sensors are sensitive to the wavefront gradient, $(\partial W / \partial x, \partial W / \partial y)$. A schlieren technique is one example. The schlieren set-up is similar to the shadowgraph set-up, where the light source is expanded into the collimated beam and passed through the flow of interest. After passing the flow, the collimated beam is focused to a point and the knife edge is placed near this point, as shown in Figure 3.4(a). Let us say that the knife edge is positioned normal to the flow direction. If the light ray passes through a flow region with a positive streamwise wavefront gradient, indicated by Path 1 in Figure 3.4(a), it will miss the knife edge and, as in the shadowgraph technique, make this part of the image brighter. However, if the light ray traverses through a region with a negative streamwise wavefront, labeled Path 2 in Figure 3.4(a), it will be deflected toward the knife edge and will be blocked by it. As a result, the corresponding part of the image, where the ray would have arrived, becomes dim. Thus, the image intensity depends on the sign of streamwise wavefront gradients caused by the flow. The sensitivity of the schlieren system can be changed by moving the knife edge closer to or farther away from the focal point. If the large area needs to be studied, large concave off-axis mirrors are often used instead of lenses to expand the light to a collimated beam and to focus it on the knife edge.

It is easy to see that the schlieren system is sensitive to the wavefront gradients normal to the knife edge. Examples of the schlieren images of a supersonic flow around a cone and a Pitot tube with two different knife-edge orientations are presented in Figure 3.4(b). When the knife edge is oriented vertically, streamwise gradients are emphasized, like the straight shocks around the Pitot tube. If the knife edge is rotated to the horizontal orientation, the schlieren image will be sensitive to the vertical (cross-stream) gradients, making the oblique shocks and the boundary layer at the bottom tunnel wall clearly visible.

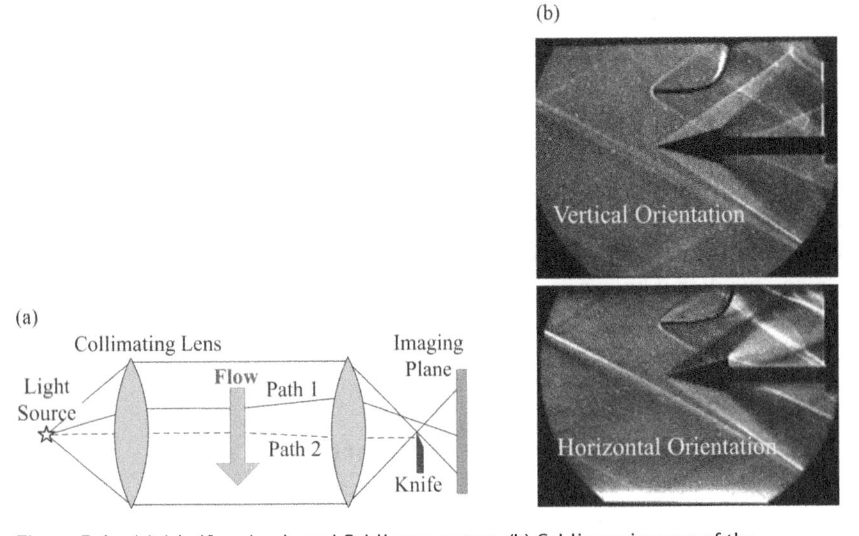

Figure 3.4 (a) A knife-edge based Schlieren system. (b) Schlieren images of the supersonic flow around a cone with a vertical and a horizontal orientation of the knife edge. *Source:* Courtesy of Stanislav Gordeyev.

Because of its relative simplicity to set-up, adjustable sensitivity to the wavefront gradients and good spatial resolution, the schlieren technique is widely used in aerospace engineering to visualize the variable-density flow around objects. Numerous variations and extensions of the traditional schlieren, like rainbow schlieren, focused schlieren and Background-Oriented Schlieren (BOS) were developed in the last few decades. The reader is referred to Settles (2001) as an excellent source on the schlieren technique and its variations. The schlieren technique is usually used as a qualitative flow-visualization technique. However, if the flow is measured with two orthogonal knife-edge orientations, the wavefront can be reconstructed from the resultant images (Geary 1995).

Another wavefront instrument relies on measuring local deflection angle in the image plane. The generic name for this class of wavefront sensors are Hartmann-type sensors. Below we will discuss the Hartmann type sensors, which the authors have used for more than 20 years and have significant experience working with.

3.3.1 Shack-Hartmann Wavefront Sensor

In 1904, a German physicist and astronomer Johannes Franz Hartmann proposed a technique to measure optical distortions of the recently built Great Refractor of Potsdam. In the initial tests it was discovered that the telescope primary mirror, 80 cm diameter and 12.5 m in focal length, had some unintended distortions. To investigate these optical distortions, Hartmann placed a plate with a regular grid of holes in front of the primary optical element (a mirror) of the telescope as a means of creating and tracing individual rays of light through the optical system. To record a resulting pattern of spots, he put a photographic plate at a given distance, as shown schematically in Figure 3.5. For an unaberrated (that is no optical distortions) optical system, the recorded dot pattern should appear regular. Any optical distortion, present in the optical element, will deflect the rays of light,

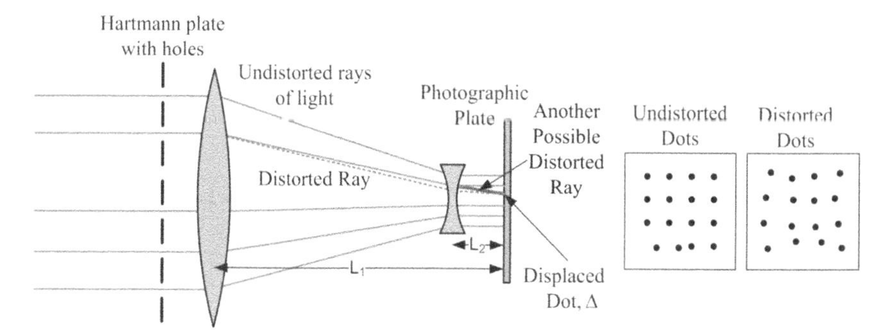

Figure 3.5 Schematic of the Hartmann test plate to quantify optical performance of a telescope.

indicated by a dotted red line in Figure 3.5, from the otherwise undistorted path. As a result, the dots will move from their undistorted locations. By measuring the displacement, Δ, of each light ray from its unaberrated position and knowing the distance, L_1, from the primary optical element to the photographic plate, it is possible to find the local deflection angles as $\theta = \Delta / L_1$ and reconstruct the wavefront distortions by integrating Equation (2.6). The results of the test, conducted by Hartmann, led to the optical components being reconfigured to alleviate the problem (Harvey and Hooker 2005).

The advantage of a Hartmann test is that it is easy to set up and possible to accurately test the entire area of the optical element, if the distance from the optical element to the plate is large. For this reason, the Hartmann plate test is still widely used by amateur astronomers. But if more than one optical element is present in the telescope, like a secondary lens in Figure 3.5, the same dot displacement can be caused by much larger distorted ray with a deflection angle, $\theta = \Delta / L_2$, but originated closer to the plate, $L_2 < L_1$, as indicated by a thick line in Figure 3.5. The situation becomes even more complicated if both or more elements distort light. In this case, the resultant dot displacement will be a combination of different deflections and lateral translations imposed on the ray. One way to remove this ambiguity from the measurements of the angular deflections is to place a lens along the beam path and measure the dot displacement at the focal plane, as shown in Figure 3.6. The resultant dot position depends only on the angular deflection of the incoming beam, and it is insensitive to the lateral position of the beam. Based on this idea, Roland Shack and Ben Platt in the late 1960s proposed to replace the Hartmann plate with an array of lenslets. They also moved the lenslet array to after a telescope, where a lenslet array can be of a much smaller size. Finally, they took advantage of the use of then-newly emerged digital CCD detectors, making the process of recording the dot pattern and its subsequent analysis much easier. The new wavefront sensor, called a Shack-Hartmann Wavefront Sensor (SHWFS), is now a standard tool in professional astronomy, adaptive optics systems, ophthalmology, and other areas, where accurate measurements of optical distortions are required.

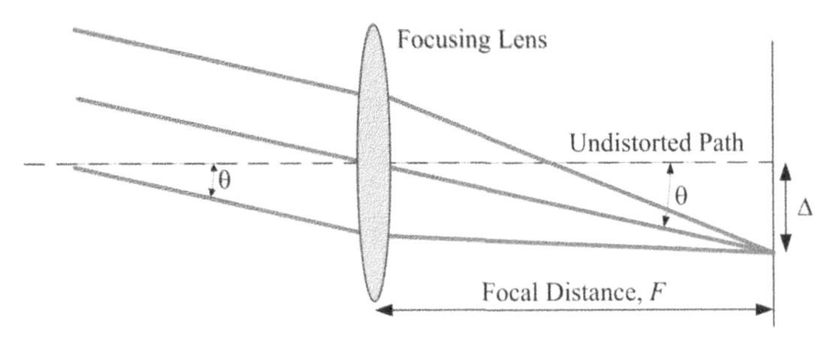

Figure 3.6 Principle of deflection angle measurements, insensitive to the lateral displacement of the small beam.

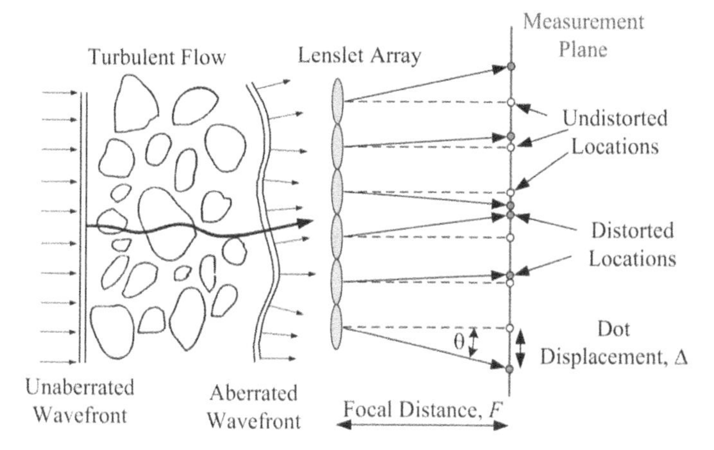

Figure 3.7 Principle of operation of the Shack-Hartmann Wavefront Sensor (SHWFS).

The principle of operation of the SHWFS is presented in Figure 3.7. It is based on the Huygens' principle, which states that a ray of light travels normally to its associated optical wavefront. The lenslet array with a focal length, F, spatially divides the incoming beam into an array of sub-apertures, where the local deflection angle, θ, over each sub-aperture represents the average wavefront local tilt of the incoming beam. An analog or digital sensor is placed at the focal plane of the lenslet array. The dot locations are recorded and the local displacements, relative to the undistorted locations, Δ, are computed. As in the Hartmann plate test, the dot displacements are used to compute the local deflection angles in both x- and y-directions as $\tan\theta_x = \Delta_x / F, \tan\theta_y = \Delta_y / F$. Knowing the deflection angles at an array of finite points, the wavefront is reconstructed by solving a discretized version of Equation (2.6), as will be described later in this section.

The spatial resolution of the SHWFS is determined by sub-aperture spacing of the lenslet array. The temporal resolution is only limited by digital image recording technology. Currently, commercially available high-speed digital cameras can provide sampling frame rates up to several million frames per second.

Once the images, like the one shown in Figure 3.8, are recorded, the spot centers can be determined. A common practice is to implement a "centroiding" approach. To do this in the case of a rectangular pattern array, the image is broken into a series of square areas of size N x N pixels, indicated as Areas of Interest (AOI) in Figure 3.8. The areas should be large enough to include the dot, but small enough to exclude other dots. Typically, the size of AOI matches the spacing, called a pitch, between the lenslets. The most prevailing method to compute the

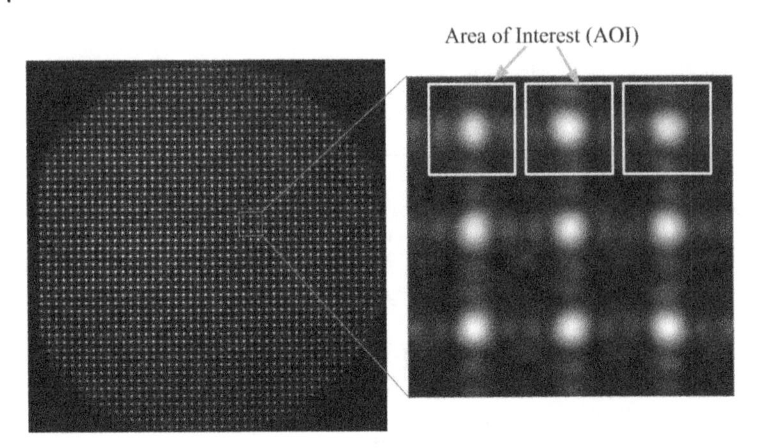

Figure 3.8 Image of a dots' pattern, measured by the Shack-Hartmann Wavefront Sensor, with a zoom-in image, with the Areas of Interest (AOI) identified for each dot.

center of each spot is to compute the "center-of-mass" or "centroid" of the dot, defined as,

$$x_C = \frac{\sum_i^N \sum_i^N x_i I_{i,j}}{\sum_i^N \sum_j^N I_{i,j}}, \quad y_C = \frac{\sum_i^N \sum_i^N y_i I_{i,j}}{\sum_i^N \sum_j^N I_{i,j}},$$

where $I_{i,j}$ is a part of the image within AOI, and (x_i, y_i) is the location of each pixel. Most of the time, this method provides a robust centroid calculation with minimal computational time. However, since the centroid is an intensity-based method, increased image noise levels would result in significant spot estimation error. In this case, various image processing approaches can be implemented to reduce the noise influence. The gamma correction is a simple means of enhancing the image's bright pixels while reducing noise. Thresholding is another common technique used to reduce noise by raising the relative "floor" of the image. Windowing similarly removes noise by isolating the relevant or bright pixels within the image. A detailed analysis of the accuracy of all these approaches is provided in Nightingale and Gordeyev (2013), and it was concluded that the 4th-order gamma correction provides the most accurate algorithm, reducing absolute error of the commonly used first moment centroid given low to moderate mean noise levels.

To compute the relative dot displacement, the location of an undistorted, or a reference dot needs to be measured. If the lenslet dimensions are known, it is a straightforward process to predict the undistorted location of the dot for each lenslet. Alternatively, a reference image without any aero-optical distortions (for instance, a no-flow case, if a wind tunnel is used) can be recorded, and the centroiding algorithm can be used to compute the undistorted locations of the dots.

The advantage of this method is that any steady optical distortions due to non-ideal optical set-up will be automatically removed from the measured wavefronts. If the reference image cannot be taken or was not taken, a time-averaged image of all the frames can be used as a reference image. In this case, only the unsteady component of the wavefront is recovered.

Another useful feature of the SHWFS is its potential robustness to the large tip/tilt (jitter) component, present in the incoming beam, also known as a beam walk. A global tip/tilt will displace all dots by the same amount but will not change their relative positions. A simple algorithm can be implemented to account for this global dot displacement by tracking average positions of some or all dots. When implemented, it makes the SHWFS a very robust wavefront sensor to collect reliable wavefront data in large industrial tunnels (Vukasinovic et al. 2013) or in-flight (Porter et al. 2013b; Morrida et al. 2017a; Gordeyev and Kalensky 2020), where a large, vibration-induced beam jitter is often present.

3.3.1.1 Wavefront Reconstruction Algorithm

An essential part of a SHWFS system is the wavefront reconstruction algorithm. Recall that the SHWFS measures the local deflection angles (θ_x, θ_y) over a discrete set of spatial points, and a discrete version of Equation (2.6),

$$\frac{\partial W}{\partial x} = -\theta_x, \quad \frac{\partial W}{\partial y} = -\theta_y$$

is used to compute the resulting wavefronts. The above system is not unique, as the wavefront is known up to an arbitrary constant. To assure a unique solution, a zero wavefront value should be imposed on one of the nodes, or one may require that the piston mode, defined is a spatial average of the wavefront, is zero,

$$\int_{Aperture} W(x,y)dxdy = 0. \qquad (3.4)$$

Different solution methods have been developed by several people (Fried 1977; Hudgin 1977; Southwell 1980) to solve the above equations. Here, we will briefly discuss the approach, similar to the one proposed by William Southwell (Southwell 1980). The grid geometry is presented in Figure 3.9(a). Here, the measured deflection angles and the estimated wavefronts lie in the corners (nodes).

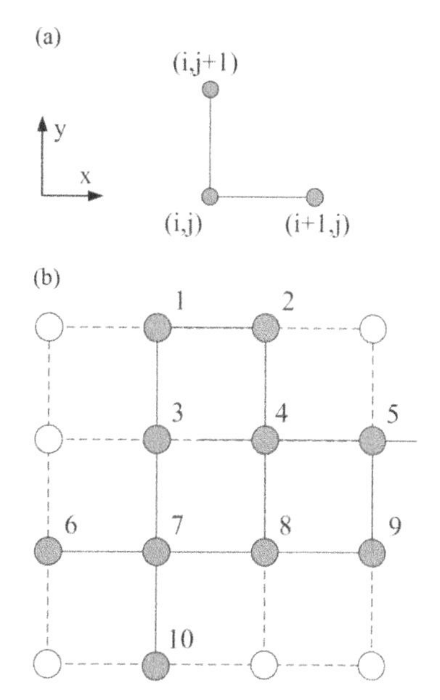

Figure 3.9 (a) Cell geometry, used by Southwell (1980). The measured deflection angles and the estimated wavefronts lie at the corners (nodes). (b) Wavefront sensor grid array.

A second-order central difference method for interior points can be used to discretize the partial derivatives in Equation (2.6),

$$\begin{aligned}\frac{W_{i+1,j}-W_{i,j}}{\Delta x} &= -0.5(\theta_{x;i+1,j}+\theta_{x;i,j})\\ \frac{W_{i,j+1}-W_{i,j}}{\Delta y} &= -0.5(\theta_{y;i,j+1}+\theta_{y;i,j}).\end{aligned} \tag{3.5}$$

Equivalently, these equations can be written for a vector of points, $\mathbf{W}=\{W_1,W_2,W_3,\ldots\}^T$, $\boldsymbol{\theta}_x=\{\theta_{x,1},\theta_{x,2},\theta_{x,3},\ldots\}^T$ and $\boldsymbol{\theta}_y=\{\theta_{y,1},\theta_{y,2},\theta_{y,3},\ldots\}^T$, where all nodes with non-zero light intensity, indicated by gray circles, are enumerated sequentially, as shown in Figure 3.9(b). This enumeration is useful when the dots' pattern is not rectangular, but for instance, a round one, with an example shown in Figure 3.8. Using matrix notations, the resulting equation becomes,

$$(\mathbf{A}_x+\mathrm{A}_y)\mathbf{W}=\mathbf{A}\boldsymbol{W}=-0.5\Delta x(\mathbf{B}_x\boldsymbol{\theta}_x+\mathbf{B}_y\boldsymbol{\theta}_y),\quad \mathbf{A}=\mathbf{A}_x+\mathrm{A}_y. \tag{3.6}$$

Here $\mathbf{A}_x$ and $\mathbf{A}_y$ are discrete derivative matrix operators of $\frac{\partial W}{\partial x}$ and $\frac{\partial W}{\partial y}$. $\mathbf{B}_x$ and $\mathbf{B}_y$ are discrete versions of the averaging operators on the right-hand side of Equation (3.6). The operators are rectangular sparse matrices. Each row or column of the matrix represents a different node. Replacing one of the rows in the matrix $\mathbf{A}$ with the discrete version of the Equation (3.4) to ensure the unique solution, Equation (3.6) can be solved numerically,

$$\mathbf{W}=-0.5\Delta x(C_x\boldsymbol{\theta}_x+C_y\boldsymbol{\theta}_y),$$

where the reconstruction matrices are $\mathbf{C}_x=\mathbf{A}^{-1}\mathbf{B}_x,\mathbf{C}_y=\mathbf{A}^{-1}\mathbf{B}_y$. The inversion of A-matrix can be performed using a computationally effective LU-factorization algorithm. Note that the reconstruction matrices depend only on the locations of the nodes and are independent of the measured deflection angles. Thus, the reconstruction matrices only need to be calculated once and then stored. After extracting the deflection angles from each frame, only a simple matrix multiplication is required to compute the wavefront.

As a final note on the reconstruction algorithm, if we apply a 2-dimensional gradient to Equation (2.6), it becomes a Poisson equation with a Neumann boundary condition,

$$\begin{cases}\nabla^2 W(x,y,t)=-\nabla\boldsymbol{\theta}(x,y,t),(x,y)\in S\\ \left.\dfrac{\partial W(x,y,t)}{\partial \mathbf{n}}\right|_{\partial S}=-\boldsymbol{\theta}(x,y,t)\end{cases}.$$

This well-studied equation provides another means to reconstruct the wavefronts from the local deflection angles.

3.3.2 Malley Probe

The Malley Probe is another optical instrument that can make direct, accurate measurements of dynamically distorting wavefronts at a fixed point, including the characteristics of the *OPD*(t). This characterization includes not only the measurement of OPD_{rms}, but also the spatial and temporal frequencies of the aberrations. The original idea of the Malley probe was proposed in 1992 by Michael Malley (Malley et al 1992). As mentioned earlier, he asserted that the aberrating structures in the flow are convecting with the flow and thus the wavefront itself is "convecting" at the same speed. The concept was fairly simple. A small aperture laser beam passes through a turbulent flow of interest and gets deflected, as shown in Figure 3.10. Using a single lens, the laser beam is then focused on a Position Sensing Device (PSD), sometimes also called Lateral Position/Displacement Detectors, which accurately measures the beam's centroid position in the flow direction $\Delta(t)$ as a function of time. Both the streamwise and spanwise deflection angles are calculated as $\theta_x(t) = \Delta_x(t) / F, \theta_y(t) = \Delta_y(t) / F$, where F is the lens focal length.

PSDs are usually made of semi-conductive photosensitive material, that provide an analog output directly proportional to the position of a light spot on the detector active area. Incident light creates a flow of free electrons on a surface of the photodiode; see Figure 3.11. These electrons are collected at cathodes at all four edges of the element as currents. For convenience, currents are converted to voltages using trans-impedance amplifiers. Beam position (X,Y) is proportional to relative differences in voltages at opposite anodes, $X = F_x \frac{V4 - V2}{V4 + V2}, Y = F_y \frac{V3 - V1}{V3 + V1}$. Constants F_x and F_y are found by performing a calibration, where the PSD is placed on a micrometer translation stage and a fixed small-aperture light source shines on it; the PSD is then traversed to different (X,Y)-positions and the individual voltages from edges are recorded.

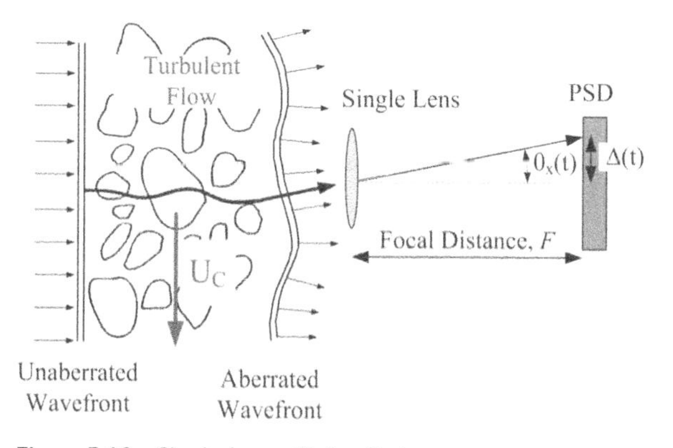

Figure 3.10 Single-beam Malley Probe.

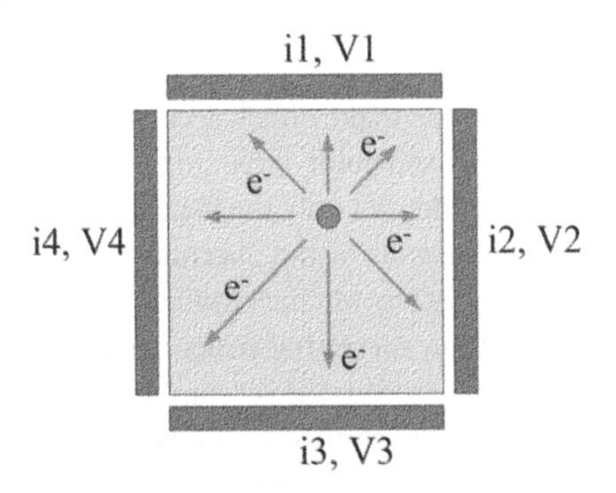

Figure 3.11 Principle of operation of a Position Sensing Device.

There are two types of PSDs. Tetra-lateral PSDs feature a common anode and four cathodes. Due to corner effects, they provide a linear response only at the center of the sensor, see Figure 3.12(a). Duo-lateral PSDs feature detectors with two separate resistive layers, one located on the top and the other at the bottom of the chip. This arrangement provides a linear response everywhere on the sensor surface, as shown in Figure 3.12(b).

Compared to digital sensors with discretized pixel areas, the PSD is a continuous analog position sensor, and offers outstanding position linearity, high analog resolution (less than few microns), fast response time (~0.5 – 1 μsec), and simple operating circuits.

Knowing the lens focal length, the measured beam positions can be converted into the deflection angles, as discussed before. From Equation (2.6), the stream-wise deflection angle of the small aperture beam is the spatial gradient of OPD at the probe-beam location,

$$\theta_x(x_0,t) = -\frac{\partial W(x_0,t)}{\partial x} = \frac{\partial OPD(x_0,t)}{\partial x}.$$

Assuming a convective nature of structures, the spatial derivative can be replaced with the temporal one as,

$$\theta_x(x_0,t) = \frac{\partial OPD(x_0,t)}{\partial x} = \frac{dt}{dx}\frac{dOPD(x_0,t)}{dt} = -\frac{1}{U_C}\frac{dOPD(x_0,t)}{dt},$$

where U_C is the convective speed of the turbulent structures. So, the OPD can be computed from the temporal evolution of the deflection angle, $\theta_x(x_0,t)$, at the fixed location as

$$OPD(x_0,t) = -U_C\int_0^t \theta_x(x_0,\tau)d\tau. \tag{3.7}$$

Finally, the time can be replaced with a pseudo-spatial coordinate,

$$OPD(x_0,t) \rightarrow OPD(x_0, x = -U_C t).$$

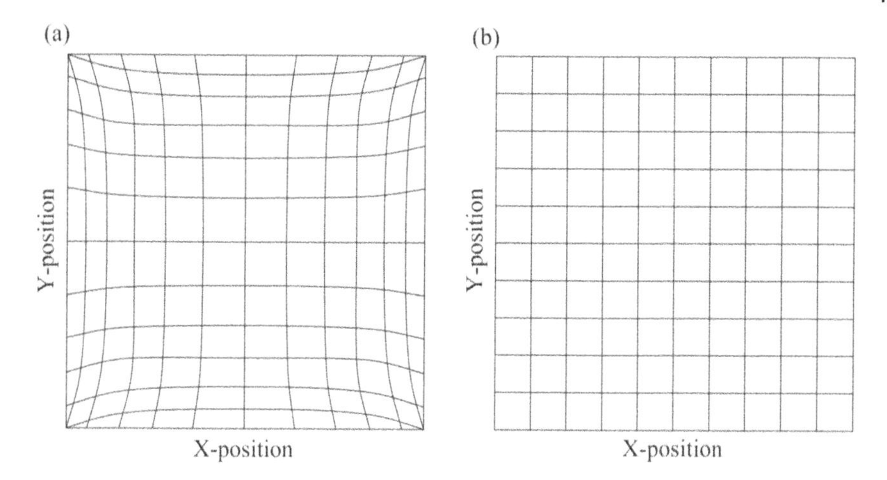

Figure 3.12 Position detectability for (a) tetra- and (b) duo-lateral types of Position Sensing Devices (PSD).

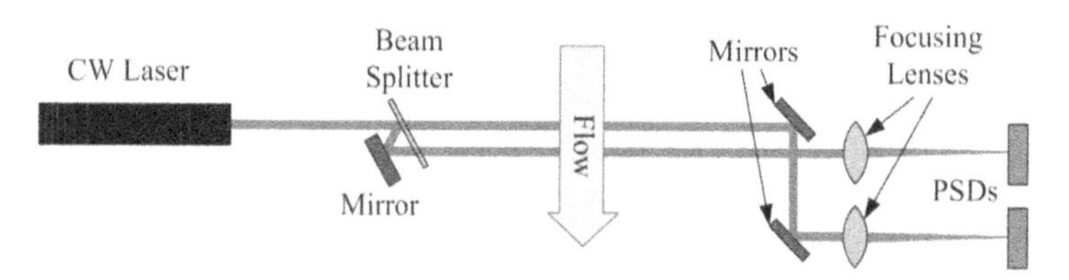

Figure 3.13 Schematic of a two-beam Malley Probe layout.

Thus, by measuring a deflection of a single beam at a fixed location, pseudo-spatial distribution of OPD in the streamwise direction can be extracted if U_C is known. Since the position-sensing device is an analog photosensitive crystal with a response time of less than 1 μsec, it allows sampling deflection angles at high sampling rates, up to hundreds of kilohertz. Consequently, it leads to very good (pseudo-) spatial resolution of the Malley probe, with $\Delta x_{pseudo} = |U_C|\Delta t = |U_C| / f_{samp}$.

The reconstruction algorithm requires the knowledge of the convective speed. Originally, it was proposed to be estimated or measured separately. In 2003, Gordeyev et al. (2003) proposed to add a second laser beam, parallel to the first one and placed some distance downstream from it, as shown in Figure 3.13. The distance should be large enough to be accurately measured, but it should be small enough to make sure that Taylor's frozen flow assumption does not break down

(Wang and Wang 2012, 2013). In real applications, this spacing is typically a few millimeters. The second beam is used to extract convection velocity information contained on the beam-deflection angles by cross correlating them and obtaining the time delay between signals. Knowing the displacement between the beams and this delay time, the convective velocity can be computed, as will be discussed in more details in Chapter 4.

One obvious drawback of the Malley probe is that it measures only slices of *OPD* in the streamwise direction. In addition, it relies on the Taylor frozen-field assumption, so it might give incorrect results for the spatially evolving flows, like shear layers. The big advantage of the Malley probe is its low cost and ease of set-up and use. The data analysis is straightforward and not computationally expensive. The extracted pseudo-spatial 1-D slices of the streamwise wavefront can be used to collect various statistics of aero-optical distortions, like OPD_{rms} for various apertures, Ap, and the wavefront spectra. It is still one of the fastest wavefront sensors, with the sampling frequency limited only by the time response of PSDs. Nowadays, high-speed cameras can also be used instead of PSDs to record the positions of dots.

3.3.3 SABT Sensor

Similar to the concept of the two-beam Malley probe is another wavefront sensor which uses the ideas of Malley et al. (1992) is termed a Small Aperture Beam Technique (SABT) (Jumper and Hugo 1995; Hugo et al. 1997). Historically it was developed before the two-beam Malley probe in an attempt to perform accurate time-resolved wavefront measurements over large regions for spatially evolving flows. The schematic of the SABT sensor is shown in Figure 3.14. It looks similar to the two-beam Malley probe set-up, but can have several parallel small-aperture beams traversing the flow. The instantaneous deflection angles of each beam are measured by either PSDs or a digital camera with sufficient temporal resolution. Unlike the Malley probe set-up, the spacing between the beams is larger in order to cover the large region. Similar to the Malley probe, the SABT sensor takes advantage of the fact that the optically aberrating flow field is

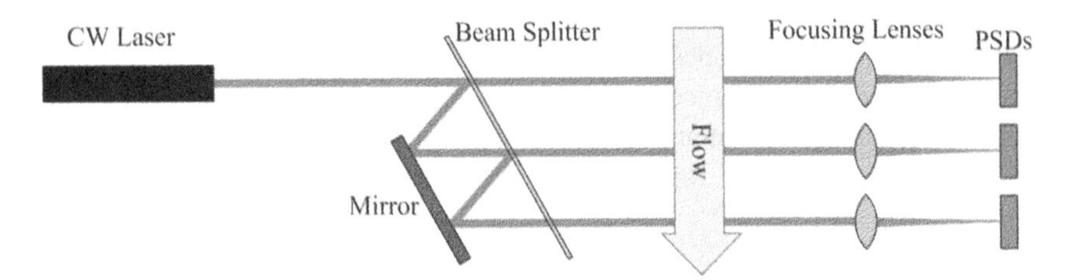

Figure 3.14 Schematic of the Small-Aperture Beam Technique (SABT) sensor.

convecting. In this case, instantaneous deflection angles are used to reconstruct the pseudo-spatial one-dimensional wavefronts in the streamwise direction at several locations, using Equation (3.7). The pseudo-spatial wavefronts between the adjacent locations are blended into one wavefront, using linear weighting functions. Blending all locations results in a spatially temporally resolved $OPD(x,t)$, which is continuous over the region of the measurements. An example of such reconstruction is shown in Figure 6.4 later in this book. It should be kept in mind that, like the Malley probe, the wavefront reconstruction requires knowledge of a convection speed. Various cross-correlation methods, described in Chapter 4, can be used to estimate the convective speed from the deflection angle data.

High sampling frequency of the deflection angles allows the capturing of small spatial features of aero-optical flows. Measurements in multiple beam locations allowed data collection for aero-optical flows, which are inhomogeneous or spatially evolving in the streamwise direction. The set-up requires only several (as few as two) small-aperture laser beams and corresponding inexpensive PSDs to simultaneously measure deflection angles. This reduced requirement on the number of sensors necessary to construct a given optical wavefront reduces the demands on the data-acquisition system, leading to the case where high temporal bandwidth (up to few hundred of kHz) can be achieved. The SABT sensor was instrumental in correctly measuring aero-optical distortions due to transonic shear and boundary layers at a sampling frequency of 75 kHz as early as 1997 (Hugo et al. 1997). In comparison, the fastest commercially available SHWFS at that time could measure the wavefronts at only about a few hundred Hz. Although the SABT is only able to construct wavefronts in the flow direction, it is still a useful sensor to study many flow fields of aero-optical interest, such as the shear layers or boundary layers over large regions where optical propagation would take place.

3.4 Typical Optical Set-Ups

In a typical set-up to measure wavefronts, a laser beam is expanded to a large-aperture beam using expanding lens sets or beam expanders, as shown in Figure 3.15. This set-up is often called a single-path set-up. After transmission through the flow of interest with a series of steering mirrors, the beam is contracted to a size required by the wavefront sensor using a separate set of receiving optics. For the SHWFS, the lenslet size is typically 0.3–0.5 mm with the total size ~20 mm, thus giving 40–60 spatial points in one direction. While straightforward to set-up, the single-path requires two optical benches, one on each side of the flow of interest, and demands a careful alignment of the optical components. The optical benches must be properly secured to

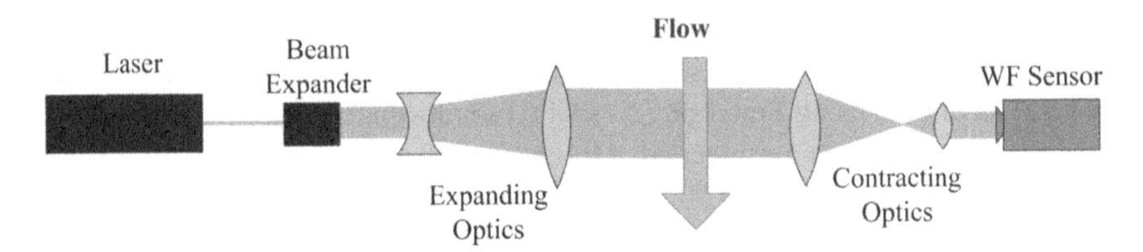

Figure 3.15 Schematic of the single-path optical set-ups.

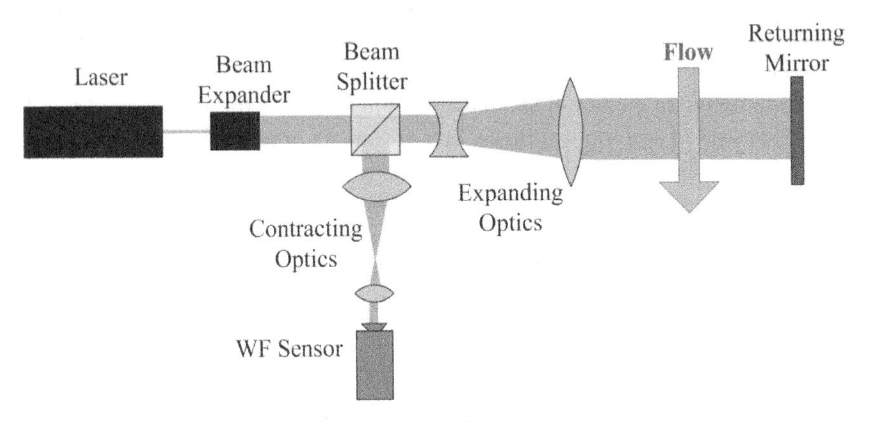

Figure 3.16 Schematic of the double-path optical set-up.

minimize the relative motion and subsequent unwanted beam motion. Finally, if the aperture of the beam passing the flow of interest needs changing, it requires changing both expanding and contracting optical sets, with a follow-up realignment.

If the laser beam can be reflected back using a large flat mirror, an alternative approach is to use a double-path set-up, shown in Figure 3.16. Essentially, this approach simplifies the optical set-up by combining transmitting and receiving optics. It uses the same expanding optics to forward the beam into the flow. After the flow, the return mirror is used to reflect the beam along the same path to return it back onto the optical bench. After being contracted using the same set used for expanding, the beam is split, using a 50/50 beam-splitter. Finally, after an additional pair of contracting lenses, the beam is sent to the wavefront sensor. There are several advantages with the double-path system. Firstly, it requires only one optical table to put all the optical elements on, with possibly an additional mount for the return mirror, greatly reducing the set-up complexity and the alignment problems. Secondly, the beam travels through the turbulent flow twice, doubling the signal-to-noise ratio of the measured aberration. Finally, if the beam size across the flow needs changing, it requires the replacement of only the expanding set of optics, without changing any other components on the table.

This approach was successfully used by the authors in different tunnels to study boundary layers and flow over different turrets, including the ones with the conformal window (Gordeyev et al. 2007). In the last case, a special optical canister was built and installed into the turret. The canister was equipped with a set of lenses and a mirror, which were designed to return a coaxial collimated beam back to the optical bench with minimal distortions. The optical canister consisted of three elements: a positive lens, a negative lens, and a flat mirror, as schematically shown in Figure 3.17(a). The positive lens had an outer radius of curvature matching the curvature of the hemispheric portion of the turret. The custom-made negative lens was mounted on an adjustable platform and was designed to compensate for aspherical aberrations from the positive lens. The collimated beam, passing through the flow, positive and negative lenses, was reflected by the flat mirror back coaxially to the optical table. Thus, the whole optical assembly works equivalent to an optical flat mirror, as demonstrated in Figure 3.17(b), while fluidically it had the same curvature, as the rest of the turret surface.

A few words are needed about the proper positioning of the wavefront sensor. Ideally, it should be placed on the plane, which reimages the region of the flow of interest to eliminate or minimize beam walking. Most of the time, a simple geometric optics analysis is sufficient to predict the location of the re-imaging plane. Recall that the thin lens equation relates the object location, L_o, and its image, L_i, as,

$$\frac{1}{L_o}+\frac{1}{L_i}=\frac{1}{f}, \tag{3.8}$$

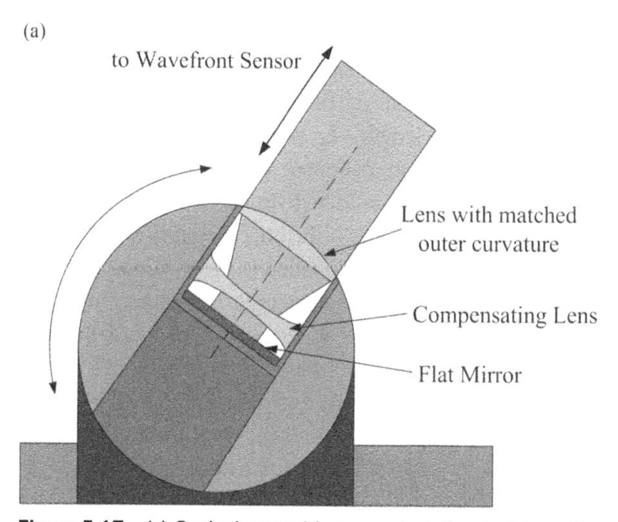

Figure 3.17 (a) Optical assembly to conduct the double-path experiments with a conformal surface turret.

(b)

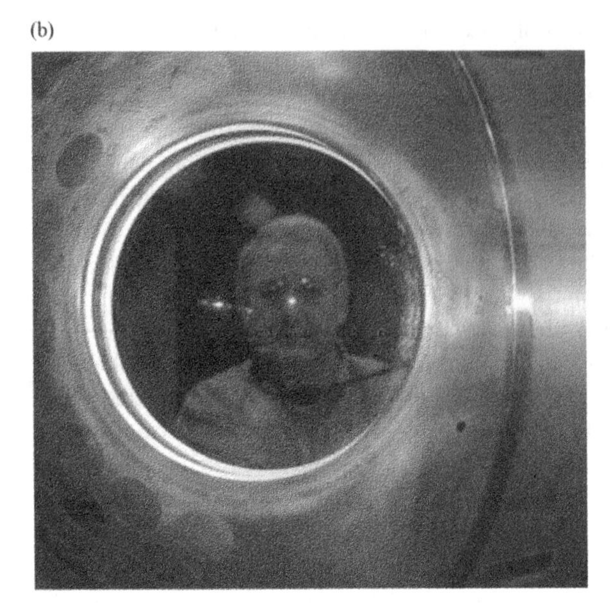

Figure 3.17 (Cont'd) (b) A view through the optical assembly, demonstrating that it works as an optical flat mirror. *Source:* Courtesy of Eric Jumper.

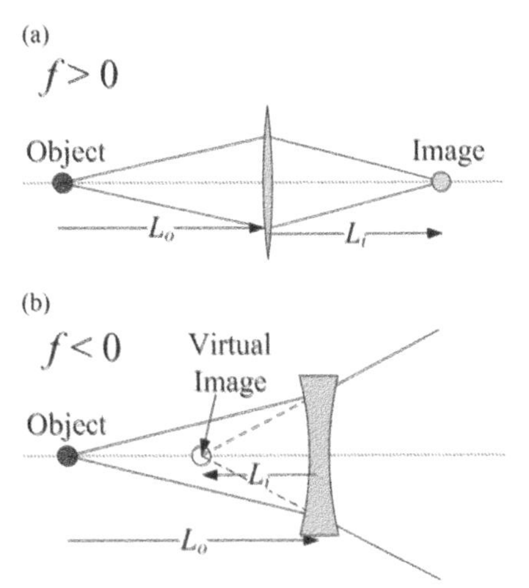

Figure 3.18 Relation between the object and image locations in case of a single lens with (a) a positive and (b) negative focal lengths.

where f is the focal distance of the lens. Focal distance can be positive or negative. For positive focal length, the image will form on the other side of the lens, as shown in Figure 3.18(a), and it can be observed by placing a screen at the location of the image. For negative focal length, L_i is negative, and the image cannot be observed. In this case, the image is considered virtual, as shown in Figure 3.18(b). If the optical system has multiple lenses, Equation (3.8) can be used to consecutively propagate or reimage the object through all lenses and compute the final image location. To physically observe or record the final image, the image cannot be virtual; another word, the image should be located after the final lens.

Often, as shown in Figure 3.15, a series of contracting or expanding optics, called telescopes, are used to reimage the flow region onto the wavefront sensor, and the beam is collimated between the telescopes. To preserve the collimated nature of the beam before and after the telescope, the distance between two optical elements should be equal to the sum of their focal lengths. In this case, it is convenient to compute the relationship between the object and the image locations for this two-lens telescope, schematically shown in Figure 3.19. Let us say that the first lens has the focal distance of f_1 and the second lens has the focal distance of f_2. Note that these focal distances can be positive or negative. If the distance from the object to the first (closest to the object) lens in the telescope is L_1, the distance to the image location, L_2, *from the second lens* is given by

$$L_2 = f_2 \cdot \frac{M+1-(L_1/f_1)}{M}, \text{ where } M = \frac{f_1}{f_2}. \tag{3.9}$$

In order to have an observable image, L_2 should be positive, and the wavefront sensor should be placed at this image location. If more than one telescope is used in the set-up, like in the set-up in Figure 3.16, the corresponding images through each telescope can be consecutively computed using Equation (3.9).

In a double-path set-up, since the beam goes through the flow twice, there are two objects and two corresponding re-imaging planes. As a compromise, the return mirror is reimaged onto the sensor. Another benefit of re-imaging the mirror is that any potential vibration of the return mirror will not result in beam walk on the wavefront sensor.

The size of the laser beam in the flow is usually dictated by the experimentalist and depends on the desired cross-sectional size of the interrogating region. The laser beam is then forwarded to the wavefront sensor, which generally requires the laser beam to be of a particular size. Thus, in general, the laser beam in the flow is different from the beam size at the sensor. This is shown schematically in

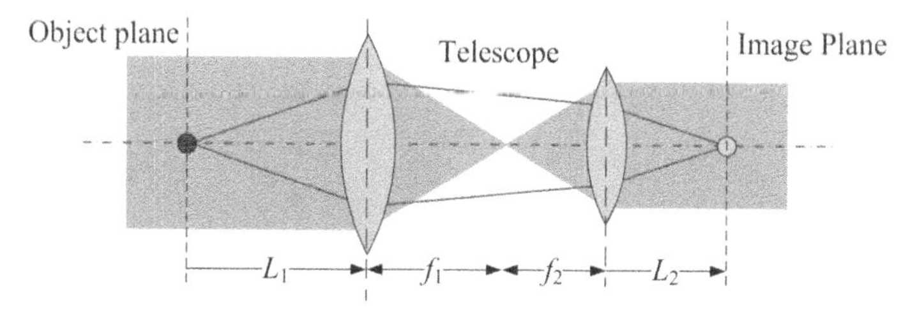

Figure 3.19 Object and image in the case of the telescope, which preserves the collimating nature of the beam.

Figure 3.20, where a simple pair of optics is used to contract the laser beam from diameter D in the flow space to the diameter, d, in the sensor space. The ratio between the beam size in the flow space (sometimes called an output space) in the sensor space (sometimes called an input space) is called a magnification ratio, $Mag = D / d$. The amplitude of the wavefront remains the same through the optical system, while its aperture size on the sensor is reduced by the Mag-ratio. Therefore, all local deflection angles on the sensor (input space) are *magnified* by the magnification ratio, hence the name. For gradient-type wavefront sensors, like the SHWFS, it would mean the increase in the sensor sensitivity, as the increased deflection angles would result in larger dot displacements on the sensor. Thus, one way to increase the sensitivity of the SHWFS is to increase Mag-ratio by having a large beam in the flow and contracting it to a small beam on the sensor. One potential issue with this approach is that the spatial resolution of the resulting wavefronts will also increase as the spacing between the measured deflection angles in *the flow (output) space* is equal to the lenslet pitch, multiplied by the Mag-ratio. Another issue with contracting the beam on the sensor is the increased sub-aperture averaging effect. While the lenslet pitch, A_{sub}, is fixed, the corresponding spatial region in the flow space is Mag-times larger, $Mag \cdot A_{sub}$. As it will be discussed in more detail in Chapter 7, due to spatial averaging over the larger spatial region, information about structure sizes below the size of this spatial region will be essentially lost. Another word, the spatial averaging works as a low-pass filter, eliminating any spatial structures, which are smaller than $Mag \cdot A_{sub}$. For large Mag-ratios, it might result in significant underestimating of optical distortions at small scales, and this effect should be recognized when planning optical set-ups.

In real experiments, both factors should be considered when designing the optical set-up. As a final comment, the magnification effect should be kept in mind, if only the deflection angles are used in the analysis, and the measured

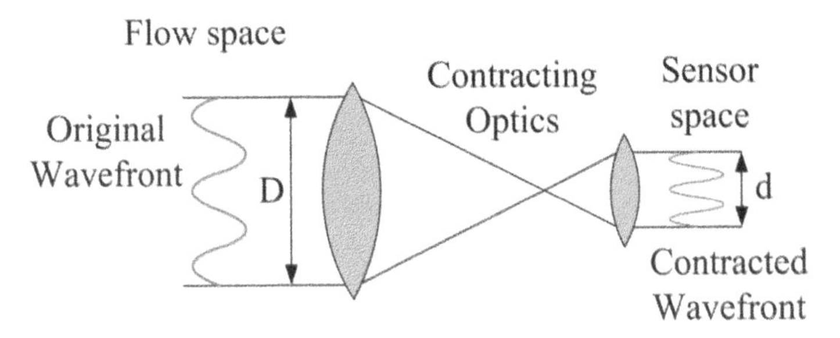

Figure 3.20 Wavefront, reimaged from the flow space into sensor space, using contracting optics, has larger deflection angles in the sensor space, compared to the angles in the flow space.

deflection angles should be *divided* by the *Mag*-ratio to restore true deflection angles in the flow.

In tunnel tests, one or more regions of aero-optical flows are often present along the laser beam. Examples include the boundary layers on the tunnel walls or aero-optical structures present in an open-jet flow. Thus, the measured *OPD* becomes a linear superposition of all flows,

$$OPD_{total}(x,y,t) = OPD_1(x,y,t) + OPD_2(x,y,t) + \dots$$

While it is not trivial to separate time-resolved wavefronts for each flow, it is still possible to estimate the level of the aero-optical distortions for the flow of interest. Assuming that all regions are statistically independent from each other, the overall level of aero-optical distortions, expressed as spatially temporally averaged $OPD^2_{total,rms}$, simply becomes a linear sum of individual OPD^2_{rms},

$$\begin{aligned}
OPD^2_{total,rms} &\equiv \overline{\left\langle \left(OPD_1(x,y,t) + OPD_2(x,y,t) + \dots\right)^2 \right\rangle_{(x,y)}} \\
&= \overline{\left\langle \left(OPD_1^2(x,y,t) + OPD_2^2(x,y,t) + 2OPD_1(x,y,t)OPD_2(x,y,t) + \dots\right)^2 \right\rangle_{(x,y)}} \\
&= \overline{\left\langle \left(OPD_1^2(x,y,t)\right)^2 \right\rangle_{(x,y)}} + \overline{\left\langle \left(OPD_2^2(x,y,t)\right)^2 \right\rangle_{(x,y)}} + 2\overline{\left\langle OPD_1(x,y,t)OPD_2(x,y,t) \right\rangle_{(x,y)}} + \dots \\
&= OPD^2_{1,rms} + OPD^2_{2,rms} + \dots
\end{aligned}$$

For instance, if the beam passes through two flows with the same OPD_{rms}, the resultant $OPD_{total,rms} = \sqrt{2}OPD_{rms}$. This equation was used to extract statistics of aero-optical distortion due to a single boundary layer, using the test section with two statistically similar boundary layers present on the opposite tunnel walls (Gordeyev et al. 2014). If the aero-optical distortions for one flow is known, it can be subtracted from the total levels of $OPD_{total,rms}$ to get an estimate of OPD_{rms} for the flow of interest. This approach was used, for instance, to study the aero-optical effects from a heated/cooled boundary layer present on one tunnel wall, with the canonical boundary layer present along the beam on the opposite tunnel wall (Gordeyev et al. 2015), discussed in more detail in Chapter 6.

4

Data Reduction and Interpretation

With the emergence of high-speed wavefront sensors, long time sequences of wavefronts are typically collected over a two-dimensional aperture at several hundred spatial points, resulting in a large amount of data, $OPD(x, y, t)$. In addition, some parameters, like the flow speed, viewing angle, etc. are commonly varied and the wavefronts are measured under different conditions. The question becomes: how to efficiently analyze these large datasets? The answer depends on the application. From the point of view of an optical-system engineer, the main objective is to quantify the changes in the far-field intensity pattern, resulting from the aberrated wavefronts. The typical far-field quantities are the instantaneous and time-averaged intensities at the target, point-spread functions (PSF) and the statistics of overall jitter imposed on the beam. The connection between the wavefront statistics and the far-field patterns were discussed earlier in Chapter 2. If detailed information about the far-field pattern is required, one needs to use the Fourier-optics equation, Equation (2.26). However, most of the time the information about the instantaneous on-axis intensity is sufficient to estimate the aero-optical effects on the target by using the Large-Aperture Approximation, Equation (2.30). In this case, only spatial root-mean-squared values, $OPD_{rms}(t)$, are needed to be extracted from the wavefront analysis. Time series of $OPD_{rms}(t)$ then can be analyzed to estimate the spectral content of the far-field intensity. In addition, probability density distributions can be extracted to be used in conjunction with Equation (2.33).

From a fluid dynamist point of view, however, the wavefronts are integrals of the density field. It is important to emphasize that the wavefronts depend only on the density field. Thus, there is a lot of information about the turbulent flow "imprinted" on the wavefronts. This information includes some important flow parameters, like local flow thickness and convective speed, turbulence levels, sizes of the dominant flow structures and their temporal content, the state of the flow (laminar, transient, or turbulent) etc. In this case, a more detailed analysis of

Aero-Optical Effects: Physics, Analysis and Mitigation, First Edition. Stanislav Gordeyev, Eric J. Jumper, and Matthew R. Whiteley.

the wavefronts, such as cross-correlations, dispersion, and modal analyses are needed to extract this type of information from the wavefronts.

Data analysis serves two important purposes. The first one, often called post-processing analysis, will be discussed later in this chapter. But another, equally important purpose, called preprocessing analysis, is to check the quality of the collected data. Various contaminating factors, such as mechanical vibrations of optical components on the optical set-up, non-ideal optics, and issues with the laser source will affect the collected data. Some of these issues can be easily spotted by inspecting the raw images with the dots from a Shack-Hartmann wavefront sensor, while others are difficult to recognize right away. The preprocessing analysis is obviously a sensor-dependent one, so it is difficult to come up with a generic preprocess analysis approach. Thus, in this chapter we will be mostly dealing with various post-process analysis techniques, with only a few examples of the preprocessed analysis.

In this Chapter, examples of wavefront data collected by sensors described in Chapter 3 are used to present some of the common methods that have been adapted and used in the aero-optical community over the last two decades. Each of the examples will make use of specific approaches and principles developed to overcome some of the contamination that is always present in the data. The main techniques can be broadly split into several categories.

- *Statistical Analysis*: Includes spatial and temporal averaging and probability density functions.
- *Spectral Analysis*: Fourier-based analysis of temporal, spatial, and spatio-temporal spectra. It also includes a dispersion analysis.
- *Modal Analysis*: Includes Zernike decomposition, Proper Orthogonal Decomposition (POD), and Dynamic Mode Decomposition (DMD).

Some of the analyses mentioned involve the combination of several techniques. For instance, Dynamic Mode Decomposition uses a Fourier-based approach to compute the modes.

As a final comment, it is assumed that the steady-lensing, and the instantaneous piston and tip/tilt components are always removed from the wavefront data before the post-process analysis (see Equation (2.35)), so only the higher-order components of the wavefront sequence are analyzed.

4.1 Statistical Analysis

4.1.1 Temporal and Spatial OPD_{rms}

As mentioned before, one would first want to characterize the overall level of wavefront aberrations over the aperture for each time frame or snapshot. A traditional way to represent aero-optical distortions over the aperture at different viewing angles is to determine time-dependent *spatial variation* of the OPD across the aperture, often called a temporal evolution of $OPD_{rms}(t)$, introduced earlier in Equation (2.15). Time evolution of $OPD_{rms}(t)$ quantifies the aberrations present in the beam at a given moment and can be used to estimate the instantaneous far-field intensity at the target, using Equation (2.30).

Another main quantity of interest used to characterize the aero-optical distortions is a time-averaged value of the time-dependent $OPD_{rms}(t)$, $OPD_{rms} = \overline{OPD_{rms}(t)}$. Recall from Chapter 2 that if this quantity is known, the time-averaged far-field effects can be estimated using the Large-Aperture Approximation, given in Equation (2.32). For this reason, OPD_{rms} is the main metric for aero-optical effects. Dependence of OPD_{rms} on flow parameters, as well as some semiempirical models for several fundamental flows, like turbulent boundary layers, shear layers, and flows around turrets will be presented in Chapter 6.

Another way to quantify aero-optical distortions is to compute the spatial distribution or a spatial map of *temporal variance* of the wavefronts at different points over the turret

$$S(x,y) = \left(\overline{OPD(x,y,t)^2}\right)^{1/2} \tag{4.1}$$

This quantity will be called the spatial map of the OPD_{rms} for brevity. It is important to distinguish between the quantities defined in Equations (2.15) and (4.1). $OPD_{rms}(t)$ describes the *spatial* variation of the aero-optical distortions over an *entire aperture* and it is directly related to the far-field intensity at the target, as shown in Equation (2.30). This quantity depends on the aperture size. The spatial map, $S(x,y)$, describes the *temporal* variation of aero-optical wavefronts at a given *point over the aperture*. It is directly related to the local strength of aero-optical flow features. However, it should be noted that the tip/tilt and piston components, removed from the wavefront before the analysis, are affected by the aperture size, so their removal does indirectly affect $S(x,y)$. Some representative examples of the spatial maps of OPD_{rms} are shown in Figure 4.1. The plot on the left reveals that the strength of aero-optical distortions is approximately the same across the aperture, indicating a homogeneous turbulent flow over the aperture. The example on the right plot, on the other hand, shows that the aero-optical distortions increase in the horizontal (streamwise) direction. This spatial evolution is typical for the spatially growing flows, like a free shear layer. The spatial map is also very useful in identifying the potential problems with the data during the preprocess analysis. Any unexpected spatial variations or sharp local peaks/troughs are usually indicative of issues with either the data collection or the data reduction.

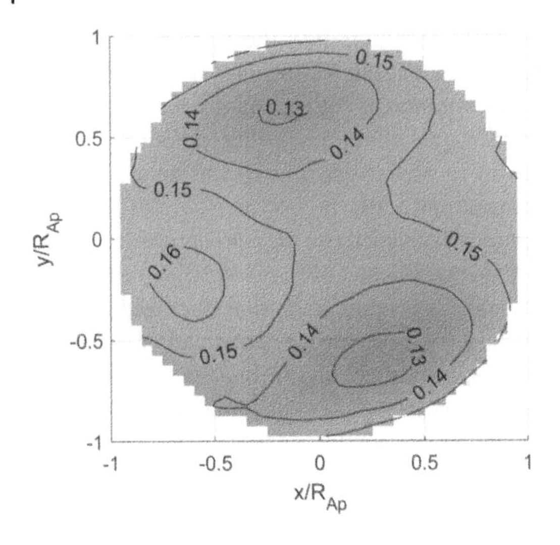

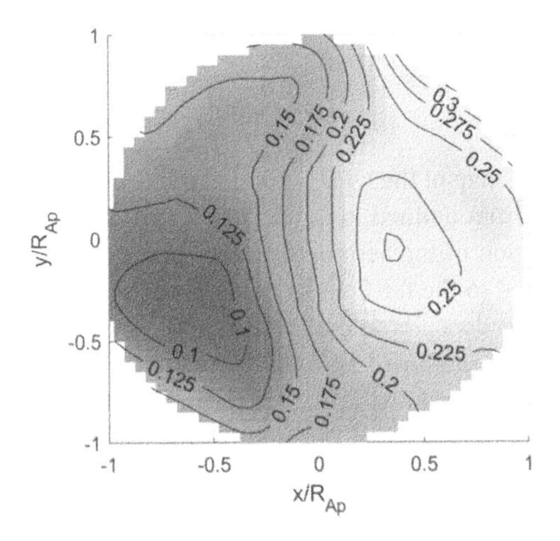

Figure 4.1 Examples of the spatial distribution of OPD_{rms} over the aperture.

4.1.2 Histograms and Higher-Moment Statistics

As discussed before, a time sequence of spatial root-mean-squared values of OPD_{rms}, defined in Equation (2.15), is the main quantity to describe the levels of aero-optical distortions. Since OPD_{rms} is not constant, but changes from one moment to another, it is useful to study the main statistical properties of the OPD_{rms} time series. The most helpful property describing the OPD_{rms} series is a probability density function, $PDF(OPD_{rms})$, also known as a histogram plot. An example of $PDF(OPD_{rms})$ is shown in Figure 4.2. This distribution provides useful

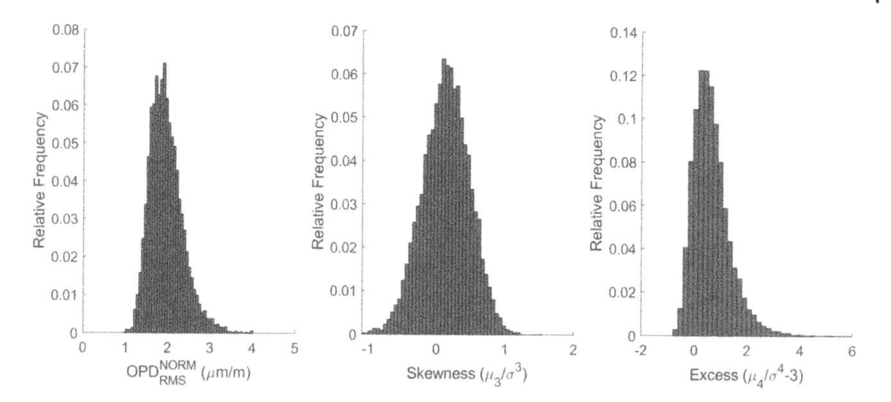

Figure 4.2 Examples of P.D.F. of spatial second, third and fourth order moments of *OPD*. *Source:* Porter et. al. (2011), figure 17 / American Institute of Aeronautics and Astronautics.

information about the range of values of OPD_{rms}, and its minimum, maximum, and time-averaged values. The probability density function can be used to directly estimate the time-averaged intensity using Equation (2.33). In addition, plotting and inspecting $PDF(OPD_{rms})$ becomes another useful diagnostic preprocessing tool. For most physical processes, $PDF(OPD_{rms})$ has a log-normal distribution, which states that the natural logarithm of OPD_{rms} has a normal distribution,

$$PDF(OPD_{rms};m,\Sigma)=\frac{1}{\sqrt{2\pi}X\Sigma}\exp\left(-\frac{(\ln(OPD_{rms})-m)^2}{2\Sigma^2}\right).$$

The distribution has two parameters, $m=\overline{\ln(OPD_{rms})}$ and $\Sigma^2=\overline{\left[\ln(OPD_{rms})\right]}^2$. The time-averaged OPD_{rms} can be computed from these parameters as $\overline{OPD_{rms}}=\overline{OPD_{rms}(t)}=\exp(m+\Sigma^2/2)$. Thus, any significant deviation of the measured $PDF(OPD_{rms})$ from the expected one usually indicates some issues with the data collection and/or processing. Some examples of these issues are long thick tails in PDF, presence of additional spikes, etc. All of these indicate that there might be a problem with the collected data.

From an optical system point of view, the primary aero-optic effect is the reduction of the far-field intensity delivered to the target as a result of these aberrations, typically quantified by the instantaneous and time-averaged normalized intensity (Strehl ratios). As mentioned in Chapter 2, if the spatial distribution of the instantaneous wavefronts over the aperture is Gaussian, the instantaneous OPD_{rms} can then be used with Equation (2.31) to predict the instantaneous Strehl ratio *exactly*. The question remains whether the wavefront data resulting from aero-optic aberrations meet this criterion of being Gaussian-distributed in space. For simplicity, let us define a new variable, X, which is the values of OPD over the aperture for *a fixed time*, $X \equiv OPD(x, y; t = fixed)$. Since we assume that the piston mode is removed from each wavefront, the averaged value of X is zero, $\overline{X} = 0$. The second moment of X is σ, where $\sigma^2 \equiv \overline{X^2} - \overline{X}^2 = \overline{X^2}$. Comparing this equation to Equation (2.15) leads to an obvious conclusion that $\sigma = \overline{OPD_{rms}}$. Similarly, higher moments are defined as, $\mu_n \equiv \overline{X^n}, n > 2$. Particularly, the third and the fourth moments, properly normalized by σ, are called the skewness γ_1, and the excess γ_2,

$$\gamma_1 = \frac{\mu_3}{\sigma^3} = \frac{\overline{X^3}}{\sigma^3}, \quad \gamma_2 = \frac{\mu_4}{\sigma^4}\text{-}3 = \frac{\overline{X^4}}{\sigma^4} - 3.$$

For Gaussian-distributed wavefronts, these quantities should be zero, $\gamma_1 = \gamma_2 = 0$. Thus, any deviations from zero indicates the degree of "non-Gaussian" spatial distribution of the wavefronts. Now let us recall that these moments are computed for every time frame. Therefore, similar to $OPD_{rms}(t)$-series, they also form the time sequences and have corresponding probability density functions, with examples shown in Figure 4.2. As seen in this figure, for this example, a large number of the instantaneous wavefronts do not meet the requirement that skewness and excess are zero, indicating a strong non-Gaussian process.

4.2 Spectral Analysis

While the statistical analysis presented in the previous section uses the time-resolved wavefronts, the main focus was on the spatial statistics of the wavefronts. The advantage of this approach is that it does not depend on the sampling frequency of the wavefront sensor. As long as the wavefront sequence is sufficiently long to ensure a convergence of the time-averaging operator, statistical methods will provide useful information. A drawback of the statistical methods is that it is difficult to identify and remove various contaminations, like mechanical vibration of mirrors or other optical components. Another issue is that with low sampling rates, any temporal information about the aero-optical distortions is lost.

On the other hand, if the wavefronts are sampled with sufficiently high sampling rates, they can be analyzed using a Fourier transformation (FT), the so-called spectral methods. FT is a well-established technique and can be computed efficiently using a Fast Fourier Transform algorithm. FT allows one to compute the spectral (frequency) distribution of the energy of the signal. FT can be performed in time (temporal spectra), space (spatial spectra) or in a space–time domain (spatiotemporal spectra). Detailed information about a proper implementation of FT can be found in Bendat and Piersol (2010).

One way to quantify the temporal behavior of wavefronts is to study the wavefront temporal spectra, $S(f;x,y)$, at every spatial point, (x,y), over the aperture. As a typical wavefront has hundreds or even thousands of spatial points, this approach is not practical. The alternative approach is to average individual spectra over all spatial points to get the aperture-averaged wavefront spectrum,

$$S(f) = \langle S(f;x,y) \rangle_{\text{over aperture}}, \text{ where}$$

$$S(f;x,y) = \frac{\left|\widehat{OPD}(f;x,y)\right|^2}{T}, \text{ and } \widehat{OPD}(f;x,y) = \int_0^T OPD(t,x,y) e^{-2\pi i f t} dt. \tag{4.2}$$

Here T defines the duration of the wavefront sequence, and the square brackets denote the spatial averaging over all spatial points. The aperture-averaged spectrum shows the spectral features, which are common to all spatial points.

The analysis of the aperture-averaged spectra is especially useful for studying parametric changes of the aero-optical environment. Figure 4.3 shows an example of such analysis, applied to the transonic aero-optical environment around a turret in-flight at different viewing angles. The viewing angle, α, is the angle between the incoming flow velocity and the direction of the outgoing laser beam, as schematically shown in Figure 6.29 in Chapter 6. The frequency is converted into

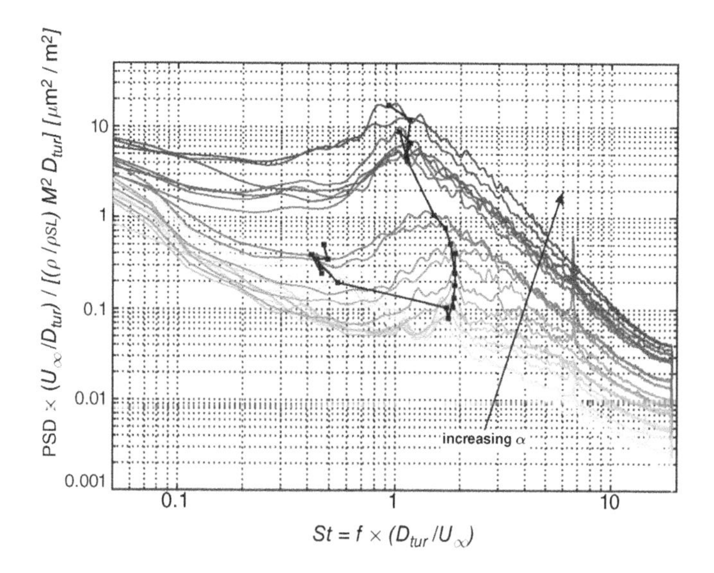

Figure 4.3 Aperture-averaged wavefront spectra, $S(f)$, plotted as a function of the normalized frequency, $St = fD_{turr} / U_\infty$ of AAOL flight test at Mach 0.65 as a function of the viewing angle, α. *Source:* Goorskey et. al. (2013), figure 4/ with permission of SPIE.

a Strouhal number, based on the turret diameter, $St = fD_{turr} / U_{\infty}$. The main peak of $St \sim 1$ in the spectra at downstream viewing angles, $\alpha > 95$ degrees, is associated with the shedding of the shear-layer-like vortical structures from top of the turret; these shear-layer structures create significant aero-optical effects, evidenced in Figure 6.29 by large spectral values; see Chapter 6 for more discussion. The frequency of the shedding decreases with the increasing viewing angle and the overall spectral energy increases; both effects are consistent with the spatial growth of the shear layer farther downstream of the turret. At viewing angles $\alpha \sim$ 90 degrees, the aero-optical distortions are dominated by a local unsteady shock, present at this speed over the aperture. The shock dynamics are coupled with the wake dynamics, resulting in a lower normalized frequency of $St \sim 0.4$. More discussion about the aero-optical distortions around the turrets can be found in Chapter 6.

4.2.1 Relation between the Deflection Angle Spectrum and the Wavefront Statistics

As discussed in Chapter 3, in most cases, a reasonably homogeneous turbulent flow (boundary layer, for instance) convects over the aperture and the frozen field assumption can be used to relate the local deflection angles and wavefront statistics. Let us assume that the flow convects along the x-direction. In this case, the streamwise component of local deflection angle at a fixed location, x_0, is related to the local wavefront as

$$\theta_x(x_0,t) = -\frac{1}{U_C}\frac{dOPD(x_0,t)}{dt},$$

where U_C is the convective speed of the turbulent structures. Applying a Fourier Transform to the above equation yields

$$\hat{\theta}_x(f) = -\frac{2\pi i f}{U_C}\widehat{OPD}(f),$$

which provides a useful relation between the auto-spectral density functions of the wavefront at a fixed point, $S_{OPD}(f)$, and the local streamwise deflection angle, $S_\theta(f)$,

$$S_{OPD}(f) = \left(\frac{U_C}{2\pi f}\right)^2 S_\theta(f). \qquad (4.3)$$

In many cases, the spectral form is more advantageous, as it allows various filters to be applied for removal of vibrations, electronic noise, etc. After proper filtering, the overall level of aero-optical distortions, OPD_{rms}, can then be calculated from the deflection angle spectrum by integrating over frequency domain,

$$OPD_{rms}^2 = 2\int_0^\infty S_{OPD}(f)df = 2U_c^2 \int_0^\infty \frac{S_\theta(f)}{(2\pi f)^2} df. \tag{4.4}$$

4.2.2 Dispersion Analysis

As aero-optical aberrations often convect with the flow, it is useful to decompose them into several traveling waves, using 3-dimensional Fourier Transform,

$$OPD(x,y,t) = \frac{1}{(2\pi)^3} \int_{-\infty}^{\infty}\int_{-\infty}^{\infty}\int_{-\infty}^{\infty} \widehat{OPD}(f,k_x,k_y) e^{i(\omega t - k_x x - k_y y)} d\omega dk_x dk_y,$$

where k_x and k_y are the spatial frequencies in the x-direction (streamwise) and y-direction (cross-stream), respectively, and $\omega = 2\pi f$ is the temporal angular frequency. From here one can compute the phase velocities in each direction, $U_{C,x}$ and $U_{C,y}$ via the dispersion relation,

$$\omega(k_x) = k_x U_{C,x}(k_x), \quad \omega(k_y) = k_y U_{C,y}(k_y).$$

To find the streamwise convective velocity, one can transform a three-dimensional wavefront data set, $OPD(x,y,t)$ into the spatial wavenumber-frequency domain using a three-dimensional Fourier Transform and compute a three-dimensional auto-spectral density function or a power spectrum, $S_{OPD}(f,k_x,k_y)$,

$$\widehat{OPD}(f,k_x,k_y) = \int_{-\infty}^{\infty}\int_{-\infty}^{\infty}\int_{-\infty}^{\infty} OPD(t,x,y) e^{-i(2\pi ft - k_x x - k_y y)}\, dtdxdy$$
$$S_{OPD}(f,k_x,k_y) = \frac{\left|\widehat{OPD}(f,k_x,k_y)\right|^2}{T \cdot L^2}. \tag{4.5}$$

Here L is the size of the aperture in one direction and T is the sampling time.

Often, the structures convect in the streamwise direction, so only 1-dimensional "slices" of the wavefront data in the streamwise direction, $OPD(x, y = const, t)$ can be used to compute a two-dimensional power spectrum, $S_{OPD}(f,k_x)$.The traveling structures correspond to lines $2\pi ft - k_x x = const$. Consequently, the traveling structures of various frequencies would appear in the 2-dimensional spectrum as branches with constant slopes. The slope indicates the value and more importantly, the sign of the convective velocity as $U_C = 2\pi f / k_x$.

An example of the dispersion analysis for the boundary layer data (Gordeyev et al. 2014) is shown in Figure 4.4. Here the 2-D spectrum is represented as a function of the normalized frequency $f\delta / U_\infty$ and the normalized spatial wavenumber, $k_x\delta / 2\pi$, where δ is the boundary layer thickness. The lower "branch" of the dispersion curve maxima corresponds to the modes moving downstream and it is related to aero-optical effects of boundary layers. For the range of aero-optically active maxima, the branch, that is the "crest" in the temporal–spatial wavefront spectrum, is linear, confirming that the frozen flow hypothesis is valid for aero-optical structures in the boundary layer. The convective speed is 0.83 of the freestream speed agrees with the measurements of 0.82 of the freestream speed from the aero-optical distortions for the subsonic boundary layer, using the Malley probe (Gordeyev et al. 2014). More information on the aero-optical properties of the boundary layers is provided in Chapter 6. A small amount of the spectral aliasing, visible in the upper right corner in Figure 4.4, suggests that the temporal sampling frequency, which was 50 kHz for this set of data, would be barely enough to properly resolve the wavefront spectra in frequency.

In addition to the lower branch, two more branches can be seen in Figure 4.4. A horizontal, $f = 0$, branch corresponds to a stationary aero-optical structure. An additional upper branch was found to correspond to modes convecting upstream in the flow that are caused by acoustic noise propagating upstream from the fan motor into the wind tunnel test section, as it has a velocity of approximately $-c + U_C$. Another indication that this branch represents low-frequency acoustic

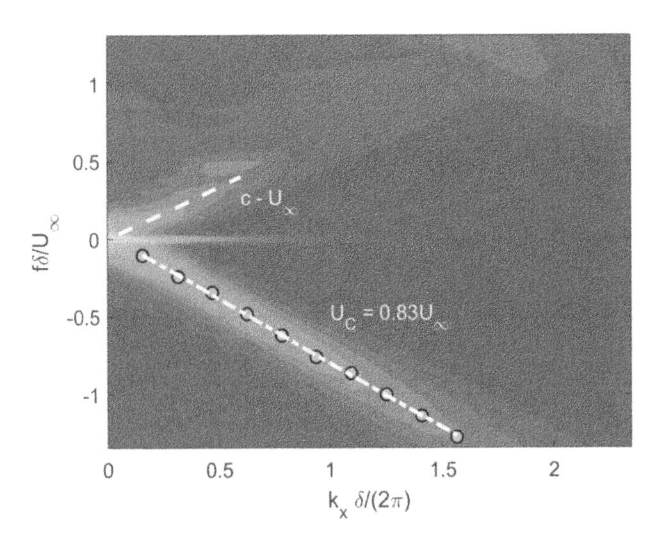

Figure 4.4 Wavefront spatial wavenumber-frequency spectra, $S_{OPD}(f, k_x)$ for $M = 0.6$ boundary layer data. *Source:* Gordeyev et. al. (2014, reproduced with permission of Cambridge University Press.

effects is that it is dominant only in the low frequency end of the spectra, consistent with the low-frequency noise emitted by the tunnel motor. In contrast, the lower branch of dispersion curve maxima shows a broadband family of downstream-propagating aero-optical structures corresponding to the turbulent boundary layer.

The dispersion analysis can be used to separate the wavefronts into the corresponding upstream and downstream moving components by performing the inverse Fourier transform using either the downstream convecting ($f > 0, k_x > 0$) harmonics, or the upstream traveling harmonics, ($f > 0, k_x < 0$) of the Fourier transform,

$$OPD(t,x,y) = OPD^{Down}(t,x,y) + OPD^{Up}(t,x,y)$$

$$OPD^{Down}(t,x,y) = \frac{2}{(2\pi)^2}\mathrm{Re}\left\{\int_0^{\infty}\int_0^{\infty}\int_{-\infty}^{\infty}\widehat{OPD}\,(f,k_x,k_y)e^{i(2\pi ft-k_x x-k_y y)}dfdk_x dk_y\right\}$$

$$OPD^{Up}(t,x,y) = \frac{2}{(2\pi)^2}\mathrm{Re}\left\{\int_{-\infty}^{0}\int_0^{\infty}\int_{-\infty}^{\infty}\widehat{OPD}\,(f,k_x,k_y)e^{i(2\pi ft-k_x x-k_y y)}dfdk_x dk_y\right\}.$$

After separating the wavefronts into downstream and upstream traveling components, they can be further analyzed, using other techniques. The dispersion analysis is a very useful technique, if the wavefronts are significantly contaminated by the acoustic noise, often present in the tunnels or in flight from engine noise propagating upstream through the beam (Gordeyev and Kalensky 2020).

When the aero-optical distortion convects over several spatial measurement points, it creates multiple versions of the temporal evolution at these points. This redundancy allows bypassing the Nyquist criterion and reconstructing the temporal spectrum of aero-optical distortions above the Nyquist frequency from the dispersion spectrum, using a stacking method (Lynch et al. 2021; Sontag 2022). The dispersion analysis was also found to be a more robust technique to extract a convective speed of the aero-optical structures in case of temporally undersampled data, compared to the two-point cross-correlation analysis (Sontag 2022).

4.3 Modal Analysis

The spectral analysis decomposes the signal into a sum of spatially periodic harmonics, multiplied by the coefficients. While it might be a useful mathematical reconstruction to study the energy distribution among harmonics (spectra), periodic functions with infinite span might not be well suited to study the spatio-temporal signal, like wavefronts, with finite structure sizes. Alternative approach is to identify and/or extract physically important spatial features, or modes, and represent spatio-temporal wavefronts as an infinite sum of spatial modes, multiplied by temporal coefficients,

$$W(x,y,t)=\sum_n a_n(t)\psi_n(x,y) \tag{4.6}$$

This so-called modal analysis allows separating and independently analyzing spatial and temporal aspects of the distortions and shedding some light on the underlying spatial features and dynamics of the wavefronts.

Often, the spatial modes are selected or defined in such a way that they are orthogonal to each other,

$$\iint_{Aperture} \psi_n(x,y)\psi_m(x,y)dxdy = C_n\delta_{nm}, \text{ where } \delta_{nm} = \begin{cases} =1, n=m \\ =0, n\neq m \end{cases}$$
$$C_n = \iint_{Aperture} \{\psi_n(x,y)\}^2 dxdy \tag{4.7}$$

and the integration is performed over the entire aperture. If $C_n = 1$, the set of modes is called an orthonormal set. If the modes are orthogonal to each other, a-coefficients can be computed directly using Equation (4.6) and the orthogonality property, Equation (4.7),

$$\iint_{Aperture} W(x,y,t)\psi_n(x,y)dxdy = \iint_{Aperture} \left\{\sum_m a_m(t)\psi_m(x,y)\right\}\psi_n(x,y)dxdy =$$
$$\sum_m \left\{a_m(t) \iint_{Aperture} \psi_m(x,y)\psi_n(x,y)dxdy\right\} = \sum_m a_m(t)C_n\delta_{nm} = a_n(t)C_n \tag{4.8}$$

or

$$a_n(t) = \iint_{Aperture} W(x,y,t)\psi_n(x,y)dxdy \,/\, C_n$$

The spatial modes can be either prescribed a priori or computed from the dataset. Below we will consider both cases: Zernike expansion, when the modes are defined analytically, and POD and DMD analysis, when modes depend on the given dataset.

4.3.1 Zernike Functions

Zernike functions or modes, introduced by Fritz Zernike in 1934 (Zernike 1934) form a complete set of two-dimensional functions, $Z_n^m(r,\varphi)$, $n \geq m$, which are orthogonal over a unit circle (Born and Wolf 1999; Mahajan 2011). Mathematically, they are defined as

$Z_n^m(r,\varphi) = R_n^m(r)\cos(m\varphi)$ - even Zernike function

$Z_n^{-m}(r,\varphi) = R_n^m(r)\sin(m\varphi)$ - odd Zernike function

where Zernike polynomial is defined as

$$R_n^m(r) = \begin{cases} \sum_{k=0}^{[n-m]/2} \frac{(-1)^k (n-k)!}{k!([n+m]/2-k)!([n-m]/2-k)!} r^{n-2k}, (n-m) - even \\ 0, (n-m) - odd \end{cases}$$

Orthogonality property

$$\int_0^1 \int_0^{2\pi} Z_n^m(r,\varphi) Z_{n'}^{m'}(r,\varphi) r dr d\varphi = \frac{\varepsilon_m \pi}{2n+2} \delta_{n,n'} \delta_{m,m'}, \quad \varepsilon_m = \begin{cases} 2, m=0 \\ 1, m \neq 0 \end{cases} \tag{4.9}$$

The first few $R_n^m(r)$:

$R_0^0(r) = 1$

$R_1^1(r) = r;\quad R_2^0(r) = 2r^2 - 1;\quad R_2^2(r) = r^2$

$R_3^1(r) = 3r^3 - 2r;\quad R_3^3(r) = r^3,$

$R_4^0(r) = 6r^4 - 6r^2 + 1;\quad R_4^2(r) = 4r^4 - 3r^2;\quad R_4^4(r) = r^4$

Several Zernike functions for $n \le 4$ are plotted in Figure 4.5, and the corresponding far-field images are given in Figure 4.6. All Zernike functions with $m = 0$ are axi-symmetric. $Z_0^0(r,\varphi)$ represent the constant or piston mode and, as was demonstrated earlier in Equation (2.36), it does not affect the far-field image. $Z_1^{-1}(r,\varphi)$ and $Z_1^1(r,\varphi)$

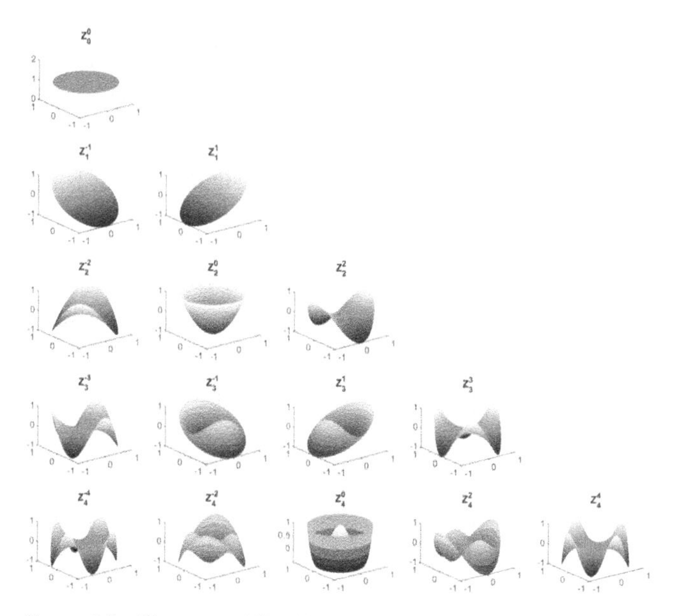

Figure 4.5 First several Zernike polynomials.

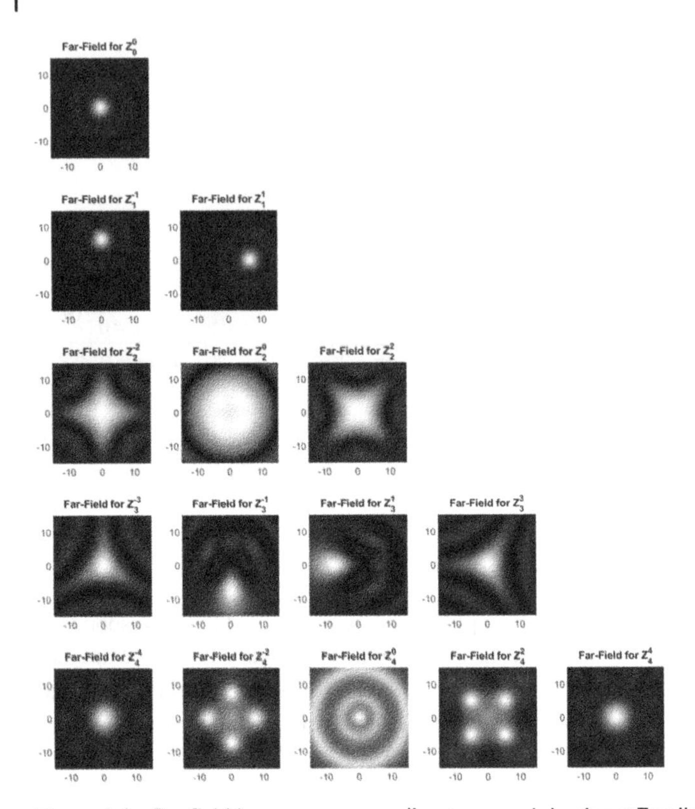

Figure 4.6 Far-field images corresponding to several dominant Zernike polynomials, computed using Equation (2.26), for $\lambda = 1$ micron.

describe the tip/tilt modes in Equation (2.35) and result in the images simply being shifted in the far-field in either the horizontal or vertical directions, as shown in Equation (2.36). $Z_2^{-2}(r,\varphi)$ and $Z_2^2(r,\varphi)$ describe amount of primary astigmatism in the wavefront, and $Z_2^0(r,\varphi)$ is related to the beam defocus. $Z_3^{-1}(r,\varphi)$ and $Z_3^1(r,\varphi)$ describe amount of coma in the wavefront, and $Z_4^0(r,\varphi)$addresses aspherical distortions. Other Zernike modes describe various high-order distortions present in the wavefront (Born and Wolf 1999; Mahajan 2011).

Any wavefront over a circular aperture can be decomposed as a series of Zernike modes, $W(r,\varphi) = \sum_{n,m} a_{nm} Z_n^m(r,\varphi)$. For time-varying wavefronts, a-coefficients become time-dependent, $W(r,\varphi,t) = \sum_{n,m} a_{nm}(t) Z_n^m(r,\varphi)$.

Using the orthogonality property in Equation (4.9), a-coefficients can be computed directly as,

$$\int_0^1\int_0^{2\pi} W(r,\varphi)Z_{n'}^{m'}(r,\varphi)rdrd\varphi = \int_0^1\int_0^{2\pi}\sum_{n,m} a_{nm}Z_n^m(r,\varphi)Z_{n'}^{m'}(r,\varphi)rdrd\varphi =$$

$$\sum_{nm} a_{nm}\int_0^1\int_0^{2\pi} Z_n^m(r,\varphi)Z_{n'}^{m'}(r,\varphi)rdrd\varphi = \sum_{nm} a_{nm}\frac{\varepsilon_m\pi}{2n+2}\delta_{n,n'}\delta_{m,m'} = a_{n'm'}\frac{\varepsilon_{m'}\pi}{2n+2}$$

Zernike-based expansion is commonly used in optical design and testing to describe the lens or beam quality. However, by construction they are defined only over circular apertures. While technically they can be used over annular, ring-type apertures, Zernike polynomials over these apertures are non-orthogonal and the computation of the a-coefficients becomes cumbersome. To overcome this issue, different definitions of the Zernike polynomials were introduced (Dai and Mahajan 2007; Huang et al. 2012; Navarro et al. 2014) to form complete and orthogonal set of Zernike modes over annular apertures. For any arbitrary aperture, Zernike modes cannot be defined.

Another case when the Zernike decomposition is ill-suited is when the signal is not spatially isotropic but has some degree of spatial anisotropy. These types of wavefronts are when aero-optical distortions are due to turbulent large-scale structures with preferred spatial orientation present in the flow, like in free shear layers and wakes. To demonstrate this, let us compute Zernike coefficients for a traveling two-dimensional wave defined as,

$$W(x,y,t) = \sin\left(2\pi\left[x/0.75 - t\right]\right) \tag{4.10}$$

The results for the first 60 Zernike modes for several time instances are presented in Figure 4.7. The distribution among the coefficients changes significantly as a function of time. At time $t = 0$, as shown in Figure 4.7(a), only 14 out 60 Zernike modes are needed to decompose the wavefront. But, when the time is changed to $t = 0.0094$, as demonstrated in Figure 4.7(b), 29 out 60 coefficients are non-zero; thus, 15 more Zernike modes must be included to decompose the signal at this time. For $t = 0.1875$, see Figure 4.7(c); the number of essentially non-zero coefficients drops down to 14.

This example demonstrates an important point. While the Zernike modes are still orthogonal and still may be used to describe any signal over the circular aperture, many of these modes might be included into the expansion to describe the signal, making the analysis difficult and time-consuming. From the considered example, it is clear that the Zernike modes defined in the polar frame of reference are not a good choice to describe this simple convecting wavefront. The convecting wave is better defined in a Cartesian frame of reference. Also, while the signal has a well-defined underlying spatial structure, this information about the structure is completely lost, when decomposed into Zernike modes. Thus, while predefining the modes might be convenient and useful for some applications, it is not well-suited to analyze a

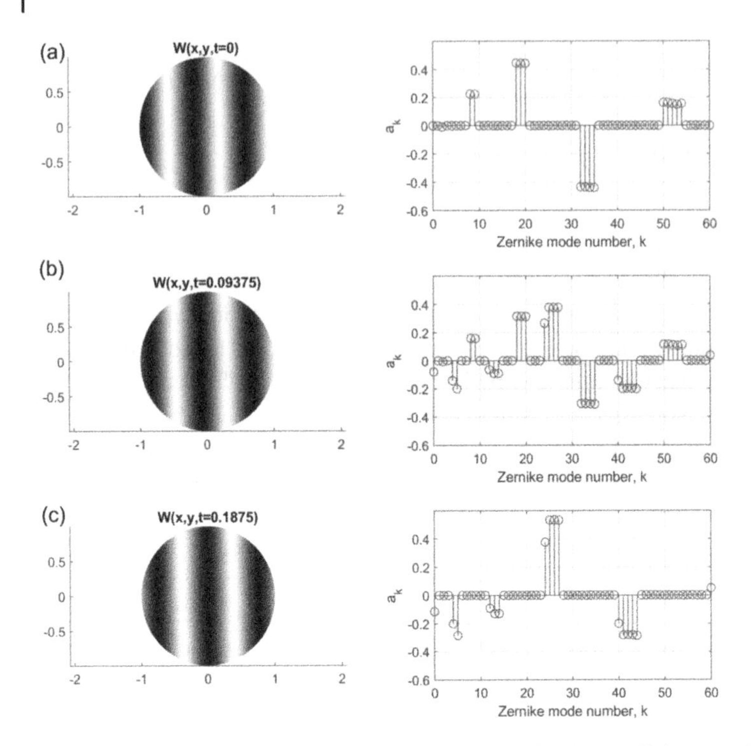

Figure 4.7 Example wavefronts, and corresponding Zernike coefficients at different time instances, (a) $t = 0$, (b) $t = 0.0094$, and (c) $t = 0.1875$.

signal with some structural components present in the turbulent flow. For these types of signals, another modal decomposition, termed POD, is often used.

4.3.2 Proper Orthogonal Decomposition (POD)

Proper Orthogonal Decomposition (POD) was introduced by John Lumley in fluid mechanics in 1970s (Berkooz et al. 1993; Holmes et al. 1996; Lumley 1970) to describe and analyze coherent structures in fluids. Currently it is one of the most popular techniques for performing a modal analysis of multi-dimensional, time-dependent data (Taira et al. 2017). The idea behind POD, also historically known as Karhunen-Loéve or KL expansion in optics, is to describe a given spatio-temporal signal, $W(x,y,t)$, with the *minimum* number of spatial modes, $\phi(x,y)$. One approach is to maximize the following functional in L^2-norm,

$$\frac{\int\limits_{Aperture} W(x,y,t)\phi(x,y)dxdy}{\int\limits_{Aperture} \{\phi(x,y)\}^2 dxdy} = \lambda \geq 0 \rightarrow \max$$

Using the classical methods of the calculus of variations, it is straightforward to demonstrate that the spatial modes, ϕ, are the solutions of the following integral equation,

$$\int R(x,y;x',y')\varphi(x',y')dx'dy' = \lambda\varphi(x,y) \qquad (4.11)$$

where $R(x,y;x',y') = \overline{W(x,y,t)W(x',y',t)}$ is the two-point spatial correlation function of the signal, and the overbar defines the temporal averaging, $\overline{f(t)} = (1/T)\int_T f(t)dt$. The solution of Equation (4.11) forms a complete set of orthogonal functions, $\{\phi_n(x,y)\}$, with associated eigenvalues, λ_n. The original signal can be represented by a series of the orthonormal functions or modes with corresponding temporal coefficients, $a_n(t)$, and the coefficients are uncorrelated with one another in time,

$$\begin{aligned}
&W(x,y,t) = \sum_{n=1}^{\infty} a_n(t)\varphi_n(x,y), \text{ where} \\
&a_n(t) = \int W(x,y,t)\varphi_n(x,y)dxdy \\
&\overline{a_n(t)a_m(t)} = \delta_{nm}\lambda_m \\
&\int_{Aperture} \varphi_n(x,y)\varphi_m(x,y)dxdy = \delta_{nm}
\end{aligned} \qquad (4.12)$$

The eigenvalues, λ_n, are the energy of the various modes, $\lambda_n = \overline{a_n^2(t)}$. It is convenient to introduce the normalized energy of the modes, $\ddot{\lambda}_n = \lambda_n / \sum_n \lambda_n$, which describes the *relative* energy contribution of the modes, and the cumulative energy of the modes, $\sigma_m = \sum_{n=1}^{m} \lambda_n / \sum_n \lambda_n$, which quantifies the relative energy contribution of the first m POD modes.

The cross-correlation function, R, can also be reconstructed from POD modes as

$$R(x,y;x',y') = \sum_{n=1}^{\infty} \lambda_n \varphi_n(x,y)\varphi_n(x',y') \qquad (4.13)$$

From here it follows that the contribution to the correlation function from a particular POD mode comes from spatial points where the mode is essentially non-zero, and the contribution is given by the mode energy. Studying the spatial distribution of dominant POD modes (and their relative energies) provides some information about the level of correlation between different spatial regions over the aperture.

From POD expansion, various statistical properties of the wavefront data can be directly extracted. For instance, the temporal variance of the wavefronts, $S(x,y)$, defined in Equation (4.1), is related to POD modes as,

$$\left[S(x,y)\right]^2 \equiv R(x,y;x,y) = \sum_{n=1}^{\infty} \lambda_n \left\{\varphi_n(x,y)\right\}^2 \tag{4.14}$$

From this equation, the temporal variance of the wavefronts is a linear combination of the (squared) spatial POD modes, and the energies indicate the relative contribution of the POD modes to $S(x,y)$. Thus, by analyzing the spatial distribution of the POD modes (where the mode is mostly non-zero) and the modal energies, various spatial regions of significant wavefront changes can be identified. Combined with the analysis of the spectra of the temporal coefficients, it provides very useful insight into underlying physical processes, contributing to aero-optical distortions. One example of POD analysis, when an unsteady shock is present over the hemispherical turret at transonic speeds will be given later in this section.

Also, POD energies are related to the time-averaged OPD_{rms}, since,

$$OPD_{rms} = \left[\frac{\int_{Aperture} \left\{S(x,y)\right\}^2 dxdy}{ApertureArea}\right]^{1/2} = \left[\frac{\sum_{n=1}^{\infty} \lambda_n \int_{Aperture} \left\{\varphi_n(x,y)\right\}^2 dxdy}{Aperture\ Area}\right]^{1/2}$$

$$= \left[\frac{\sum_{n=1}^{\infty} \lambda_n}{Aperure\ Area}\right]^{1/2}.$$

It is important to point out that to compute POD modes and their energies, only the two-point correlation function, $R(x,y;x',y')$, is needed; see Equation (4.11). Thus, POD analysis can be performed even when only a collection of wavefronts, uncorrelated or sampled at frequencies smaller than typical frequencies of the flow, is given, if the total number of wavefronts is sufficient to guarantee the statistical convergence of time averaging when calculating the R-function.

Since the modes are determined by maximizing the amount of energy, λ, in each mode, the series in Equation (4.12) converges rapidly. This means that it gives rise to an optimal set of basis functions or modes from *all possible* sets. On practical grounds, the expansion Equation (4.12) is usually represented only in terms of a finite or truncated series,

$$W(x,y,t) \approx \sum_{n=1}^{L} a_n(t)\varphi_n(x,y)$$

The optimality (fastest-convergence property) of the POD modes reduces the amount of information required to represent the spatial field to a minimum. This

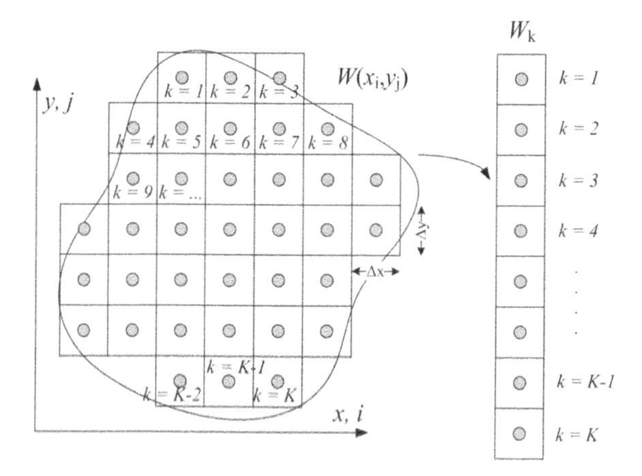

Figure 4.8 Re-numerating the wavefronts in discrete spatial points over an arbitrary aperture.

crucial feature explains the wide usage of POD in the process of analyzing data in the fields of detection, estimation, pattern recognition, and image processing. It is also an efficient tool to store large amounts of data and is particularly beneficial to control systems, with some examples presented in Chapter 8.

Optical wavefronts are often given at a finite number of spatial points over the aperture; see Figure 4.8. It is convenient to convert these points to a single column of data, as shown in Figure 4.8, by sequentially re-numerating the spatial points; symbolically, it can be denoted as, $\mathbf{W} = \{W_k\}$, with the number of spatial points of K. If the wavefronts are given in M discrete time instances, the data can be represented as a large matrix $\mathbf{B} = \{\mathbf{W}(t_1), \mathbf{W}(t_2), \ldots, \mathbf{W}(t_M)\}$ of size $[K \times M]$, where K is a total number of the spatial points. Once the data matrix is formed, POD modes and the coefficients can be computed by applying either the direct or the snapshot methods.

4.3.2.1 Direct Method

By replacing all the integrals in Equations (4.11) and (4.12) with the first-order discrete approximations, the two-point correlation function, R, can be approximated with the discrete correlation matrix, $\mathbf{R} = (\mathbf{B}\mathbf{B}^T)/M$ of the size $[K \times K]$ and the calculation of the POD modes is reduced to solving the eigenvalue problem,

$$\left[(\mathbf{B}\mathbf{B}^T)\Delta x \Delta y / M\right]\varphi_n = \lambda_n \varphi_n, \tag{4.15}$$

where φ_n of the size $[1 \times K]$ are often called eigenvectors.

The second equation, Equation (4.11), can be rewritten in the discrete matrix form as

$$\mathbf{R} = \mathbf{\Phi}\mathbf{\Lambda}\mathbf{\Phi}^T, \tag{4.16}$$

where $\mathbf{\Phi} = \{\varphi_n\}$ is a matrix, containing all eigenvectors as columns and $\mathbf{\Lambda} = diag\{\lambda_n\}$. On the other hand, applying a Singular Value Decomposition (SVD) to the data matrix $\mathbf{B}$ gives, $\mathbf{B} = \mathbf{U\Sigma V}^T$, where $\mathbf{U}$ and $\mathbf{V}$ are called left and right singular vectors and $\mathbf{\Sigma}$ has non-zero values along its diagonal. For more details on SVD, see Taira et al. (2017). Recalling that $\mathbf{R} = (\mathbf{BB}^T)/M = (\mathbf{U\Sigma V}^T)(\mathbf{U\Sigma V}^T)^T//M$. and recognizing that $\mathbf{VV}^T = \mathbf{I}$, we finally obtain

$$\mathbf{R} = (\mathbf{BB}^T)/M = (\mathbf{U\Sigma}^2\mathbf{U}^T)//M. \tag{4.17}$$

Comparing Equation (4.16) with Equation (4.17) shows that the eigenvectors φ_n are (up to a scaling factor of M) simply columns of the left-hand matrix $\mathbf{U}$ and the eigenvalues, λ_n, are squared diagonal values of $\mathbf{\Sigma}$. Once the eigenvectors are known, the temporal coefficients can be computed, as $\mathbf{A} = \mathbf{B}^T\mathbf{\Phi}$, where the temporal coefficients $\mathbf{a} = \{a(t_1), a(t_2), \ldots, a(t_M)\}$ are the columns of $\mathbf{A} = \{\mathbf{a}_1, \mathbf{a}_2, \ldots, \mathbf{a}_M\}$. So, SVD provides a convenient way of computing POD modes and energies, using the `svd`-function in MATLAB, for instance. Also, eigenvectors can be computed by directly solving the eigenvalue problem, Equation 4.15, using the `eig`-function in MATLAB.

4.3.2.2 Snapshot Method

In the case of a large number of the spatial points $K \gg 1$, the size of the second-order correlation matrix **R1** becomes very large, $[K \times K]$, and the direct method of finding the eigenmodes becomes computationally memory and time consuming. Sirovich (1987) pointed out that the temporal correlation matrix, $\mathbf{C}$, will yield the same dominant spatial modes, while often giving rise to a much smaller and computationally more tractable eigenproblem – the method of snapshots. Mathematically, for the discrete data, $\mathbf{B}$, instead of finding a spatial two-point correlation matrix, $\mathbf{R}$, and solving Equation (4.15) with $[K \times K]$-matrix, where K is a number of spatial points, one can compute a temporal correlation $[M \times M]$ -matrix $\mathbf{C} = \{C_{mn}\}$ over *spatial* averaging,

$$C_{mn} = \frac{1}{M} \int_{Aperture} W(x, y, t_m) W(x, y, t_n) dxdy,$$

where M is number of temporal snapshots and calculate $\phi_n(x, y)$ as $\phi_n(x, y) = \frac{1}{M}\sum_{m=1}^{M} b_{n,m} W(x, y, t_m)$, where $\mathbf{b} = \{b_{n,m}\}$ is the solution of the equation $\mathbf{Cb} = \lambda\mathbf{b}$.

If $M \ll N$, the computational cost of finding ϕ's can be reduced dramatically. The method of snapshots also overcomes the difficulties associated with the large data sets that accompany more than one dimension.

To demonstrate the optimality of POD, let's apply it to the same example data, given in Equation (4.10), previously used to demonstrate Zernike expansion. The wavefronts were spatially discretized with $\Delta x = \Delta y = 0.01$ over a circular aperture, resulting in $K = 31{,}397$ spatial points, over the time interval of the full period, with $M = 500$ time steps. As $M \ll N$, the snapshot version of POD was used to compute POD modes, coefficients, and modal energies, and the results are presented in Figure 4.9. For this example, modal energies, shown in Figure 4.9(a) are non-zero only for the first two POD eigenvalues and each mode contains

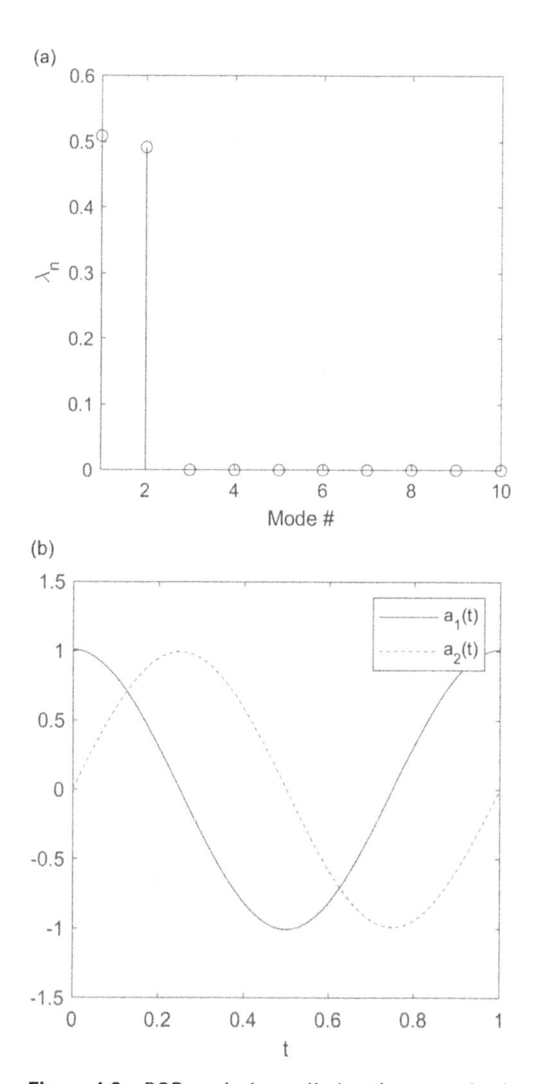

Figure 4.9 POD analysis, applied to the example signal, Equation (4.10). (a) Relative modal energies, (b) temporal POD coefficients, and (c) dominant spatial POD modes.

(c)

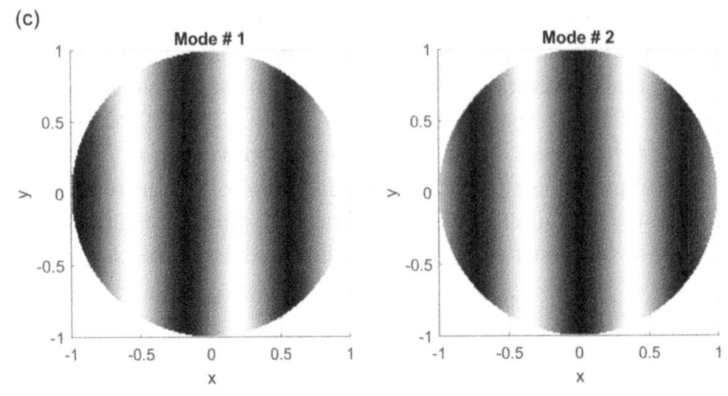

Figure 4.9 (Continued)

about 50% of the total energy. The corresponding temporal coefficients for the first two POD modes are plotted in Figure 4.9(b) and the modes are presented in Figure 4.9(c). Thus, the signal can be represented as a sum of only two POD modes, $W(x,y,t) \approx \sum_{n=1}^{L=2} a_n(t)\varphi_n(x,y)$. In contrast, it took almost 30 Zernike modes to represent the same signal at different times, clearly demonstrating the optimality of POD approach. Further, this decomposition has a simple functional pattern. In aero-optics almost all the fluid structures are moving, and the underlying spatial modes are more complicated. In some early work, we first explored Zernike decomposition and, even with 50 modes, the data was completely misleading. In our opinion, Zernike modes *should not be used* in aero-optics. On the other hand, Zernike decomposition can still be useful to capture lower-order modes like piston, tip/tilt, spherical and maybe a few others, and remove them from the wavefronts before performing POD decomposition.

From Figure 4.9, one can see that both modes look very similar, but spatially shifted relative to each other; the temporal coefficients are also shifted in time. To explain why there are two similar-looking POD modes present in POD analysis for the given example wavefront, it is straightforward to analytically compute POD modes and the coefficients, as

$$\begin{aligned} W(x,y,t) &= \sin\left[2\pi\left(\frac{x}{0.75}-t\right)\right] \\ &= \cos(2\pi t)\sin\left(2\pi\frac{x}{0.75}\right) - \sin(2\pi t)\cos\left(2\pi\frac{x}{0.75}\right). \\ &= \sum_{n=1}^{N=2} a_n(t)\varphi_n(x), \\ \lambda_1 &= \lambda_1 = 0.5 \end{aligned} \tag{4.18}$$

This simple exercise demonstrates that when using POD analysis, at least in the present form, traveling structures cannot be represented as a single mode. In general, traveling structures are represented by a *pair* of stationary POD modes, which are similar in spatial distribution but appear to be shifted in the advection direction. Indeed, for the given example, a pair of spatial modes, $\varphi_1(x.y) = \sin\left(2\pi\frac{x}{0.75}\right)$ and $\varphi_2(x.y) = \cos\left(2\pi\frac{x}{0.75}\right) = \sin\left(2\pi\frac{x}{0.75} - \pi/2\right)$, are shifted by a phase of $\pi/2$ in the x-direction; the similar shift in time is present in the temporal coefficients, $a_1(t) = \cos(2\pi t)$ and $a_2(t) = -\sin(2\pi t) = \cos(2\pi t + \pi/2)$. Both of these shifts can be seen in Figure 4.9.

Another example of POD analysis is given in Figure 4.10, where the wavefront data collected in-flight around a hemispheric turret at Mach number of$M = 0.8$ were used for POD analysis (flight and data collection details are reported in Morrida et al. 2017). The wavefront data set had 627 spatial points, collected over 21,000 time steps, so the direct method was used to calculate POD modes. For reference, the temporal variance of the wavefronts, $S(x, y)$, is presented in the upper left corner of Figure 4.10. At this angle, an unsteady shock forms over the aperture, creating large localized aero-optical distortions. The shock location is evident by a narrow spatial region of increased distortions. The upper middle plot shows the normalized and cumulative energies of POD modes. The first POD mode contains more than 30% of the total energy, the first 6 POD modes contain about 70% of the energy, and the first 20 modes contain 90% of the energy. The first six POD modes are presented in the bottom two rows in Figure 4.10. The first POD mode has a similar sharp discontinuity related to the presence of the unsteady shock. At this flight speed, the shock is strong and creates large unsteady distortions in the wavefronts, so the first POD modes primarily reflect the shock-related optical distortions. The shock presence is also evident in higher POD modes at these speeds. In addition to the shock-related non-zero values in the region where the shock is present, higher POD modes, specifically Modes #2, #4, and #6, also have non-zero values over the downstream portion of aperture, over which the flow separates and ultimately creates a separation region. As discussed before, it indicates that there is a correlation between the shock region and the separation region, a valuable insight into the dynamics of the separated region, influenced by the shock.

POD Modes #3 and #5 indicate the presence of the unsteady focus in the data, related to a slowly varying distance between the aircraft, since the laser beam defocus is a function of the distance. The different physics of the shock-related and defocus-related modes are also demonstrated in Figure 4.10, upper right plot, where the temporal coefficients for the mode #1 and the mode #5 are plotted. Mode #1 changes rapidly due to the fast shock motion, while mode #5 exhibits only slow variation, consistent with slow varying distance between the aircraft. So, POD analysis is useful in identifying and quantifying different sources of aero-optical distortions.

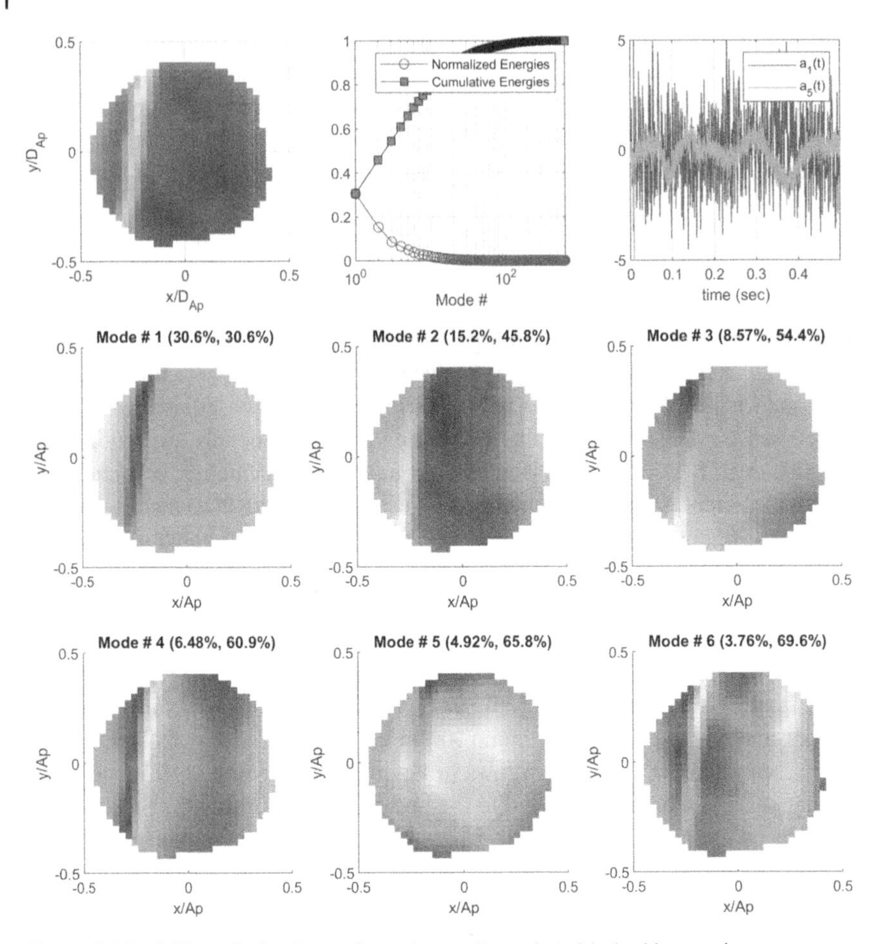

Figure 4.10 POD analysis of wavefront data, collected at side-looking angle over a turret at transonic speed of $M = 0.8$. Upper left: the temporal variance of the wavefronts, $S(x, y)$, over the aperture. Upper middle: normalized and cumulative energies of POD modes. Upper right; Temporal coefficients for the POD mode # 1 (shock-related) and Mode # 5 (related to unsteady defocus). Bottom two rows: Dominant POD modes. The corresponding normalized and the cumulative energies for each POD mode are given as a first and a second number in parenthesis. Flow goes from left to right.

The above example can also be used to demonstrate the storage efficiency, using POD series. The first 40 modes contain about 95% of the energy. If the POD series is truncated at the first 40 modes, it will result in only a 5% loss in the total energy. In this case, the original wavefront data of the size [627 x 21,000] can be replaced with storing only 40 POD modes at 627 spatial points and 40 temporal coefficients 21,000 long, with the total size of 40 x (627 + 21,000) numbers, resulting in a 15-fold reduction in data storage. If 10% of the energy losses is acceptable, only the first 20 POD modes and their coefficients need to be used to store the signal, resulting in a 30-fold reduction in storage.

POD analysis uses the two-point correlation function, R, to decompose the signal into the POD modes. In other words, the POD technique arranges modes by amount of cross-correlation or "order," so mostly incoherent noise was present in the very high-order modes. This POD property is often used to remove incoherent noise from the data by truncating POD series down to dominant POD modes. There are no well-defined criteria on where to truncate POD series, as it depends on the data. If the level of noise is small, one way to choose how many truncated modes are needed is to inspect the cumulative energy distribution of the modes to see how many modes are needed to resolve, let's say, 80% or 90% of the total energy. Another approach is to inspect higher-order POD modes. If, starting from some modes, they look like mostly incoherent spatial noise, they can be ignored in the truncated series. The noise-reduction property of POD is demonstrated in Figure 4.11, where relative and cumulative modal energies of the POD analysis of the unsteady pressure field over a turret (Gordeyev et al. 2014) are shown. The full POD set for this data had 4,651 modes, but the first POD mode contained almost 30% of the total energy of the pressure fluctuations, the first 10 modes had more than 60% of the energy, the first 100 modes held more than 80% of the total pressure-fluctuating energy, and the first 1,000 modes captured virtually all pressure "energy" of the flow.

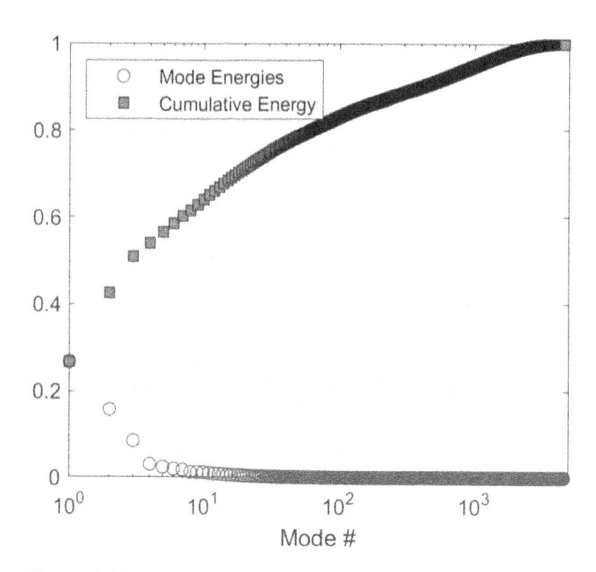

Figure 4.11 The normalized and cumulative mode energies for an unsteady pressure field. *Source:* Gordeyev et. al., (2014b), figure 8, with permission of Springer Nature.

To illustrate the effect of truncating POD series on the noise reduction, the pressure field over a turret at a given moment was reconstructed using the first 10, 100, and 1,000 POD modes and results are presented in Figure 4.12. The original pressure field is also presented in Figure 4.12(a). As few as 10 POD modes, Figure 4.12(b), were able to capture most of the essential pressure features and all features were properly recovered using 100 modes, Figure 4.12(c). The addition of higher-order modes, Figure 4.12(d), simply added more noise to the pressure field. Thus, the POD technique, in addition to providing an optimal framework to investigate the flow, also presents an efficient way to remove the noise present in the experimental data, while preserving the structures' features.

A final example of using POD analysis to extract and analyze different physical features deals with isolating and studying the acoustic noise, present in the wavefronts (De Lucca et al. 2018; Gordeyev and Kalensky 2020). Acoustic waves are pressure fluctuations which affect density, which in turn, create additional optical distortions. Acoustic disturbances often appear in experimental measurements, both in wind tunnels due to the tunnel motor and in-flight tests due to aircraft engines (De Lucca et al. 2018; Gordeyev and Kalensky 2020). For this reason, it is important to identify and remove these contaminating optical effects from the

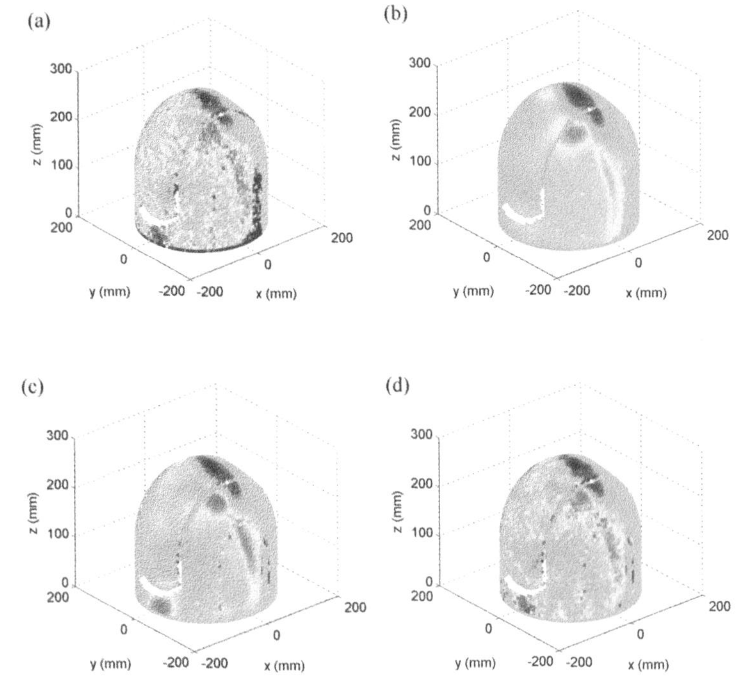

Figure 4.12 The representative instantaneous pressure field (a) and POD reconstruction using the first 10 (b), 100 (c) and 1000 (d) POD modes. *Source:* Gordeyev et. al., (2014b), reproduced with permission of Springer Nature.

overall optical distortions. In case of a tonal acoustic noise, the density field has well-defined spatial periodicity, and POD is a very good technique to identify the associated spatially periodic optical distortions. POD analysis was performed on the aero-optical data collected in-flight at wall-normal viewing angles (Gordeyev and Kalensky 2020). The first 3 POD modes, containing about 55% of the total energy, are shown in Figure 4.13, top row. These dominant modes do not show any regular spatial distortions and represent various spatial features of the turbulent flow over the aperture. On the contrary, higher-order modes 20, 22, and 23, shown in Figure 4.13, bottom row, have very different spatial distribution, characterized by predominantly streamwise-periodic and almost spanwise-uniform structure with a similar spatial wavelength. These modes just differ in phase, which indicates, as discussed before, a presence of a traveling structure. Such large, regular spatially periodic structures are not expected from the turbulent-flow aero-optic distortions. By computing spatial wavelengths for these modes and studying the corresponding temporal coefficients, their convective speed and direction was calculated. These modes were found to travel *upstream*, consistent with the upstream-propagating acoustic waves of the speed of sound minus the freestream speed. Ultimately, these regular upstream-propagating distortions were attributed to acoustic waves originating from the aircraft engine downstream of the pupil. Note that these higher-order modes contain only a small fraction of the total energy, about 0.5% each. Nevertheless, POD analysis was able to identify them due to their distinctly different spatial organization, compared to the turbulent structure.

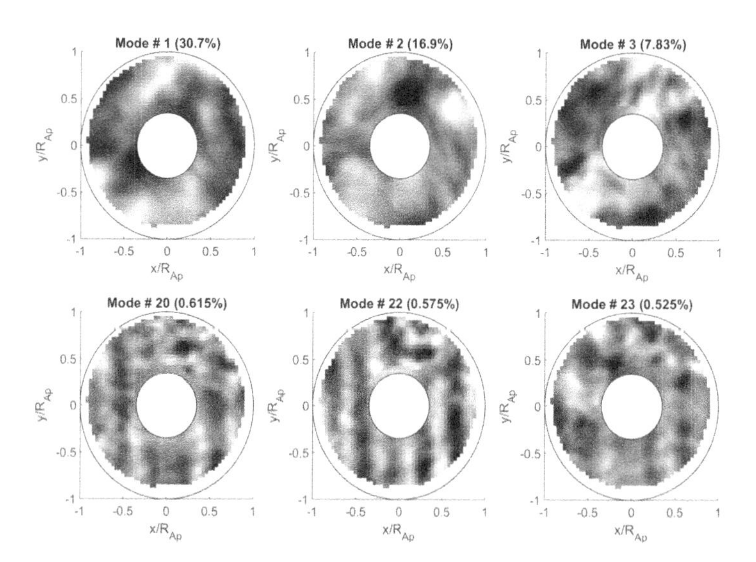

Figure 4.13 Selected POD modes 1–3, 20, 22, and 23 collected at side-looking angles in-flight. Flow goes from left to right. Number in parenthesis indicates the relative amount of energy of the mode.

Let us summarize some properties of the POD analysis.

- Provides the linear expansion of the data without any prior knowledge of the field. This expansion is the fastest-converging set of orthogonal modes for the given set of data.
- Provides spatial distribution of energetically dominant modes with their temporal evolution.
- Gives an orthogonal frame of reference with a minimal dimension to describe the dynamics of the system.
- Very useful for pattern recognition, image processing, and for optimally storing/compressing large databases.
- Useful to remove the incoherent noise from the dataset by simply removing high-order modes from the expansion.
- Relatively simple to compute using either direct or snapshot methods.
- For traveling structures, POD needs to have at least two stationary structures to describe it.

4.3.3 Dynamic Mode Decomposition (DMD)

Alternative to POD, Dynamic Mode Decomposition or DMD provides a means of decomposing time-resolved data into modes, with each mode having a single characteristic frequency of oscillation and growth/decay rate (Schmid 2010; Taira et al. 2017). DMD decomposes the series of N sequential snapshots of wavefronts, $W(s,t_k), t_k = k\Delta t, k = 1 \ldots N$, separated by a fixed time delay, $\Delta t = t_{k+1} - t_k$, into a set of spatial modes, $\varphi_n(s)$, corresponding amplitudes, c_n, and eigenvalues, λ_n, to represent the frequency and exponential decay rate of the modes,

$$W(s,t_k) = \sum_{n=1}^{N} c_n \exp(\lambda_n t_k) \phi_n(s). \tag{4.19}$$

where the spatial point $s = (x, y)$ is introduced for simplicity. Note that the modes, ϕ, and the amplitudes, c, are complex quantities. The decomposition is computed by breaking the data set into two data sets, W_1 and W_2, offset by a single time step. That is, if the data set has N total snapshots, the matrix W_1 contains the first $N-1$ snapshots as columns and W_2 contains the last $N-1$ snapshots given by

$$W_1 = \{W(s,t_1), W(s,t_2), \ldots, W(s,t_{N-1})\}$$
$$W_2 = \{W(s,t_2), W(s,t_3), \ldots, W(s,t_N)\}$$

The size of both data sets is $[K \times (N-1)]$, where K is a number of spatial points. The second data set then can be written as a linear combination of the first data set, $W_2 = AW_1$. Alternatively, it can be written as $W_2 = W_1 S$, where S is called a companion matrix to A. The matrices A and S are similar in that they share the same eigenvalues and eigenvectors. The DMD eigenvalues and modes can be calculated from the eigenvalues and eigenvectors of the matrix A or S.

It is common that the number of snapshots is smaller than the number of spatial points of each snapshot, $N \ll K$. In this case, it is not efficient to compute A or S explicitly, and the actual computation of the DMD modes is as follows (Tu et al. 2014):

(1) Perform SVD of the first data set, W_1,

$$W_1 = U\Sigma V^T.$$

During this step, the noise can be removed from the data by truncating SVD-series down to the first L-modes. This can be done by including only first L columns of U and V, and the first L rows and columns of Σ.

(2) Using the above relationship between W_1, W_2 and S, substitute SVD of the first data to obtain $W_2 = U\Sigma V^T S$, and to derive the expression for S,

$$S = U^T W_2 V \Sigma^{-1}.$$

(3) Calculate the eigenvalues, μ, and eigenvectors, Y, of the matrix S,

$$SY = \mu Y.$$

(4) The DMD modes and eigenvalues can be computed from the eigenvectors and eigenvalues of S as

$$\varphi = UY, \ \lambda = \ln(\mu) / \Delta t.$$

The frequency, ω_n, of each mode is given by the imaginary part of the eigenvalue and the growth/decay rate is given by the real part,

$$e^{\lambda_n t} = e^{\mathrm{Re}(\lambda_n)t + i Im(\lambda_n)t} = e^{\mathrm{Re}(\lambda_n)t} e^{i\omega_n t}.$$

(5) c-coefficients in Equation (4.19) can be calculated as,

$$c = \mu Y^{-1}.$$

Each DMD mode pair corresponds to a single temporal frequency. In addition, DMD mode pair gives the spatial distribution at a given frequency.

For stationary zero-mean data, it is straightforward to show (Chen et al. 2012) that DMD modes have no growth/decay rates and DMD is reduced to a Discrete Fourier Transform, with only the imaginary eigenvalues, $\lambda_n = i\omega_n$,

$$W(s,t_k) = \sum_{n=1}^{N} \exp(i\omega_n t_k)\hat{\varphi}(s;\omega_n),$$

$$\text{where } \hat{\varphi}(s;\omega_n) = DFT\{W(s,t_k)\} = \frac{1}{N}\sum_{k=1}^{N} W(s,t_k)\exp(-i\omega_n t_k), \tag{4.20}$$

$$\omega_n = (n-1)/(N\Delta t), \quad n = 1 \ldots N$$

Like POD, DMD modes and coefficients are data-driven, that is they will depend on the data set. Unlike POD, DMD modes are not orthogonal modes. Each mode is associated with a single frequency and both the modes and the coefficients are complex quantities. DMD captures both frequency and growth/decay aspects of flow.

As a demonstration, DMD was implemented to the example signal from Equation (4.10). Results are shown in Figure 4.14. From a spectral distribution of DMD coefficients, shown in Figure 14.4(a), it is possible to identify that only one frequency (1 Hz) is present in the signal. The spatial modes, both real and imaginary components, corresponding to the 1 Hz, are plotted in Figure 4.14(b). The components look very similar, but are simply spatially shifted relative to each other. In fact, DMD pair looks very similar to pair of POD modes, shown in Figure

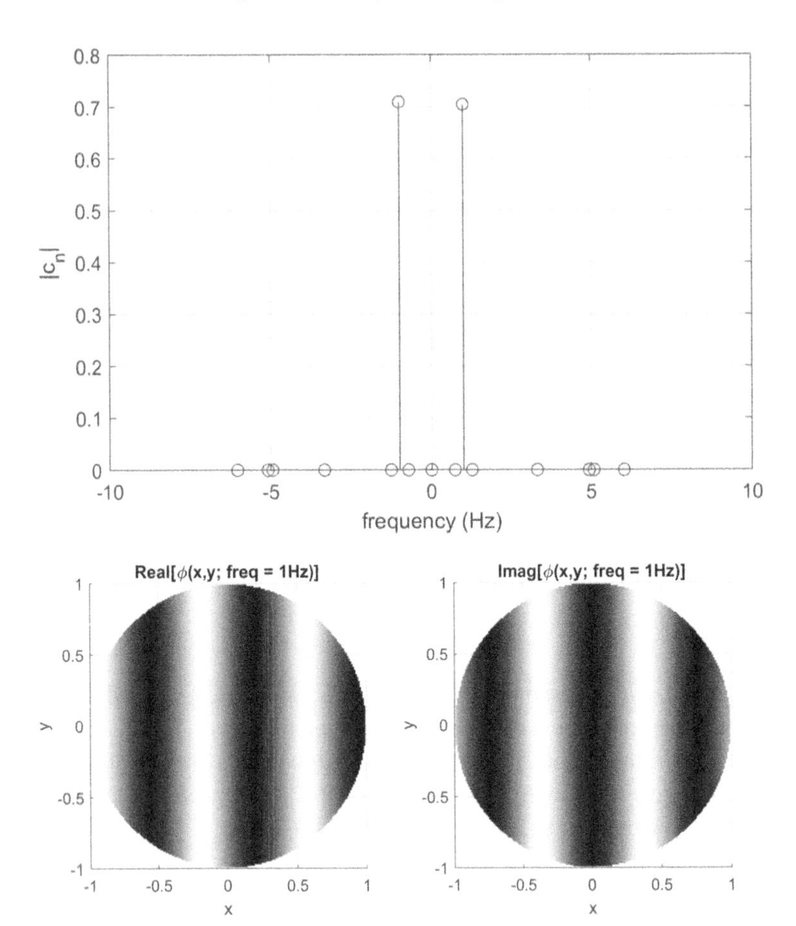

Figure 4.14 DMD analysis of the example signal from Equation (4.10): (a) amplitudes of c_n as a function of frequencies, f_n, (b) real and imaginary parts of DMD mode, corresponding to $f = 1$ Hz.

4.9(c), indicating a traveling structure. It is straightforward to demonstrate that in case of only one DMD mode, Equation (4.19) becomes,

$$W(s,t) = c_1 \exp(i\omega_1 t)\varphi_1(s) + c.c. = c_1 2\,\mathrm{Re}[\exp(\omega_1 t)\varphi_1(s)]$$
$$= 2c_1 \{\mathrm{Re}[\varphi_1(s)]\cos(\omega_1 t) - Im[\varphi_1(s)]\sin(\omega_1 t)\}.$$

This equation has the same structure as the POD expansion for this example signal, see Equation (4.18). Thus, both POD and DMD can identify traveling structures. The advantage of DMD is that it automatically contains the pair of modes as real and imaginary parts of the complex DMD mode. Knowing the wavelength of the spatial mode and the corresponding frequency, the convective speed of the DMD mode can be computed.

Like POD, the DMD can be used to identify coherent structures in the wavefront data. Figure 4.15(a), shows the DMD amplitude spectrum for the wavefront dataset, collected in-flight (Gordeyev and Kalensky 2020). The spectrum shows a strong spectrum peak at 4,663 Hz, which corresponds to the acoustic waves at the blade-passing frequency of the engine. The corresponding DMD mode, presented in Figure 4.15(b), clearly shows spanwise-uniform and streamwise-periodic structure, associated with the acoustic waves from the engine. Dynamic modes at other frequencies inside the broad-band part of the spectrum, like for 737 Hz in Figure 4.15(c), do not exhibit any distinct spanwise-uniform features, as this frequency range is associated with the structures inside the turbulent flow over the aperture.

Because the DMD organizes the modes based on frequency, the DMD is a very good technique to extract the spatial distribution of the underlying structure, if the structure has an associated fixed frequency. The POD, on the other hand, arranges the modes according to their relative energy, which might not be a good criterion to identify the structure. It can be observed in Figure 4.13, where POD modes were computed from the same dataset, used to compute DMD modes in Figure 4.15. Both techniques were able to identify the spanwise-uniform acoustic-related structure, but the modal shape is much cleaner for the DMD analysis.

4.4 Cross-correlation-based Techniques

4.4.1 Local Convective Speeds

The measurements of the convective speed are based on the assertion made by Malley et al. 1992, that the aberrating structures in the flow are convecting with the flow and thus the wavefront itself is “convecting” at the same speed, U_C. This assertion, used with Taylor’s “Frozen Field” hypothesis, gives the basis for measuring convective speeds of the wavefront structures.

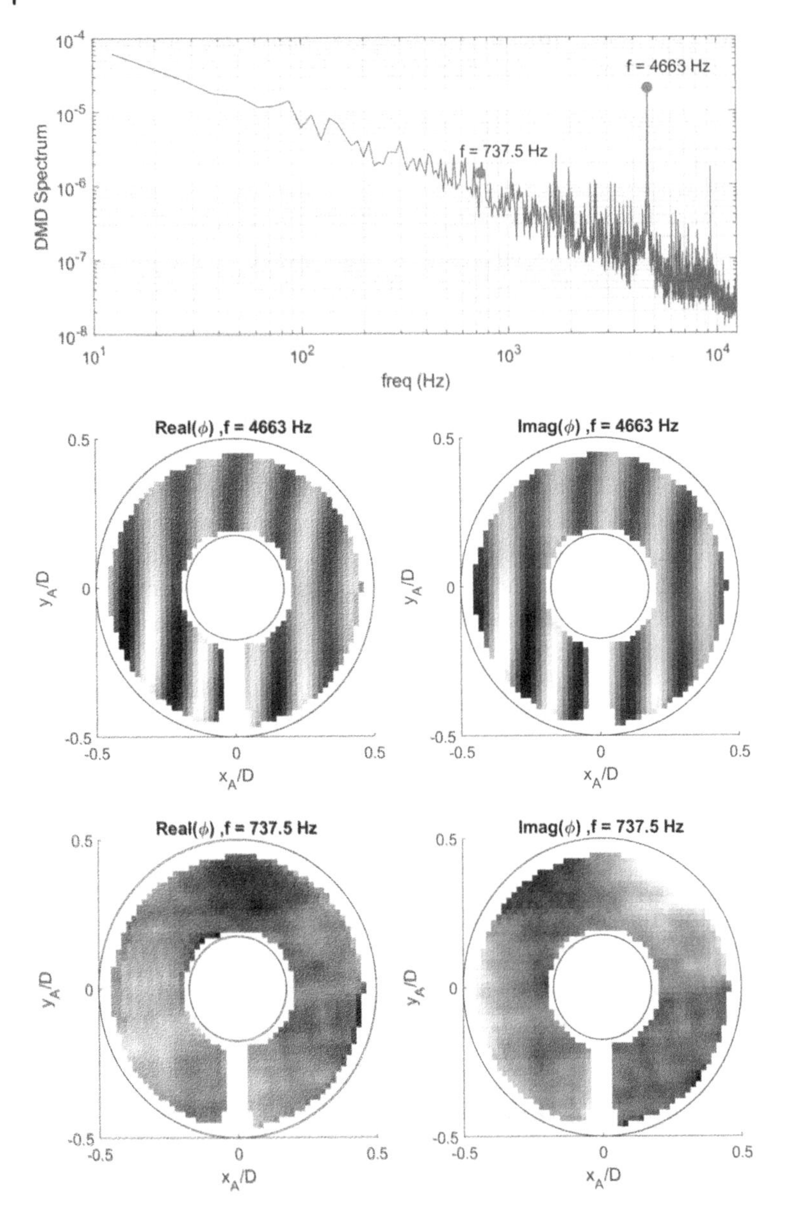

Figure 4.15 (a) DMD spectrum of for in-flight data at Mach 0.7. (b) The real and imaginary parts of the DMD modes for two selected frequencies, indicated by solid symbols in (a).

As discussed in Chapter 3, the Malley probe uses two parallel laser beams, separated in the streamwise direction. Imagine that an optically aberrating structure moves across both beams, as illustrated in Figure 4.16 (a). Initially, it will result in changes of the deflection angle of the upstream beam. If the beams are spatially sufficiently close, almost the same changes will happen in later time in the deflection angle of the second, downstream beam. This time is a "flyby" time it takes for the structure to convect from the first to the second beam. If the "flyby" time or the time delay, τ, between the two beams positions

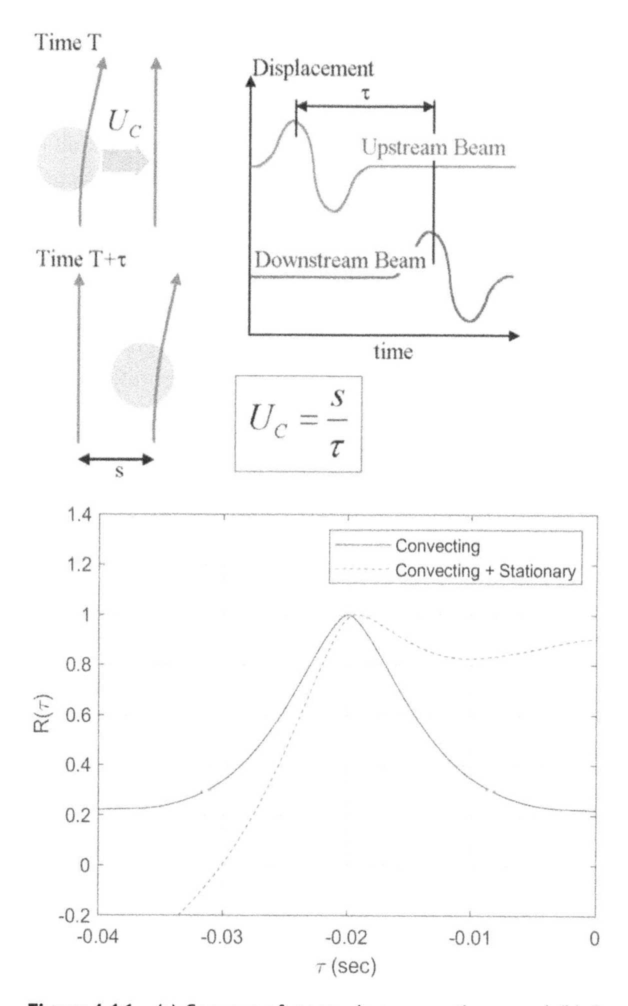

Figure 4.16 (a) Concept of measuring convective speed. (b) Cross-correlation function, R(t) for both the convecting-only distortions and a combination of convecting and stationary distortions.

and the distance between the beams, s, are known, the convective speed can be directly computed as $U_C = s/\tau$.

To compute the convective speed, a time series of the deflection angles from both beams, denoted by $\theta_1(t)$ and $\theta_2(t)$, can be cross-correlated to find the time delay between the signals. One way to do this is to compute a time-delayed correlation function $R(\tau) = E\{\theta_1(t)\theta_2(t+\tau)\}$ between the two signals from the two probe beams. This method is called a *direct method*. A convective speed U_C is computed knowing a separation s between beams and a time delay of the highest correlation, τ, where $R(\tau) = \max$.

The direct method is simple to implement. The problem with the direct approach is that it is sensitive to signal contamination, like stationary mechanical vibrations of optical components in the beam train and electronic noise, for instance; often the contamination is present in both signals. To illustrate that, consider a case, when an optical distortion simply convects across both beams, and it takes 20 msec for the distortion to travel between both beams. In this case, a signal at the second beam is the same as at the first beam, but shifted by 20 msec. The cross-correlation function is plotted as a solid line in Figure 4.16(b). The peak in the time-delayed correlation is at $\tau = -20$ msec, as expected. A negative time indicates that the second signal is delayed relative to the first one. But if a stationary single-frequency component, cos(20πt), is added to both signals by say the natural frequency of the mirror post off of which both beams are reflected, it will introduce additional correlations near time delays of zero. As a result, it will change the peak location in the correlation function, indicated by a dashed line, forcing it to move to an apparently smaller time shift of $\tau = -19.4$ msec. To avoid this contaminating effect, all vibrations or stationary components should be removed or filtered out from both signals.

Another way of computing the time delay is to use a *spectral method*. In this approach, the time delay to computed by analyzing a spectral cross-correlation function,

$$S(\omega) = \int R(\tau)\exp(-i\omega\tau)d\tau = \frac{1}{T}\langle\hat{\theta}_1(\omega)\hat{\theta}_2^*(\omega)\rangle, \tag{4.21}$$

where T is a block sampling time, and the brackets denote an ensemble average. In the case of a pure convecting structure, the signal downstream, θ_2, is just a time-delayed signal of the upstream signal, θ_1, $\theta_2(t) = \theta_1(t-\tau)$ and the Fourier transform of θ_2 becomes $\hat{\theta}_2(\omega) = \hat{\theta}_1(\omega)\exp(-i\omega\tau)$. Using this relation, the expression for the spectral correlation becomes

$$S(\omega) = 1/T\langle\hat{\theta}_1(\omega)\hat{\theta}_2^*(\omega)\rangle = 1/T\langle\hat{\theta}_1(\omega)\left[\hat{\theta}_1(\omega)\exp(-i\omega\tau)\right]^*\rangle = A(\omega)\exp(i\omega\tau),$$

where $A(\omega) = 1/T\langle\hat{\theta}_1(\omega)\hat{\theta}_1^*(\omega)\rangle$ is *a real function* of the angular frequency, ω. The phase or the argument of the spectral cross-correlation function is $Arg\left[S(\omega)\right] = i\omega\tau$, and it is a linear function of τ. Thus, by analyzing the slope of the phase of the spectral cross-correlation function, one can find the time delay, τ, $\tau = \frac{1}{2\pi}\frac{d}{df}\arg[S(f)]$.

The advantage of the spectral method is that it allows distinguishing between the convecting and stationary distortions, as the stationary distortions will have a zero-time delay, while the convecting ones will have a linearly changing phase. To demonstrate this, the same combination of convecting and stationary signals, used to illustrate the direct method in Figure 4.16(a), was used to compute the spectral cross-correlation function, $S(f)$. The plot of the phase is shown in Figure 4.17(a). It reflects the linear dependence of the phase, except for the region near 10 Hz, where the phase is nearly zero. It is an indication that the signals are most probably contaminated by the stationary distortions at this frequency. If the frequency range is excluded, it is straightforward to compute the time delay by performing a linear fit. For this example, the linear fit correctly recovers the time delay of 20 msec.

Another useful feature of the spectral method is that it allows one performing convective velocity measurement than the convective speed is a function of frequency, $U_C(f) = s/\tau(f)$. This information is very useful, for instance, when the beams encounter the flow with different optically active regions, and it is still possible to measure convective speeds for *each* region using the spectral method, as long as optical structures associated with each region have distinct frequency bands. An example of one such phase plot is presented in Figure 4.17(b), where the measurements were conducted through the test section with both the shear layer in the middle of the section and the boundary layer on one of the tunnel walls (Seigenthaler et al. 2003). There is a clear split in the phase slopes for the lower-frequency (corresponding to a large structure), shear-layer convecting structures and the higher-frequency, smaller structures associated with the attached turbulent boundary layer. The shear-layer structures travel at a lower shear-layer convection velocity, while the boundary layers convect at a higher velocity.

4.4.2 Multi-point Malley Probe Analysis

Cross-correlation analysis is also useful for Shack-Hartmann sensors, where the deflection angle measurements are performed at multiple points, $\theta(x_k,t), k = 1\ldots K$. Several methods, based on the analysis of the multiple points in space, will be presented here.

Locally, deflection angles, θ, can be assumed to be a combination of a stationary, θ_S, (for instance, mechanical jitter) and traveling, θ_T, (aero-optical structure) modes,

$$\hat{\theta}(x,f) = FT\{\theta(x,t)\} = \hat{\theta}_S(f) + \hat{\theta}_T(f)\exp(2\pi i f(t - x/U_C)). \tag{4.22}$$

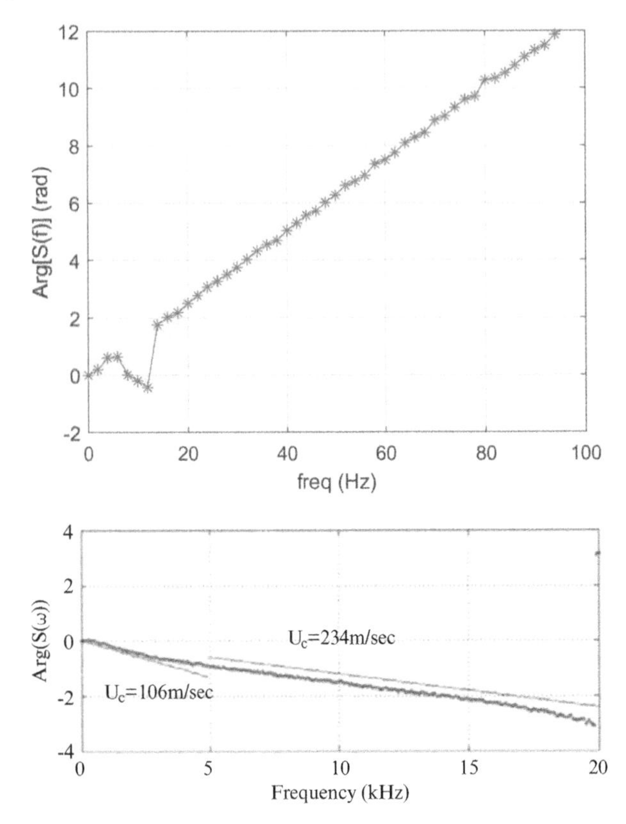

Figure 4.17 (a) Phase plot of the spectral correlation between two synthetic signals, with inclusion of the stationary time-dependent component at 10 Hz. (b) Example of the phase dependence versus frequency, indicating the presence of two linear slopes, a steeper one at low frequency below 3 kHz, and a shallower one at higher frequency above 5 kHz. The corresponding convective speeds for each frequency range are also indicated. *Source:* Seigenthaler et. al. (2003), figure 15 / American Institute of Aeronautics and Astronautics.

If the stationary and the traveling components are statistically independent, the cross-spectral correlation, S, can be computed between deflection angles for different separations,

$$S(\Delta x; f) = \langle \hat{\theta}(x,f)\hat{\theta}^*(x+\Delta x, f)\rangle = \left|\hat{\theta}_S(f)\right|^2 + \left|\hat{\theta}_T(f)\right|^2 \exp(2\pi i f \Delta x / U_C). \quad (4.23)$$

If the cross-correlation function is known for more than two different separations, the above equation is over-determined, and the convective speeds and the stationary and the traveling spectra can be found using the least-squares estimation.

To use this multi-point spectral cross-correlation technique, the convective speed of the traveling mode should be known. As two sub-apertures, adjacent in the streamwise direction, can be treated as a Malley probe, the convective speed can be measured by performing the spectral cross-correlation between two beams, as discussed before. Alternatively, the dispersion analysis can be used to measure the convective speed.

Once the convecting speed is known, the stationary and traveling modes can be extracted from the original signal, solving a system of equations, given in Equation (4.23), for all possible Δx's. A demonstration of this technique is shown in Figure 4.18, where a deflection angle spectrum from hypersonic boundary layer experiments (Gordeyev and Juliano 2016) and the extracted stationary and the traveling spectra are calculated for $M = 5.8$. The stationary spectrum, related to mechanical vibrations of the tunnel and the optical components dominates the total spectrum at low frequencies, below 1 kHz. Above 1 kHz, the traveling spectrum, related to the convecting aero-optical boundary layer structures, is the leading source of the total spectrum. The noise-reducing feature of the cross-correlation technique is clearly visible at the high end of the spectrum, above 200 kHz, where the traveling spectrum has a steeper slope, compared to the total spectrum. The slope of the traveling spectrum agrees quite well with the theoretically predicted slope of $\sim f^{-4/3}$ (Gordeyev et al. 2014), shown as a dashed line.

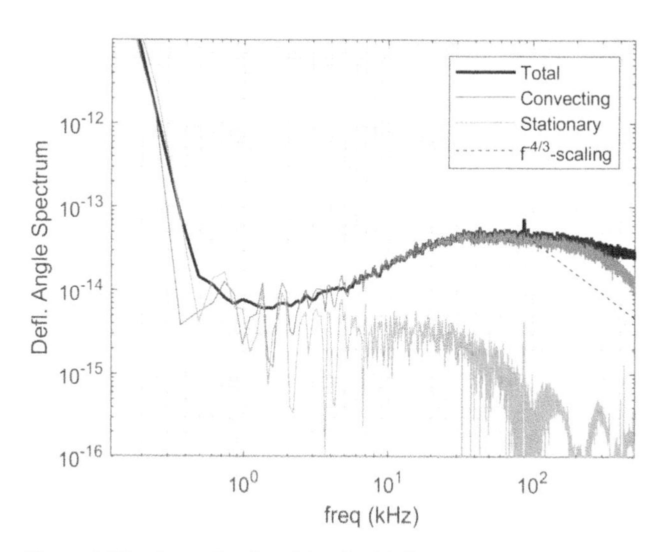

Figure 4.18 Example of multi-point Malley analysis to separate the original deflection angle spectrum into the convecting and the stationary components.

4.4.3 Spatially Varying 2-D Convective Velocity

As discussed earlier, the Malley probe principle of operation is based on the assumption that the optical aberrations are "frozen" and at least sufficient knowledge of the convection direction is known to align the two beams in the streamwise direction. In general, the convecting speed and direction might not be known or different at different locations over the aperture. These complications can be overcome by using a two-dimensional array of detectors to investigate a large aperture. Rather than using individual beams, a Shack-Hartmann wavefront sensor can be used by treating each sensor's lenslet as a separate beam of the subaperture size, where the deflection angle over each subaperture represents the average wavefront local tilt. Based on this, each beam's subaperture represents a "probe-beam" which is propagating through the flow-field, hence, a 2-D array of Malley probes. The main advantage of this multiple-beam instrument is that with an entire array of probes, the evolution of a structure as it convects across adjacent subapertures will be sufficiently negligible that it can be treated as "frozen" between the adjacent subapertures. Furthermore, different convection velocities will be obtained at the same location depending on which two adjacent subapertures are used. The true magnitude and direction of the aberration convection can then be found by vector summing of the inferred velocity from each pair in a square array if the subapertures are laid out in an equidistant rectilinear array; the lenslet arrays used most Shack-Hartmann wavefront sensors are of this type. In this array pattern, the x- and y-directions define the array so that adjacent correlations in the x- and y-directions will yield local velocity components in the x-direction, U_x, and in the y-direction, U_y, respectively. Then, both local velocity components may be used to calculate a two-dimensional convective velocity field, $\mathbf{U}_C(x,y) = (U_x, U_y)$.

Consider a structure which convects at the velocity, $\mathbf{U}_C$, between two detector points, separated by a distance $\mathbf{d}$; see Figure 4.19(a). The unit vector in the direction of convection can be represented as

$$\mathbf{n} = \frac{\mathbf{U}_C}{|\mathbf{U}_C|}.$$

The distance along the direction of convection is a scaler product between the unit velocity vector and the distance, $\mathbf{n} \cdot \mathbf{d}$. The convective time it takes to travel this distance as,

$$\tau = \frac{\mathbf{n} \cdot \mathbf{d}}{|\mathbf{U}_C|} = \frac{\mathbf{U}_C \cdot \mathbf{d}}{|\mathbf{U}_C|^2}.$$

Defining two parameters u and v as

$$u = \frac{U_x}{U_x^2 + U_y^2}, \quad v = \frac{U_y}{U_x^2 + U_y^2},$$

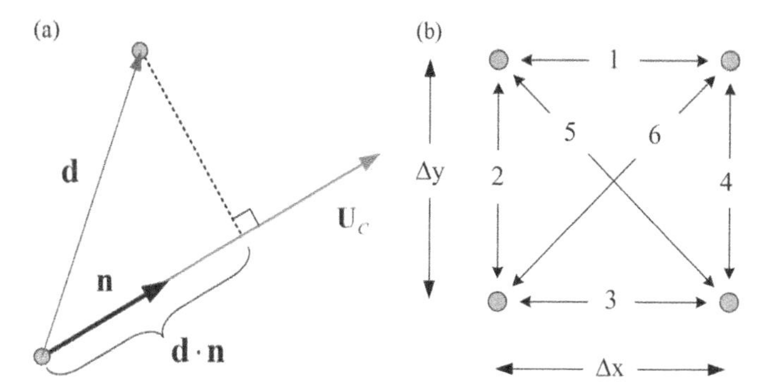

Figure 4.19 (a) Schematic of the relationship between the distance vector, **d**, and the convective velocity vector, $\mathbf{U}_C$. (b) Six possible cross-correlations for a four-beam cell.

For a rectangular lenslet array, four adjacent subapertures can be used to form a four-beam cell. An example layout of such a cell is shown in Figure 4.19(b). Here, four probe beams at the corners of the cell define a four-beam Malley probe instrument. Rather than just one correlation pair for the two-beam Malley probe, the four-beam Malley probe has six possible cross-correlation combinations. Then, for all six possible two-point separations, an over-determined system of equations with six equations and two unknowns can be defined,

$$\begin{bmatrix} \Delta x & 0 \\ 0 & -\Delta y \\ \Delta x & -\Delta y \\ -\Delta x & -\Delta y \\ 0 & -\Delta y \\ \Delta x & 0 \end{bmatrix} \cdot \begin{bmatrix} u \\ v \end{bmatrix} = \begin{bmatrix} \tau_1 \\ \tau_2 \\ \tau_3 \\ \tau_4 \\ \tau_5 \\ \tau_6 \end{bmatrix} \quad or \quad \mathbf{Ax} = \mathbf{b} \tag{4.24}$$

Here Δx and Δy are the spacings between the subapertures in the x- and y-directions and $\tau_i, i = 1 \ldots 6$, are the time delays of all Malley probe combination, which can be calculated based on either the direct or the spectral methods described in the previous section. This over-determined system of equations given above can be solved in a least-squares sense to find u and v as $\mathbf{x} = (\mathbf{A}^T\mathbf{A})^{-1}\mathbf{A}^T\mathbf{b}$. The local velocity components in the x- and y-directions can be finally expressed as

$$U_x = \frac{u}{u^2 + v^2}, \quad U_y = \frac{v}{u^2 + v^2}$$

To achieve a spatial mapping of the local convective velocities across the aperture, the analysis described above can be repeated at every possible combination of four subapertures across the entire beam's aperture.

It should be emphasized that the solution of Equation (4.24) is not an exact solution, but rather it is a solution that finds $\mathbf{x}$ which minimizes the length of $\mathbf{Ax} - \mathbf{b}$ as much as possible. The over-determined system of equations in Equation (4.24) allows us to minimize corrupting effects due to various sources of errors in computing time delays and reduce experimental noise.

An example of the calculation of the 2-D convective velocity is shown in Figure 4.20, where wavefront data collected in-flight at Mach number of 0.4 were used (Abado et al. 2013). For the dataset presented here, 498 possible four-beam Malley probes were defined to map the local phase convective velocity across the beam's aperture. The spectral method of computing time delays was used over a range of frequencies below 3 kHz, to eliminate noise present at higher frequencies. The flow is moving in a near-uniform direction at some angle. For this viewing angle, the flow-field is dominated by large, convecting, coherent structures that form in the shear layer. These dominant coherent structures explain the uniformity of the velocity mapping. At other viewing angles when local structures might exist in the flow-field, such as local separation bubbles, this mapping uniformity breaks down and regions of different flow directions appear in the velocity mapping. For the case in Figure 4.20, a mean convective speed of the entire aperture was calculated to be about 106 m/sec.

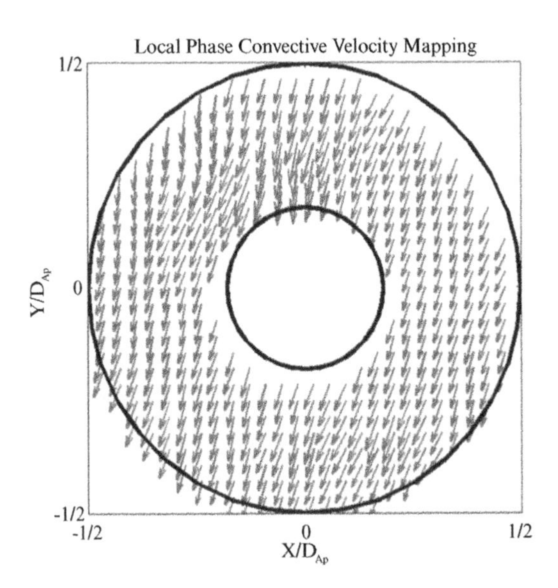

Figure 4.20 Local phase convective velocity mapping for in-flight data. *Source:* Abado et al. (2013), Figure 7. Reproduced with permission of SPIE.

The local phase convective velocity mapping has inaccuracies. These inaccuracies are mainly due to the limited number of data points available to perform ensemble averaging when calculating the power spectra, in addition to the sensitivity of the phase convective velocity calculation to the frequency range at which a linear line can be fitted. Here, the ensemble average was performed over 35 data blocks of 28 data points each. These values were determined to be the optimal values for the trade-off between sufficient temporal frequency resolution and noise rejection. In addition, it was assumed here that the flow was convecting at the same frequency range in all six possible cross-correlations and in all 498 possible four-beam Malley probe combinations; this is not always the case.

5

Aperture Effects

A true wavefront is a property of the turbulent field, as it is proportional to the integral of the three-dimensional density field, $OPD(x,y,t) = K_{GD}\int \rho'(x,y,z)dz$. However, this quantity is often known only over a finite aperture. Later in this chapter we will call it an apertured wavefront for simplicity. As discussed before, the instantaneous tip/tilt components are often contaminated by mechanical vibrations, and it is often assumed that some kind of a closed-loop system using a fast-steering mirror will be implemented to mitigate pointing error in the system. These factors prompt the investigation of how the statistics of the apertured wavefronts with removed piston and tip/tilt components depend on the size and the shape of the beam aperture.

First, let us derive the explicit equations for the global tip/tilt and piston components in Equation 2.35. One way to derive the equations is by minimizing of the wavefront's L_2-norm in the least-squares sense over the aperture area, S, at each time instant,

$$\min_{A,B_1,B_2} \left\| W(x,y,t) \right\|_2 = \int_S \left\{ W(x,y,t) - \left[A(t) + B_1(t)x + B_2(t)y \right] \right\}^2 dxdy \rightarrow \min.$$

This definition of the global tip/tilt is called Z-tilt (Sasiela 2007; Tyson and Frasier 2012). Using the least-square analysis, the above equation can be solved, resulting in a system of linear equations for A, B_1 and B_2,

$$\begin{bmatrix} \int_S dxdy & \int_S xdxdy & \int_S ydxdy \\ \int_S xdxdy & \int_S x^2 dxdy & \int_S xydxdy \\ \int_S ydxdy & \int_S xydxdy & \int_S y^2 dxdy \end{bmatrix} \begin{bmatrix} A(t) \\ B_1(t) \\ B_2(t) \end{bmatrix} = \begin{bmatrix} \int_S W(x,y,t)dxdy \\ \int_S xW(x,y,t)dxdy \\ \int_S yW(x,y,t)dxdy \end{bmatrix} \tag{5.1}$$

Aero-Optical Effects: Physics, Analysis and Mitigation, First Edition. Stanislav Gordeyev, Eric J. Jumper, and Matthew R. Whiteley.

From here, it follows that the values for the piston and the tip/tilt components depend on the aperture size and geometry. For symmetric apertures, like circular or rectangular ones, with the center of the aperture at $x = 0, y = 0$, is it straightforward to show that all off-diagonal terms in Equation (5.1) are zero. In this case, the expressions for the unsteady piston and tip/tilt terms become,

$$A(t) = \frac{\int_S W(x,y,t)dxdy}{\int_S dxdy}, \quad B_1(t) = \frac{\int_S xW(x,y,t)dxdy}{\int_S x^2 dxdy},$$
$$B_2(t) = \frac{\int_S yW(x,y,t)dxdy}{\int_S y^2 dxdy} \tag{5.2}$$

After the instantaneous components of piston and tip/tilt are computed, they can be removed from the instantaneous wavefronts,

$$W(x,y,t;Aperture) = W(x,y,t) - (A(t) + B_1(t)x + B_2(t)y), \tag{5.3}$$

and the statistics of the apertured wavefronts, such as $OPD_{rms}(Aperture)$ and the power spectra, $S_W(f;Aperture)$ can be calculated. Note that the aperture, piston/tilt-removed wavefronts exactly equal the higher-order component of the unsteady wavefront in Equation (2.35). In this chapter we will use the notation $W(x,y,t;Aperture)$to emphasize the aperture dependence on the resulting wavefronts.

Figure 5.1 shows two wavefronts in the form of a periodic sine wave, with the same amplitude, over the same size of aperture. In Figure 5.1(a), the wavelength of the optical distortion is small and equal to one-third of the aperture size. In this case, piston and tilt components, indicated by a dashed line, are fairly small, and their removal does not significantly affect the piston/tilt-removed wavefront, plotted as a thick solid line. But when the wavelength of the optical distortion is larger than the aperture, with the example of the wavelength to be twice larger that the aperture shown in Figure 5.1(b), the apertured wavefront shows a significant degree of the piston and tilt. Removing these piston/tilt components from the original wavefront significantly reduces the amplitude of the variations in the resulting piston/tilt-removed wavefront, indicated by a thick solid line. Therefore, a finite aperture of the beam acts as a spatial filter, separating the distortions caused by larger-scale structures from those caused by smaller-scale structures. If $Aperture / \Lambda \ll 1$, then piston and tip/tilt components become the dominant part of the distortion. When these dominant components are removed, the resultant

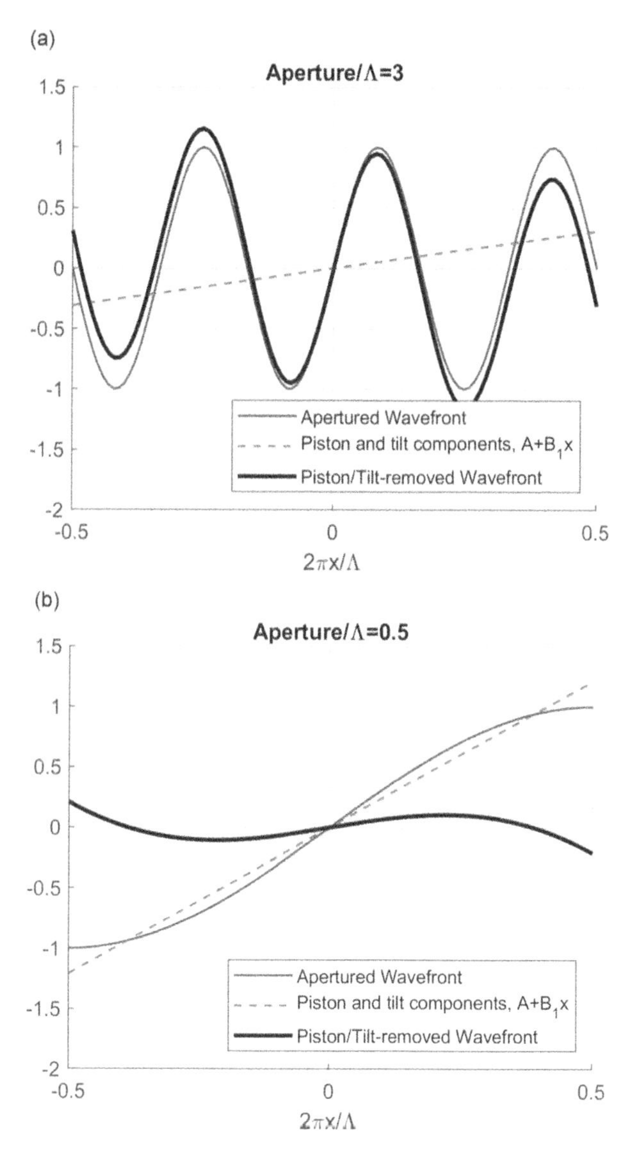

Figure 5.1 Examples of the aperture effect produced by the piston and tilt-removal for (a) $Aperture / \Lambda = 3$ and (b) for $Aperture / \Lambda = 0.5$.

levels of the piston/tilt-removed wavefront, $OPD_{rms}(Aperture)$, will be much smaller than the level of the original optical distortion.

It is useful to treat the wavefront as a sum of periodic harmonics. If the aperture effect on each harmonic is determined, the resulting $OPD_{rms}(Aperture)$ and $S_W(f;Aperture)$ can be found. Consider a simple one-dimensional harmonic wavefront, $W(x;\Lambda,\phi) = \sin(2\pi x / \Lambda + \phi)$. This wavefront represents a periodic spatial disturbance with a fixed wavelength, Λ, and some arbitrary phase offset, ϕ. Substituting this wavefront into Equation (5.2), we can get equations for piston and tip/tilt components. Knowing these components, we can compute the ratio $G(\Lambda; Aperture) = \frac{W_{rms}^2(Aperture)}{W_{rms}^2}$. This function, called an aperture function, can be treated as a filter, which attenuates the energy of a harmonic at a given wavelength, Λ. As the phase, ϕ, can be arbitrary, one way to compute the aperture function is to phase-average the piston/tilt removed harmonic over all phases,

$$G(\Lambda; Aperture) = \frac{\left\langle \int_S \left\{W(x;\Lambda,\phi) - \left[A(\Lambda,\phi) + xB_1(\Lambda,\phi)\right]\right\}^2 dxdy \right\rangle_\phi}{(1/2)\int_S dxdy}, \tag{5.4}$$

where the square brackets denote the phase averaging, and the factor (1/2) in the denominator accounts for the energy of the unbounded wavefront. Below we provide solutions for two important cases, a one-dimensional aperture and a circular aperture.

One-dimensional aperture with the aperture size, Ap

The equations for piston and tip/tilt components becomes,

$$A(\Lambda,\phi) = \frac{\int_{-Ap/2}^{Ap/2} \sin\left(\frac{2\pi x}{\Lambda} + \varphi\right) dx}{\int_{-Ap/2}^{Ap/2} dx} = \frac{\Lambda}{2\pi Ap}\left[\cos\left(\frac{\pi x}{\Lambda} + \varphi\right)\right]\Bigg|_{-Ap/2}^{Ap/2}$$

$$= \frac{\Lambda}{\pi Ap}\sin\left(\frac{\pi Ap}{\Lambda}\right)\sin(\phi) = \frac{\sin z}{z}\sin(\phi),$$

$$B_1(\Lambda,\phi) = \frac{\int_{-Ap/2}^{Ap/2} x \cdot \sin\left(\frac{2\pi x}{\Lambda} + \varphi\right) dx}{\int_{-Ap/2}^{Ap/2} x^2 dx} \tag{5.5}$$

$$= \frac{3}{2(Ap/2)^3}\left(\frac{\Lambda}{2\pi}\right)^2\left[\sin\left(\frac{2\pi x}{\Lambda} + \varphi\right) - \frac{2\pi x}{\Lambda}\cdot\cos\left(\frac{2\pi x}{\Lambda} + \varphi\right)\right]\Bigg|_{-Ap/2}^{Ap/2}$$

$$= \left(\frac{\Lambda}{\pi Ap}\right)^2 \frac{6}{Ap}\left[\sin\left(\frac{\pi Ap}{\Lambda}\right)\cos(\phi) - \frac{\pi Ap}{\Lambda}\cdot\cos\left(\frac{\pi Ap}{\Lambda}\right)\cos(\phi)\right]$$

$$= \frac{6\left[\sin z - z\cdot\cos z\right]}{Ap\cdot z^2}\cos(\phi)$$

where $z = \pi Ap / \Lambda$.

Substituting the values of the piston and the tilt components into Equation (5.4) and solving the integrals, the aperture transfer function, $G(z)$, becomes

$$G(z) = \frac{\frac{1}{2\pi}\int_0^{2\pi}\int_{-Ap/2}^{Ap/2}\left\{\sin\left(\frac{2\pi x}{\Lambda}+\varphi\right)-\left[A(\Lambda,\phi)+xB_1(\Lambda,\phi)\right]\right\}^2 dx d\varphi}{(1/2)\int_{-Ap/2}^{Ap/2} dx}$$

$$= 1 - \frac{2\cos^2(z)+1}{z^2} + 6\frac{\cos(z)\sin(z)}{z^3} + 3\frac{\cos^2(z)-1}{z^4}$$

For small z, $G(z) \sim \frac{\pi^4}{45} z^4$.

Circular aperture with the diameter, D

Writing Equation (5.2) in the polar frame of reference (r,α), so that $x = r\cos(\alpha)$, and using integral identities for the Bessel functions, $J_0(t) = \frac{1}{2\pi}\int_0^{2\pi} \exp(it\cos(\theta)d\theta$ and $\int_0^u tJ_0(t)dt = uJ_1(u)$, (Source: https://mathworld.wolfram.com/BesselFunctionoftheFirstKind.html) the equations for the piston and tilt components become,

$$A(\Lambda,\phi) = \frac{\int_0^{2\pi}\int_0^{D/2}\sin\left(\frac{2\pi r\cos(\alpha)}{\Lambda}+\varphi\right) r dr d\alpha}{\int_0^{2\pi}\int_0^{D/2} r dr d\alpha}$$

$$= \frac{\int_0^{D/2}\mathrm{Imag}\left\{\int_0^{2\pi}\exp\left(i\frac{2\pi r\cos(\alpha)}{\Lambda}\right)\exp(i\varphi)d\alpha\right\} r dr}{\pi(D/2)^2}$$

$$= \frac{2\pi\int_0^{D/2}\mathrm{Imag}\left\{J_0\left(\frac{2\pi r}{\Lambda}\right)\exp(i\varphi)\right\} r dr}{\pi(D/2)^2} = \frac{8\sin(\varphi)\int_0^{D/2} J_0\left(\frac{2\pi r}{\Lambda}\right) r dr}{D^2}$$

$$= \frac{2\Lambda}{\pi D}J_1\left(\frac{\pi D}{\Lambda}\right)\sin(\varphi) = \frac{2J_1(z)}{z}\sin(\phi),$$

$$B_1(\Lambda,\phi)=\frac{\int_0^{2\pi}\int_0^{D/2} r\cos(\alpha)\sin\left(\frac{2\pi r\cos(\alpha)}{\Lambda}+\varphi\right)rdrd\alpha}{\int_0^{2\pi}\int_0^{D/2}\left[r\cos(\alpha)\right]^2 rdrd\alpha} = \frac{16\left[2J_1(z)-z\cdot J_0(z)\right]}{Dz^2}\cos(\phi) \tag{5.6}$$

where $z=\pi D/\Lambda$, and J_0 and J_1 are Bessel functions of the zeroth and the first order, respectively.

Substituting the values of the piston and the tilt components into Equation (5.4) and solving the integrals, after considerable algebra the aperture transfer function, $G(z)$, becomes

$$G(z)=\frac{\frac{1}{2\pi}\int_0^{2\pi}\int_0^{2\pi}\int_0^{D/2}\left\{\sin\left(\frac{2\pi r\cos(\alpha)}{\Lambda}+\varphi\right)-\left[A(\Lambda,\phi)+r\cos(\alpha)B_1(\Lambda,\phi)\right]\right\}^2 rdrd\alpha d\varphi}{(1/2)\int_0^{2\pi}\int_0^{D/2} rdrd\alpha}$$

$$=1-\frac{16J_0^2(z)+4J_1^2(z)}{z^2}+64\frac{J_0(z)J_1(z)}{z^3}-64\frac{J_1^2(z)}{z^4}.$$

For small z, $G(z)\sim\frac{\pi^4}{64}z^4$.

Plots of the aperture transfer functions for both the 1-D and the circular apertures are shown in Figure 5.2. The transfer function is effectively a fourth-order high-pass filter, primarily attenuating the large wavelengths. For a fixed aperture, wavefront harmonics with large wavelengths, $\Lambda \geq Ap$ (or $z \leq \pi$), are significantly suppressed, while for small wavelengths, $\Lambda < Ap/2$ (or $z > 2\pi$), the energy of the piston-tilt-removed wavefronts is mostly unchanged.

Even when the tip/tilt is kept in the wavefronts, removing only the piston component will also reduce the overall levels of aero-optical distortions, although not quite as dramatically as piston/tilt removal. It is important to note that removing the piston component from the wavefronts *will not* affect the far-field intensity distribution, as demonstrated in Equation 2.36. Repeating the analysis with the removal of the piston component only,

$$G_{Piston}(\Lambda;Aperture)=\frac{\left\langle\int_S\left\{W(x;\Lambda,\phi)-A(\Lambda,\phi)\right\}^2 dxdy\right\rangle_\phi}{(1/2)\int_S dxdy}$$

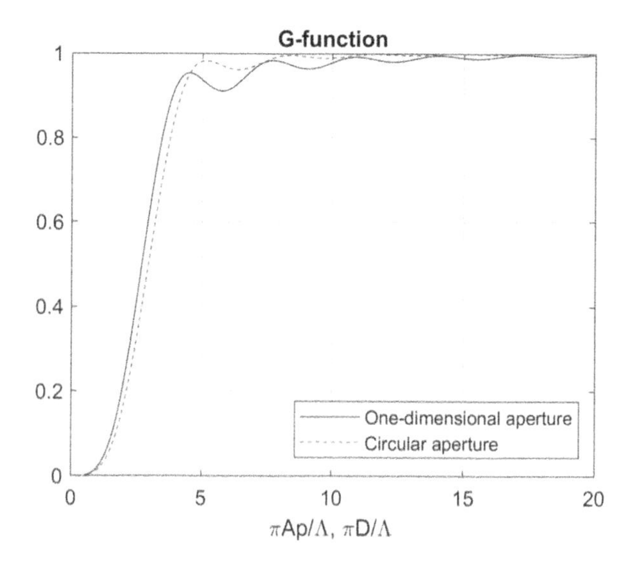

Figure 5.2 Aperture transfer function, *G*, for one-dimensional and circular apertures.

gives the following aperture transfer functions:

$$\textit{One-dimensional aperture of size, } Ap\text{: } G_{Piston}(z) = 1 + \frac{\cos^2(z) - 1}{z^2}, \quad z = \pi Ap / \Lambda.$$

$$\textit{Circular aperture of diameter, } D\text{: } G_{Piston}(z) = 1 - \frac{4J_1^2(z)}{z^2}, \quad z = \pi D / \Lambda.$$

For both cases, for small z, $G_{Piston}(z) \sim z^2$, so the piston-only-removed aperture function is a second-order high-pass filter.

Plots of the piston-only-removed aperture transfer functions for both the 1-D and the circular apertures are shown in Figure 5.3. Like the piston/tilt removed aperture function, G_{Piston}-function also behaves as a high-pass filter, although only a second-order filter. Removal of only the piston component affects the harmonics with $z < 2$, or with wavelengths $\Lambda > \pi Ap / 2$.

Once the aperture transfer function is established, the overall levels of aero-optical distortions over a finite-size aperture can be estimated, if the second-dimensional wavefront spatial spectrum, $S_{OPD}(\mathbf{k})$, is known, by multiplying the spectrum by the filter-type aperture function and integrating over the spatial wavenumber domain,

$$OPD_{rms}^2(Aperture) = \frac{1}{2\pi} \int_{\vec{k}} G(z = |\mathbf{k}| Ap / 2) S_{OPD}(\vec{k}) d\vec{k}.$$

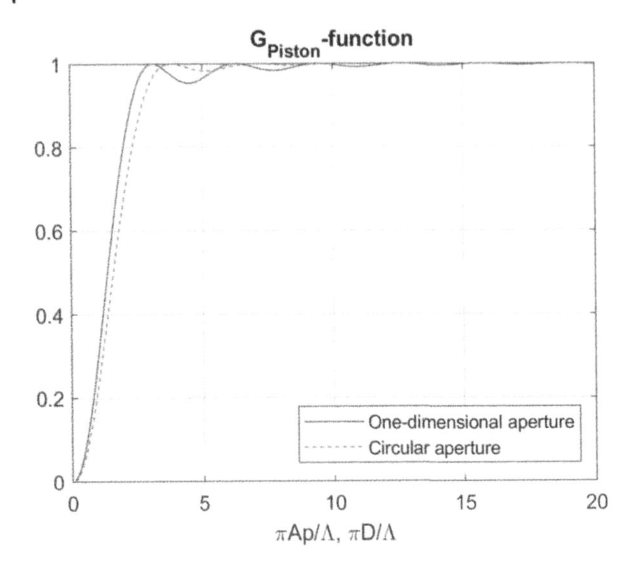

Figure 5.3 The only-piston-removed aperture function, G_{Piston}, for 1-D and circular apertures.

Quite often, only the temporal wavefront spectrum in the streamwise direction, $S_{OPD}(f)$, is available. Assuming frozen flow with a known convective speed, U_C, the z-argument in the aperture transfer becomes $z = \pi Ap / \Lambda = \pi fAp / U_C$ for one-dimensional aperture. In this case, the overall levels of aero-optical distortions over a finite aperture becomes,

$$OPD_{rms}^2(Ap) = 2\int_0^\infty G(z = \pi fAp / U_C) S_{OPD}(f) df \tag{5.7}$$

Both a Shack-Hartmann sensor and a Malley probe provide direct measurements of the streamwise deflection angles. For convective wavefronts, it has been shown in Equation (4.3), that the wavefront and the deflection angle spectra are related as, $S_\theta(f) = \left(\frac{2\pi f}{U_C}\right)^2 S_{OPD}(f)$. Plugging this relation into Equation (5.7) provides a very useful equation to estimate the aperture effects on the wavefront statistics directly from the deflection angle spectra,

$$OPD_{rms}^2(Ap) = 2U_C^2 \int_0^\infty G(z = \pi fAp / U_C) \frac{S_\theta(f)}{(2\pi f)^2} df \tag{5.8}$$

6

Typical Aero-Optical Flows

6.1 Scaling Arguments

To properly analyze experimental results, it is important to understand the scaling of these unsteady aberrations. Recall from Equation (2.14) that Optical Path Difference, *OPD*, is proportional to the integral of the density variations through the turbulent field, $OPD(x,y,t) \sim \int \rho'(x,y,z)dz$. For isentropic flows, $p'/p_\infty \sim \gamma\rho'/\rho_\infty$, where ρ_∞ and p_∞ are a freestream density and pressure, respectively. Noting that the pressure drop inside a vortical structure of characteristic size θ with a characteristic velocity u is $\Delta p \sim \rho_\infty u^2 \sim u^2 p_\infty / T_\infty \sim p_\infty (u/a)^2 \sim p_\infty M^2$, we get the following scaling for the variations in OPD_{rms},

$$OPD_{rms} \sim \theta\rho_\infty M^2, \tag{6.1}$$

where M is the convective Mach number and a is a freestream speed of sound. Thus, optical aberrations due to the vortical structures depend linearly on the freestream density and the typical structure size and the square of the convective Mach number. Later in this chapter we will demonstrate that this "ρM^2"-dependence of the levels of optical distortions is correct for most turbulent flows at subsonic speeds.

In general, optical aberrations are caused by density fluctuations inside the turbulent flow around various geometries. It can be a boundary layer on a skin of an aircraft, a separated shear layer, or a wake behind a beam steering configuration, like a turret. We can assume that these aero-optical distortions depend on the incoming-boundary layer thickness, δ, a typical vortical structure size/scale θ, a typical size of the turret, *D*, the freestream density, ρ_∞, the freestream Mach number, M, the Reynolds number, Re, the viewing angles, α and β, defined in Figure 6.29 later in this chapter. Finally, the flow and related aero-optical distortions might depend on geometric features of the turret, such as a flat or a

Aero-Optical Effects: Physics, Analysis and Mitigation, First Edition. Stanislav Gordeyev, Eric J. Jumper, and Matthew R. Whiteley.

conformal window, or a presence of various surface slope discontinuities, like cut-outs or gaps (Gordeyev et al. 2014).

$$OPD_{rms} = f(\delta, \theta, D, \rho_{\infty}, M, \mathrm{Re}, \alpha, \beta, geometry) \tag{6.2}$$

Using Equation (6.2) and applying dimensional analysis, we can get the following relationship for the OPD_{rms},

$$\frac{OPD_{rms}}{D} = \frac{\theta}{D}\frac{\rho_0}{\rho_{SL}} G(M) g\left(\frac{\delta}{D}, \alpha, \beta, \mathrm{Rc}; geometry\right) \tag{6.3}$$

From Equation (6.1), we can conclude that the Mach number dependent function, $G(M) \sim M^2$ at subsonic speeds. More details about this Mach number dependent function for some fundamental flows will be given later in this chapter.

Note that the aperture effects, discussed in the previous chapter, are not included in the provided scaling laws. The aperture effects depend on the relative ratio of the beam size and the characteristic spatial scale of aero-optical distortions. They also depend on whether only piston or both piston and tip/tilt components are removed from the optical data. Thus, they depend on the particular type of the adaptive optics corrective system. If the aperture size, Ap, is larger than the characteristic scale, Λ, of the optical distortions, the aperture-related reduction in levels of aero-optical distortions is generally small. Only when the beam size is smaller than Λ, the aperture effects should be properly accounted for.

6.2 Free Shear Layers

6.2.1 Shear-Layer Physics

Flows around typical geometries used for airborne optical systems create free shear layers. Some examples of separated flows are those downstream of turret apexes and flows over cavities. Figure 6.1 shows several examples of the shear-layer dominant flows.

Shear layers are formed between the adjacent regions of different streamwise speeds and densities, with the high speed U_1, and the low speed, U_2, $U_1 > U_2$, as shown schematically in Figure 6.2. Because of the speed mismatch, the velocity profile inside the shear layer has an inflection point, where the second derivative of the velocity profile is positive. The inflection point is convectively unstable and leads to the exponentially growing initial Kelvin-Helmholtz-type instabilities, schematically shown in Figure 6.2. These initial instabilities will eventually enter a nonlinear instability growth regime, where they start pairing and ultimately

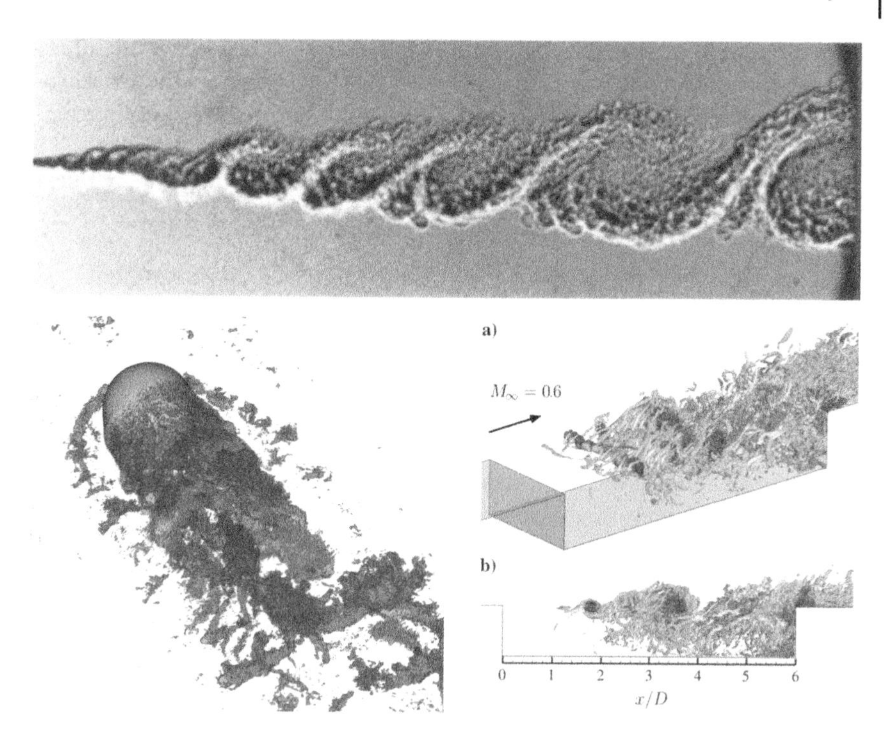

Figure 6.1 Examples of shear-layer flows: (a) A shadowgraph two-dimensional shear layer. *Source:* Brown et al., (1974), Reproduced with permission of Cambridge University Press. (b) Numerical simulations of separated shear-layer-dominant flow downstream of a turret. *Source:* Mathews et. al. (2016), figure 6 / American Institute of Aeronautics and Astronautics. (c) Numerical simulations of a flow over a cavity. *Source:* Sun et al. (2019), figure 3. Reproduced with the permission of the American Institute of Aeronautics and Astronautics, Inc.

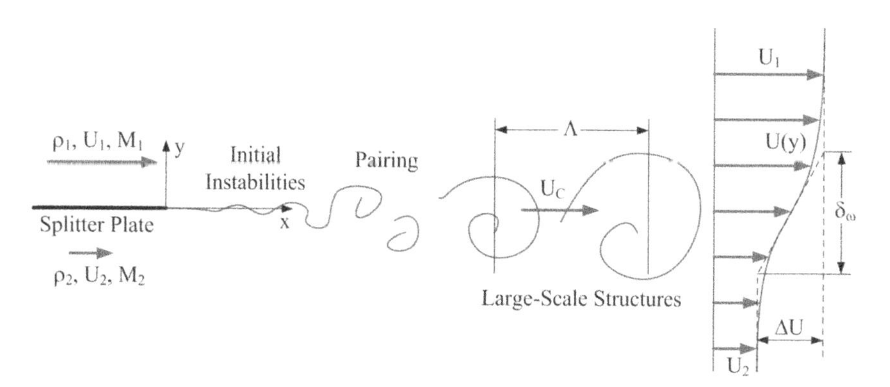

Figure 6.2 Schematic of the shear layer, with the definitions of the main parameters.

form large-scale vortical structures, moving at a constant convective speed, U_C (Drazin and Reid 1981; Huerre and Rossi 1998). The large-scale structures continue to pair, resulting in a continuously thickening shear layer. At any streamwise location, the shear layer is characterized by the mean velocity profile, $U(y)$. The local shear-layer thickness is typically characterized by the vorticity thickness, δ_ω,

$$\delta_\omega = \frac{\Delta U}{(dU/dy)_{max}}, \quad \Delta U = U_1 - U_2 \tag{6.4}$$

Graphically, the vorticity thickness is presented in Figure 6.2. Alternatively, shear-layer thickness can be quantified by the visual thickness, δ_{viz}; it is usually assumed that $\delta_{viz} = 2\delta_\omega$ (Papamoschou and Roshko 1988). Another important parameter to describe the large-scale structure is the average distance between consecutive structures, Λ, also shown in Figure 6.2. Various studies (Brown and Roshko 1974; Dimotakis 1986) have found that $\Lambda / \delta_{viz} = 1.5 - 2$.

As mentioned above, the shear layer grows in the streamwise direction. For the turbulent shear layer, it has been shown (Smits and Dussauge 1996) that the shear layer grows linearly with the distance. Using experimental results from shear layers with different gases and velocities, Papamoschou and Roshko (1988) showed that

$$\frac{d\delta_\omega}{dx} = 0.0875 \frac{(1-r)(1+\sqrt{s})}{(1+r\sqrt{s})} \tag{6.5}$$

where $r = U_2 / U_1$ is the velocity ratio and $s = \rho_2 / \rho_1$ is the density ratio across the shear layer. In case of the incompressible shear layer, $s = 1$, the growth rate is only a function of the velocity ratio, r. Consequently, the average distance between the large-scale structures also increases further downstream, with

$$\frac{d\Lambda}{dx} \approx 0.255 \frac{(1-r)(1+\sqrt{s})}{(1+r\sqrt{s})} \tag{6.6}$$

In many practical cases, the gas on both sides of the shear layer is air. In addition, most of the time the flow can be treated as an adiabatic one, that is, the total temperature on both sides is the same. Finally, for a steady (in the mean sense) shear layer, the static pressure is constant across the shear layer. Under these general conditions, the density ratio becomes a function of the Mach numbers of the high and the low-speed sides of the shear layer,

$$s = \frac{\rho_2}{\rho_1} = \frac{T_1}{T_2} = \left(\frac{1+0.2M_2^2}{1+0.2M_1^2} \right) \tag{6.7}$$

As the structures convect at the same convective speed, the dynamics of the shear layer depend on the velocity *relative* to this convective speed, $U(y)-U_C$. In particular, the dynamics depends on the convective Mach numbers, which are based on the velocity differences on both sides of the shear layer and the corresponding speeds of sound, $M_{C1}=(U_1-U_C)/a_1$ and $M_{C1}=(U_C-U_2)/a_2$, (Papamoschou and Roshko 1988). In the case of air and isentropic flow, the convective Mach numbers are the same, $M_{C1}=M_{C2}=M_C$, and is defined as

$$M_C=\frac{U_1-U_2}{a_1+a_2} \tag{6.8}$$

Finally, the convective speed can be estimated as follows,

$$U_C=\frac{a_2U_1+a_1U_2}{a_1+a_2} \tag{6.9}$$

This equation provides a good approximation (within a few percent) of the convective speed, if the convective Mach number, M_C is less than 0.4 (Barre et al. 1997).

6.2.2 Aero-Optical Effects

Due to their importance, canonical free shear layers were one of first fundamental flows to be experimentally studied to quantify their aero-optical effects; however, these early measurements used indirect, hot-wire inferred OPD_{rms}. This indirect method led to a presumption that shear-layer flows were essentially benign. The hot-wire-based methods incorrectly estimated the density field from velocity data by assuming a relationship between the velocity and the density based on the Strong Reynolds Analogy, SRA. We will further discuss SRA later in this chapter; however, this assumption and the further presumption that the unsteady pressure in shear layers is negligible, led early researchers to significantly underestimate free-shear-layer aberrations. Another method was to use dissimilar index of refraction gases and extract instantaneous index of refraction field via Rayleigh scattering technique (see a summary in Fitzgerald and Jumper 2004).

In the 1970s when incorrect experimental methods were used to attempt to measure the aberrations, there was another obstacle to properly estimate the extent of the aberrations imposed by a shear layer – the immaturity of computational fluid mechanics. The computation of separated flows made use of turbulence models to fill in the mixing in shear layers. The turbulence models attempted to predict the mean velocity profiles in the shear layer. These turbulence models gave an estimate on what they presumed was the unsteady velocity field in the shear layer, from which the same method of predicting density fluctuations used in the measurements, which presumed that the unsteady pressure in the layer was negligible. Computational methods making these assumptions appeared as late as 2006 (Pond and Sutton 2006). In a book published in the late 1990s, a statement was made that by not having information about pressure fluctuations in shear layers, it is presumed that they are negligible (Smits and Dussauge 1996).

By the early 1990s, numerical methods began to be able to routinely compute incompressible shear layers. These methods showed vortical structures like those seen experimentally as shown in Figure 6.1(a). However, they also failed to correctly predict the aero-optical distortions, caused by the shear layers.

6.2.3 Historical Shear Layer Measurements in AEDC

In order to perform direct measurements of aero-optical effects caused by transonic shear layers, a Mach 0.8 shear-layer facility was constructed in the Acoustic Research Tunnel (ART Facility) at the Arnold Engineering and Development Center (AEDC) in the early 1990s. The ART compressible shear-layer facility, as schematically in Figure 6.3, was designed to create a variety of tunnel pressure-altitudes and Mach numbers for the high- and low-speed sides of the shear layer. The tunnel had Schlieren-quality windows in the side walls of the facility, labeled #1, #2, and #3, which were centered approximately at the end of the splitter plate, half, and one meter downstream of the splitter plate, respectively. The sidewall windows were 30 cm in diameter; however, the windows centered at the same locations but on the top and bottom of the tunnel had only a 5 cm clear aperture.

The original intention was to use an interferometry technique, which was a standard technique of measuring density variations at the time; however, they found that interferometry was unable to reliably measure optical distortions

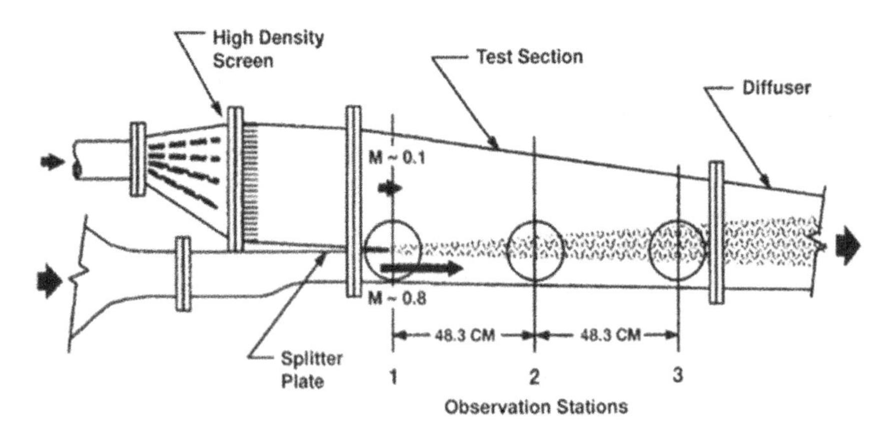

Figure 6.3 AEDC ART Facility modified to a Mach 0.8 Compressible Shear Layer Facility, with a Test Section Pressure Altitude of ~13,500 ft. *Source:* Havener and Heltsley (1994), figure 4. Reproduced with permission of the American Institute of Aeronautics and Astronautics, Inc.

because of the intense vibration environment of the facility. At about the same time, the SABT sensor, described in Chapter 3, was developed by researchers at the University of Notre Dame (Jumper and Hugo 1992). Using a heated jet facility at the University of Notre Dame, the SABT sensor was demonstrated to measure time-resolved one-dimensional slices of the wavefronts in the streamwise direction. More importantly, the sensor was largely insensitive to fairly large amounts of jitter, imposed on the sensor. Both papers on the preliminary interferometry measurements in ART facility and on the SABT technique were presented in AIAA meeting in 1992, resulting in a serendipitous meeting between Dr. Jumper and the lead engineer from the ATR facility. This meeting led to an invitation to implement this new sensor at the ART facility at AEDC.

The SABT sensor for the AEDC experiment used only two small-aperture beams separated by 2.5 cm, from which a time series of wavefronts were constructed over the full 5 cm aperture. In addition to optical data, the experiment collected concurrent accelerometer data. Indeed, the vibration environment in the tunnel was excessive, which showed up in the reconstructed one-dimensional wavefronts as a large streamwise tilt. Based on the measured accelerometer data, the vibration corruption was below 1 kHz. In order to remove this vibration-related corruption, the temporal deflection angle data used to compute wavefronts were high-pass filtered at 2.5 kHz before reconstructing the wavefronts. Wavefronts for the first and second beam locations were constructed, using Equation (3.7), and blended using a weighted average between small-aperture beams to reconstruct the time series of one-dimensional wavefronts over the 5 cm apertures.

It should be noted that the tip/tilt was not formally removed from the resulting wavefronts; however, the frame-by-frame tip/tilt was effectively removed by the high-pass spatial filter. The filter served as tip/tilt removal over the 5 cm aperture now well understood as an aperture effect; aperture effects will be discussed in more detail in Chapter 5. Essentially, for a fixed aperture size, Ap, the aero-optical distortions from structures with streamwise size, Λ, larger than the aperture size are now routinely removed from the resulting wavefronts. For convective structures, traveling at the speed, U_C, the relation between the spatial aperture size and the corresponding time, T, it takes to convect over the aperture is $T = Ap / U_C$. Consequently, the aperture effects, applied to the temporal deflection angle data, can be interpreted as a high-pass temporal filter, with the cutoff frequency, $f_{cut-off} \approx 1/T = U_C / Ap$. But if one wants to keep aero-optical structures with larger than the aperture size, i.e. $\Lambda > Ap$, the cutoff frequency should be chosen to be larger than

$$f_{cut-off} \approx U_C / \Lambda$$

High-pass filtering of the temporal deflection angle data removes not only vibrations with frequencies less than the cutoff frequency, but also any aero-optical structures with size larger than the aperture. By lowering the cutoff frequency to one that is larger than the vibration frequencies but still less than the frequencies caused by the larger structures passing by the aperture, the wavefronts will be retained uncorrupted by vibration; these structures larger than the aperture will show up as full-aperture tilt. In fact, if an approximation of structures larger than the aperture is desired, then the data from the first small-aperture beam can be extended upstream by using negative time and downstream by using future time to extend it downstream, as described in Chapter 3 in the discussion of the Malley Probe. In the original paper describing the AEDC testing (Hugo et al. 1997), the filtering initially not only removed vibration but also removed tilt over the aperture; the initially applied cut-off frequency removed full-aperture tilt consistent with the 5 cm aperture. In a later paper, Fitzgerald and Jumper (2002), the data were filtered to remove vibration, but retain tilt from aero-optical structures of wavelengths ~ 20 cm, i.e., $f_{cut-off} = 750\,Hz$. This value of the cut-off frequency allowed minimizing vibrational noise in the data without removing frequencies corresponding to flow structure sizes larger than the 5 cm aperture. It should be noted that extending the effective wavefront length up and downstream from the measurement location has its limitations. While providing an excellent estimation of the large-scale structures, it does not take into account evolution of the structures in the upstream and downstream region from the measurement location.

The data for AEDC Station 2, filtered at 750 Hz, is shown in Figure 6.4. The wavefronts are clearly dominated by large-scale structures, passing through the 5 cm aperture. The extracted OPD_{rms} over the 5 cm aperture including the tilt was 0.242 μm, which was several times larger than theoretical estimates from the models available prior to that time.

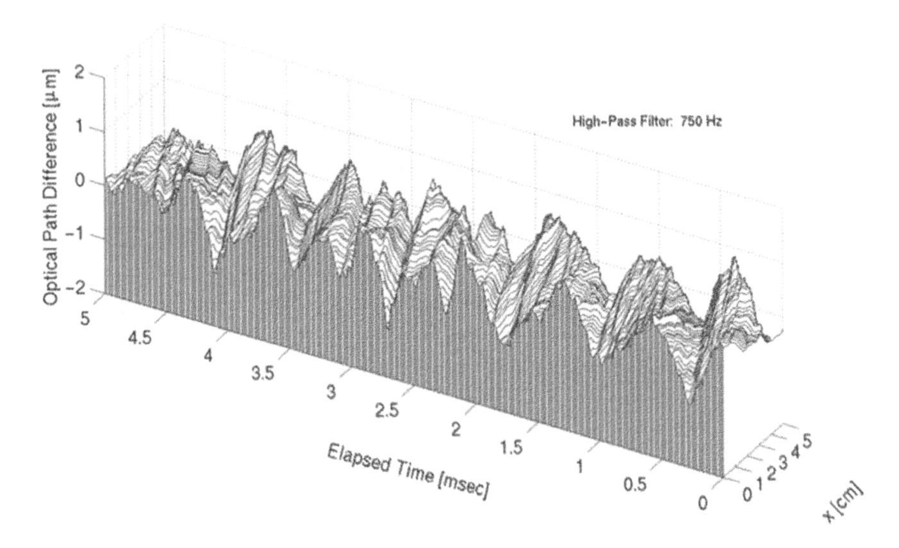

Figure 6.4 AEDC wavefront data, with high-pass filter set at 750 Hz. *Source:* Courtesy of Eric Jumper.

As mentioned earlier, a piston component is also introduced as the wavefront passes by the aperture. After removing vibration by high-pass filtering at 750 Hz, the filtered data over the 5 cm aperture still contains the increasing and decreasing piston component. This piston component was removed from each frame of the wavefront; the amount of piston removed is shown in Figure 6.4. Over a larger aperture, this removed optical piston component would result in a significant increase in the overall aero-optical levels. As a further demonstration that a large, cyclic wavefront is passing by the 5 cm aperture, it is of interest to examine the removed piston component. Figure 6.5 plots temporal evolution of the experimentally extracted piston component. From this Figure, a typical period of the associated large-scale structure was estimated to be 0.72 msec, which gives an estimated structure size $\Lambda \approx 11.3\text{cm}$ (Fitzgerald and Jumper 2002).

Using this estimated size of the large-scale structure, the visual thickness of the shear layer, δ_{viz}, could also be estimated for comparison with a numerical simulation implementing a discrete vortex method. The results of the simulations are also presented in Figure 6.5, and compare well in amplitude and phase with the

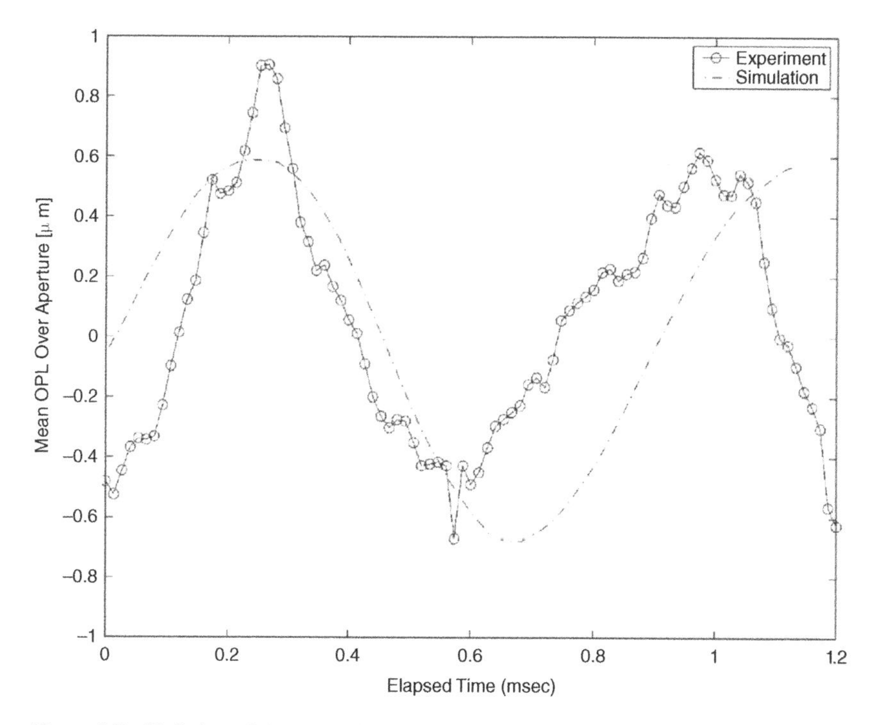

Figure 6.5 Variation of the mean *OPD* removed in *OPD* calculation with time for reconstruction of the wavefronts with a 750 Hz high-pass filter applied. The dashed line is a numerical simulation of the AEDC conditions at AEDC Station 2. *Source:* Fitzgerald and Jumper (2002), figure 4. Reproduced with permission of the American Institute of Aeronautics and Astronautics, Inc.

experimental results. The numerical simulations explicitly demonstrated the presence of the large-scale flow structure. The underlying simulation model is called the Weakly Compressible Model, which will be discussed in the next section.

The presented and the follow-up analysis of the AEDC data clearly demonstrated, for the first time, that aero-optical effects due to shear layers are significant (Fitzgerald and Jumper 2002b) and should thereafter be taken into account when designing airborne laser systems. It led to the re-thinking of many fundamental flow assumptions used to estimate aero-optical effects and the design of airborne systems (Jumper and Fitzgerald 2001). The results from the AEDC experiments ultimately resulted in the renewal of funded research in aero-optics, which continues to this day.

As a final note, the SABT results were also extended beyond the physical aperture of 5 cm using the frozen-flow assumption discussed earlier, extending the aperture to an equivalent of 20 cm aperture. As unexpected as the OPD_{rms} of 0.242 μm was for the 5 cm aperture, extending the aperture to 20 cm increased the OPD_{rms} to 0.43 μm. For spatially evolving shear layers, this frozen-flow assumption would lead to underestimating the aero-optical effects at larger apertures. To address this issue, a Compressible Shear Layer facility with large optical windows was constructed at the University of Notre Dame. It allowed directly measuring aero-optical distortions of natural and forced shear layers at transonic speeds over apertures as large as 25 cm. Analysis of these data helped determine the proper method of scaling aero-optical distortions for shear-layer dominant flows, like flows over turrets for geometrically similar turrets of different size, Mach number, and altitude. The results collected in this facility agreed quite well with the AEDC results, confirming the unexpectedly large values of aero-optical distortions by shear layer.

6.2.4 Weakly Compressible Model

As a result of experiments in AECD facilities using the SABT sensor, it was unequivocally demonstrated that the presumption of insignificant aero-optical distortions by shear layers was incorrect. However, models of that day were unable to explain these large experimentally observed optical distortions, underpredicting the values of OPD_{rms} by as much as a factor of five (Fitzgerald and Jumper 2004). The reason for this large discrepancy can be traced to the assumption of the negligible pressure fluctuations in turbulent flows. As we will discuss later in this chapter, it is a good assumption for turbulent boundary layers, where a wide range of structures with different scales is present. On the other hand, because of the inviscid instability mechanism, the main feature of shear layers is a formation of large-scale vortical structures, clearly observed in Figure 6.1. These coherent vortical, tornado-like structures are characterized by a strong radial pressure gradient accompanied by significant pressure fluctuations. Based on this fact, Jumper and Hugo (1992) and Fitzgerald and Jumper (2002, 2004) developed a physics-based numerical procedure for approximating the density field, and thus the refractive index, from known two-dimensional velocity-field data for total-temperature-matched, weakly compressible, two-dimensional shear layers. Briefly, neglecting viscosity effects, the velocity, density, and pressure fields are connected by the two-dimensional Euler equations,

$$\begin{cases} \rho\left(\dfrac{\partial u}{\partial t}+u\dfrac{\partial u}{\partial x}+v\dfrac{\partial u}{\partial y}\right)=-\dfrac{\partial P}{\partial x} \\ \rho\left(\dfrac{\partial v}{\partial t}+u\dfrac{\partial v}{\partial x}+v\dfrac{\partial v}{\partial y}\right)=-\dfrac{\partial P}{\partial y} \end{cases}$$

For a given velocity field, the Euler equations become a Laplace equation for the pressure field,

$$\nabla^2 P=-\frac{\partial}{\partial x}\left[\rho\left(\frac{\partial u}{\partial t}+u\frac{\partial u}{\partial x}+v\frac{\partial u}{\partial y}\right)\right]-\frac{\partial}{\partial y}\left[\rho\left(\frac{\partial v}{\partial t}+u\frac{\partial v}{\partial x}+v\frac{\partial v}{\partial y}\right)\right]$$

Once the pressure field is computed, the simplified isentropic equation can be used to estimate the temperature field,

$$\frac{T(x,y,t)}{T_{ad}(x,y,t)}=\left(\frac{P(x,y,t)}{P_\infty}\right)^{\frac{\gamma-1}{\gamma}},\ T_{ad}(x,y,t)=T_0-\frac{u^2(x,y,t)+v^2(x,y,t)}{2C_p}$$

Here it is assumed that the total temperature is constant inside the shear layer. Finally, using the ideal gas law, the density field can be estimated. The estimated density field can be plugged back into the equation for the pressure field and the process is repeated until the final density field is converged. This model, called the Weakly Compressible Model, takes into account the radial pressure gradients required to sustain the vortical flow and curved path lines. Figure 6.6 compares

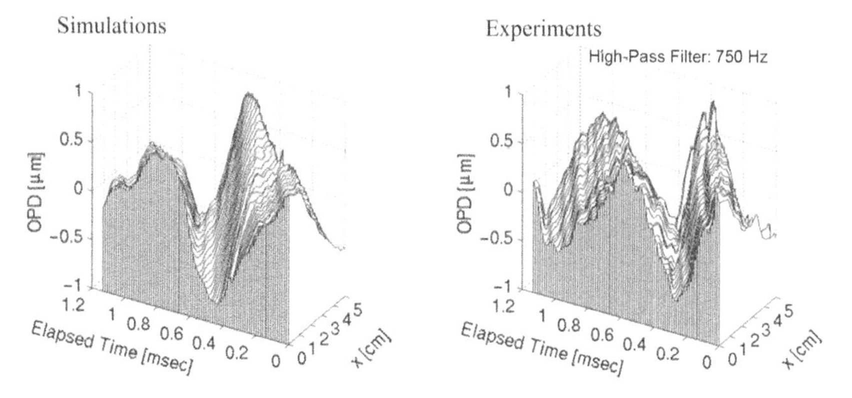

Figure 6.6 Comparison of the simulation to the experimentally reconstructed wavefront as viewed passing by a 5 cm aperture. *Source:* Courtesy of Eric Jumper.

the simulation, performed using the Weakly Compressible Model and reconstructed wavefronts from the AECD experiment passing by a 5.0 cm aperture. The model predicted reasonably well the experimentally observed aero-optical aberrations for a high subsonic shear layer, when other shear-layer models failed to do so. It also validates the use of the temporal filter and its relationship with tilt removal for the early AEDC experiments.

The Weakly Compressible Model was instrumental in understanding the physical mechanism of the aero-optical distortions by shear layers. However, the model requires knowledge of the spatially/temporally resolved velocity field to compute aero-optical distortions. Below we will provide physics-based scaling arguments to develop a scaling law for the aero-optical effects of shear layers.

To reiterate, the main source of aero-optical distortions in the subsonic and transonic shear layers are regions of lower pressure, referred to as pressure "wells," inside the large-scale coherent vortical structures rotating flow in the streamwise/wall-normal plane (Fitzgerald and Jumper 2004). The pressure gradient of these "wells" balance the centripetal acceleration of the rotating flow which keeps the vortical structures together

$$\frac{\partial P(r)}{\partial r} = \rho \frac{u_\theta^2(r)}{r} \quad (6.10)$$

Here r is the radial distance from the center of the vortical structure and u_θ is the azimuthal velocity inside the vortical structure. As the shear layer is formed by the velocity mismatch, the azimuthal velocity is proportional to the difference between the convective speed of the vortical structure, Equation (6.9), and either the higher or the lower speeds outside the shear layer,

$$u_\theta \sim (U_1 - U_c) = (U_c - U_2) = \frac{a_1 U_1 - a_2 U_2}{a_1 + a_2} \quad (6.11)$$

Combining Equations (6.10) and (6.11), it follows that the pressure drop, ΔP, inside the vortical structure is

$$\Delta P \sim \rho_{ref} (U_1 - U_c)^2 \quad (6.12)$$

where ρ_{ref} is some reference density. Assuming isentropic flow and small density/pressure variations, we can relate the pressure and the density fluctuations as

$$\frac{\Delta\rho}{\rho_{ref}} = \frac{1}{\gamma}\left(\frac{\Delta P}{P_\infty}\right) \quad (6.13)$$

Substituting Equations (6.12) into (6.13), the density variations become

$$\Delta\rho = \frac{1}{\gamma P_\infty}\left((U_1 - U_c)\rho_{ref}\right)^2 = \frac{1}{\gamma P_\infty}\left((U_1 - U_c)\rho_{ref}\right)^2 \quad (6.14)$$

Often, the shear layers are created when the flow separates and the high speed of the resulted shear layer is the freestream speed, $U_1 = U_\infty$; in this case, it is useful to use the freestream density as a reference density, $\rho_{ref} = \rho_\infty$. Using the equation of state and the definition of freestream Mach number, $M_\infty = \sqrt{\gamma R T_\infty}$, Equation (6.14) can be written as

$$\Delta\rho = \rho_\infty \left(\frac{U_1 - U_c}{a_1} \right)^2 = \rho_\infty \left(\frac{U_1 - U_c}{a_1 + a_2} \frac{a_1 + a_2}{a_1} \right)^2 = \rho_\infty M_c^2 \left(1 + \frac{a_2}{a_1} \right)^2$$

where the definition of a convective Mach number from Equation (6.8) is used. For subsonic and transonic speeds, $M_\infty < 1$, $\frac{a_2}{a_1} = \left(\frac{T_2}{T_1} \right)^{1/2} = \left(\frac{1 + 0.2M_1^2}{1 + 0.2M_2^2} \right)^{1/2} < (1 + 0.2M_\infty^2)^{1/2}$ and $1 < a_2 / a_1 < 1.1$. Thus, we can assume that $a_2 / a_1 = 1$. Using the definition of OPD from Equation (2.14) and assuming that the spatial variation of OPD, OPD_{rms} is proportional to OPD, we can finally get the scaling for the aero-optical distortions of the shear layers,

$$OPD_{rms} = AK_{GD} \rho_\infty M_c^2 L \tag{6.15}$$

where L is a characteristic thickness of the shear layer along the beam propagation direction.

Assuming the laser beam is sent through the shear layer normal to the layer, the shear layer vorticity thickness, δ_ω, can be used for L. Alternatively, the streamwise spacing between the vortical structures, Λ, can also be used for L, as $\Lambda \sim \delta_\omega$. The latter one can be directly measured in practice by cross-correlating wavefronts in the streamwise direction. Figure 6.7 shows the aero-optical distortions of shear layers measured both in tunnels (Kemnetz 2019) and in-flight (Kalensky et al. 2019), demonstrating that the scaling in Equation (6.15) does an excellent job collapsing the experimental data.

In Equation (6.1), it was argued that the levels of aero-optical distortions due to the turbulent structures at subsonic speeds should be proportional to the freestream density, the freestream Mach number squared and the characteristic size of the optically aberrating structure scales as, $OPD_{rms} \sim \rho_\infty M_\infty^2 \Lambda$, the so-called ρM^2-scaling. Using in-flight measurements (Kalensky et al. 2019), indeed it was observed that aero-optical distortions due to a shear layer are proportional to $(\rho_\infty / \rho_{SL}) M_\infty^2 \Lambda$. However, it is easy to see that the "ρM^2"-scaling is valid only in cases when the lower Mach number is proportional to the freestream Mach number and, therefore, the convective Mach number becomes proportional to the freestream Mach number. On the other hand, scaling in Equation (6.15), is more general, as it is valid for any combination of M_1 and M_2.

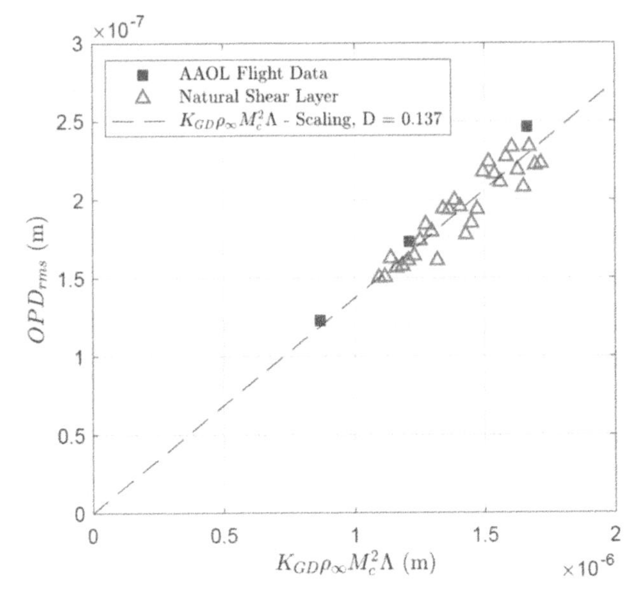

Figure 6.7 Scaled aero-optical distortions for the shear layers, collected in wind tunnel and flight experiments. *Source:* Kemnetz (2019), figure 8.7. Courtesy of Matthew Kenmetz.

6.3 Boundary Layers

Historically, attached turbulent boundary layers were the first aero-optic flow studied. Hans Liepmann conducted the first theoretical study of optical distortions caused by compressible boundary layers, published as a Douglas Aircraft Company Technical Report (Liepmann 1952). He studied a jitter angle of a thin beam of light as it traveled through a compressible boundary layer on the sides of high-speed wind tunnels to quantify the crispness of Schlieren photographs. A significant piece of work was done in 1956 by Stine and Winovich (1956). They performed photometric measurements of the time-averaged radiation field at the focal plane of a receiving telescope. Work on the turbulent boundary layer intensified in the late 1960s and through the following decade due to an interest in placing lasers on aircraft. Rose (1979) conducted the most extensive (at that time) experimental studies of optical aberrations caused by a turbulent boundary layer. He conducted hot-wire measurements in turbulent boundary layers in order to indirectly obtain their density fluctuations, $\rho_{rms}(y)$, and associated correlation lengths, $\Lambda_\rho(y)$, assuming that pressure fluctuations inside the boundary layer were zero, consistent with the Strong Reynolds Analogy, (Morkovin 1962). Unlike shear layers, where pressure variations play a dominant role in creating aero-optical distortions, pressure fluctuations inside boundary layers are indeed smaller, as will be discussed below. The density fluctuations and correlation lengths were used to estimate wavefront aberrations that would be imprinted on a laser beam propagated through the same turbulent boundary layer assuming homogeneous turbulence. The on-average wavefront aberrations, in the form of OPD_{rms}, were estimated using Sutton's linking equation, Equation (2.19). Rose (1979) empirically found OPD_{rms} to be proportional to dynamic pressure, q, and boundary layer thickness, δ, such that $OPD_{rms} \sim q\delta$.

These aircraft hot-wire measurements were complemented by the work of Gilbert (1982), who performed interferometer measurements. In this work the interferometer used a double-pulse technique, which measured the difference in the wavefront from one pulse to another, rather than the distorted wavefront at a given instant. Only a limited number of these measurements were made. Gilbert (1982) reported that the interferometry *generally* supported the hot-wire, integral-method estimations of the OPD_{rms}; however, based on his work it was concluded that the square of the OPD_{rms} depended linearly on the dynamic pressure, $OPD_{rms}^2 \sim q$.

A review of the major publication in aero-optics (Gilbert & Otten 1982) from the 1970s demonstrated that work up until 1982 focused on the measurement of the time-averaged, spatial, near-field optical distortion, OPD_{rms}, either by direct optically based methods, or assessed indirectly using fluid-mechanic measurements via the linking equation, Equation 2.19. Optical methods applied at that time to the measurement of the near-field time-averaged phase variance included direct interferometry, pulsed interferometry, and shearing interferometry. These interferometric methods provided a time-averaged assessment of the optical phase variance over the aperture; however, these methods provided no information concerning either temporal frequencies or wavefront spectra.

Masson et al. (1994) revisited the Gilbert (1982) and Rose (1979) data and concluded that after removing systematic errors from Gilbert's data, $OPD_{rms} \sim (\rho M^2)^{1.16}$. Also, Masson et al. (1994) found that there appeared to be a systematic difference between direct and indirect wavefront error measurements, with the interferometric estimates consistently yielding higher estimates of the OPD_{rms} than the hot-wire estimates but could not offer a reasonable explanation as to why optical and hot-wire data did not agree in magnitude.

These early measurements, due primarily to a lack of necessary accuracy, had conflicting conclusions about the OPD_{rms} dependence on the boundary-layer parameters. As a reminder, Rose (1979) suggested that $OPD_{rms} \sim q\delta$, while Gilbert (1982) concluded that $OPD_{rms} \sim \sqrt{q}$; finally, Masson et al. (1994) proposed that $OPD_{rms} \sim (\rho M^2)^{1.16}$. The last two dependencies contradict the general, similarity-based scaling derived in Equations (6.1) or (6.3), which showed that OPD_{rms} should be proportional to the freestream density, the boundary layer thickness and the square of Mach number.

This accuracy problem was greatly improved with the introduction of the Malley probe, which has high spatial and temporal resolution. The pioneering use of this new wavefront-sensing device in making optical measurements in turbulent subsonic boundary layers (Gordeyev et al. 2003) has shown that it yields the most accurate and highly time-resolved information about optical distortions for turbulent boundary layers with bandwidths >100 kHz. Analysis of early Malley-probe optical data (Gordeyev et al. 2003; Wittich et al. 2007) has further confirmed Rose's findings and shown that optical distortions are indeed proportional to the boundary layer thickness, the freestream density and the square of the freestream Mach number, consistent with the scaling in Equation (6.1). The Malley probe also provided first direct measurements of the convective speed of the aero-optical distortions. In subsonic boundary layers, optical distortions convect at a constant speed of 0.82–0.85 of the freestream speed, which suggests that optically active structures reside in the outer portion of the boundary layer.

In addition to experimental studies, few studies to numerically compute optical aberrations inside turbulent boundary layers have been made. Truman and Lee (1990) and Truman (1992) used a DNS spectral method to calculate time-dependent optical distortions for a low-Reynolds-number boundary layer, where the density fluctuations were computed from temperature fluctuations under the assumption of constant pressure. Large-scale streamwise elongated regions of highly correlated optical distortions were found and the link between highly anisotropic hairpin vortical structures leading to the optical distortions were observed. Also it was found that the optical distortions were anisotropic and vary significantly with the inclination angle to the wall. Tromeur et al. (2002, 2003, 2006) calculated optical aberrations by a compressible turbulent boundary layer at subsonic ($M = 0.9$) and supersonic ($M = 2.3$) speeds for $\mathrm{Re}_\theta = 2917$ using large-eddy simulations, which compared favorably with some limited experimental data by Deron et al. (2002). They found that the optical aberrations traveled at 0.8 of the freestream speed and were dominated by large-scale structures residing in the outer portion of the boundary layer. More recent high-fidelity numerical studies of aero-optical effects in turbulent boundary layers were performed by White & Visbal (2012, 2013), Wang and Wang (2013), and Kamel et al. (2016), which also confirmed that the optically active structures reside in the outer portion of the boundary layer.

Equation (6.1), which states that $OPD_{rms} \sim \rho\delta M^2$, was derived using simple scaling arguments. Later in this chapter, a more general expression for OPD_{rms} will be derived in Equation (6.22), showing that $OPD_{rms} \sim \rho\delta M^2\sqrt{C_f}$ for subsonic speeds. Some of the early aero-optical measurements, mentioned above and re-scaled using this expression are shown in Figure 6.9. The data generally agree well with the functional dependence, with larger deviations at low $\rho\delta$ values, where OPD_{rms}

Figure 6.8 Large-scale vortical structures in a turbulent boundary layer, flow goes from left to right. Courtesy of Thomas Corke, reproduced from Van Dyke (1982).

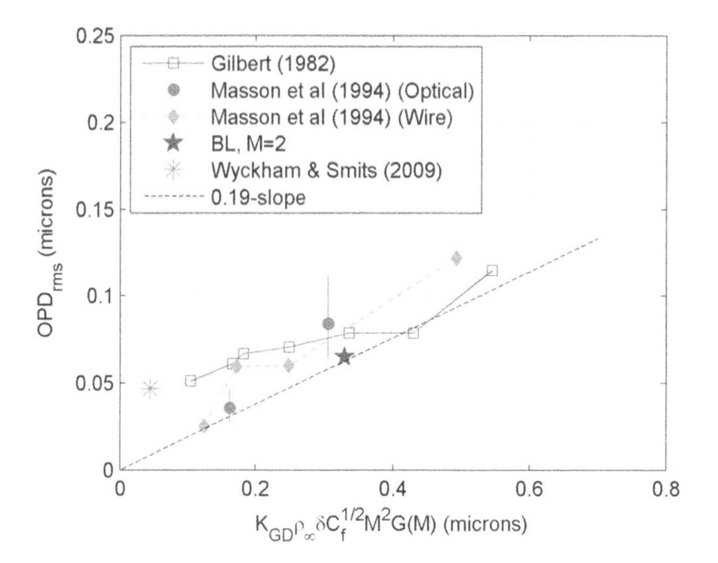

Figure 6.9 Some of early OPD_{rms} measurements in subsonic boundary layers, plotted versus the dependence given in Equation 6.22. *Source:* Gordeyev et al. (2014), figure 14 Reproduced with the permission of Cambridge University Press.

are small; these deviations are primarily due to lack of accurate optical measurements for these low OPD_{rms} values.

Wyckham and Smits (2009) pioneered the use of a high-speed digital camera for Shack-Hartmann wavefront sensing to measure aero-optical distortions in both subsonic and supersonic boundary layers. Their experimentally obtained OPD_{rms} for $M = 0.78$, shown in Figure 6.9 as an asterisk, is a factor of 3.5 higher than it was measured in the other studies, which raises some concerns about the quality and possible contamination in measurements. However, the model proposed by them, with a properly adjusted constant, agreed favorably with the experimental results and a more rigorous model, presented in the next section.

6.3.1 Model of Aero-optical Distortions for Boundary Layers with Adiabatic Walls

Unlike shear layers, where the pressure fluctuations are the dominant cause of the aero-optical distortions, numerous experimental (Spina et al. 1994; Smits and Dussauge 1996) and numerical (Gaviglio 1987; Guarini et al. 2000; Duan et al. 2010; Wang and Wang 2013) studies in boundary layers have shown that the time-averaged pressure fluctuations are several times smaller than temperature fluctuations. Thus, these pressure fluctuations can be neglected and the temperature fluctuations can be estimated using the Strong Reynolds Analogy, SRA, (Spina et al. 1994; Smits and Dussauge 1996; Wyckham and Smits 2009), which also presumes that p' is negligible,

$$\frac{T_{rms}(y)}{T_\infty} = r(\gamma - 1)M_\infty^2 \frac{u_{rms}(y)}{U_\infty}\frac{U(y)}{U_\infty} \tag{6.16}$$

Here $U(y)$ is the mean local streamwise velocity profile, the subscript "rms" denotes the temporal root-mean square values and "∞" denotes freestream values. The recovery factor, r, is defined as $r = (T_r - T_\infty)/(T_0 - T_\infty)$, where T_0 and T are the total and recovery temperature at the wall, respectively; in turbulent boundary layers with air as the fluid in motion, r is typically about 0.89. The SRA has been empirically verified for Mach numbers up to three (Spina et al. 1994).

Thus, for negligible pressure fluctuations, the equation of state, $p = \rho RT$, can be used to compute the density fluctuations,

$$\begin{aligned}\frac{\rho_{rms}(y)}{\rho(y)} &= \frac{T_{rms}(y)}{T(y)},\\ \rho_{rms}(y) &= T_{rms}(y)\frac{\rho_\infty}{T_\infty}\frac{\rho(y)/\rho_\infty}{T(y)/T_\infty} = \rho_\infty \frac{T_{rms}(y)}{T_\infty}\frac{1}{(T(y)/T_\infty)^2}\end{aligned} \tag{6.17}$$

Assuming the self-similarity of the mean velocity profile, $U(y)/U_\infty = f(y/\delta)$, the adiabatic relation between the static temperature and the velocity becomes,

$$\frac{T(y)}{T_\infty} = \left(1 + \frac{(\gamma-1)}{2}M^2\left[1 - \left(\frac{U(y)}{U_\infty}\right)^2\right]\right) = \left(1 + \frac{(\gamma-1)}{2}M^2\left[1 - f^2(y/\delta)\right]\right), \tag{6.18}$$

From a definition for the compressible skin friction coefficient, $C_f=(\rho_W u_\tau^2)/(0.5\rho_\infty U_\infty^2)$, it follows that $u_\tau = U_\infty\sqrt{\frac{\rho_\infty}{\rho_W}}\sqrt{C_f/2}$. Using this expression and the mixed scaling for compressible boundary layers (Morkovin 1962), $\sqrt{\frac{\rho(y)}{\rho_W}}\frac{u_{rms}(y)}{u_\tau} = g(y/\delta)$, where u_τ is the skin friction velocity and ρ_W is the density near the wall, the fluctuating velocity profile can be obtained as,

$$u_{rms}(y) = U_\infty\sqrt{\frac{\rho_\infty}{\rho(y)}}\sqrt{C_f/2}\cdot g(y/\delta) = U_\infty\sqrt{\frac{T(y)}{T_\infty}}\sqrt{C_f/2}\cdot g(y/\delta) \tag{6.19}$$

Substituting Equations (6.16), (6.18), and (6.19) into Equation (6.17), we can get the following expression for density fluctuations across the boundary layer,

$$\begin{aligned} \rho_{rms}(y) &= \rho_\infty\frac{T_{rms}(y)}{T_\infty}\frac{1}{\left(T(y)/T_\infty\right)^2} \\ &= \rho_\infty(\gamma-1)rM_\infty^2\sqrt{\frac{T(y)}{T_\infty}}\sqrt{C_f/2}\cdot g(y/\delta)f(y/\delta)\frac{1}{\left(T(y)/T_\infty\right)^2} \\ &= \rho_\infty(\gamma-1)rM_\infty^2\sqrt{C_f/2}\frac{f(y/\delta)g(y/\delta)}{\left(1+\frac{(\gamma-1)}{2}M^2\left[1-f^2(y/\delta)\right]\right)^{3/2}} \end{aligned} \tag{6.20}$$

Finally, substituting the estimated density fluctuations, Equation (6.20), into the linking equation, Equation (2.19), the equation for OPD_{rms} becomes,

$$OPD_{rms} = \sqrt{2}K_{GD}\rho_\infty\delta(\gamma-1)rM^2\sqrt{C_f/2}\times \left(\int_0^\infty\left[\frac{f(y/\delta)g(y/\delta)}{\left(1+\frac{(\gamma-1)}{2}M^2\left[1-f^2(y/\delta)\right]\right)^{3/2}}\right]^2\frac{\Lambda_y}{\delta}(y/\delta)d(y/\delta)\right)^{1/2} \tag{6.21}$$

or,

$$OPD_{rms} = BK_{GD}\rho_\infty M^2\delta\sqrt{C_f}F(M) \tag{6.22}$$

where

$F(M)=C(M)/C(0)$, with

$$C(M)=(\gamma-1)r\left(\int_0^\infty\left[\frac{f(y/\delta)g(y/\delta)}{\left(1+\frac{(\gamma-1)}{2}M^2\left[1-f^2(y/\delta)\right]\right)^{3/2}}\right]^2\frac{\Lambda_y}{\delta}(y/\delta)d(y/\delta)\right)^{1/2} \tag{6.23}$$

$B = C(0)$, and $\Lambda_y(y/\delta)$ is a wall-normal density correlation length. Compared with Equation (6.3), the G-function in that equation can be written as $G(M) = M^2 \cdot F(M)$. Various investigations provided some estimates for the density correlation lengths and are summarized in Figure 6.10. Using commonly accepted velocity profiles, $f(y/\delta)$ and $g(y/\delta)$, shown in Figure 6.19, the resulted Mach dependent functions, $F(M)$, were calculated for different $\Lambda_y(y/\delta)$. The results, plotted in Figure 6.11, show that the $F(M)$-function is not very sensitive to a particular choice of the density correlation function. The F-function is approximately constant for subsonic speeds, but monotonically decays at supersonic and hypersonic speeds.

Wyckham and Smits (2009) proposed another model to predict aero-optical boundary-layer distortions, where they used a bulk-flow approach to derive OPD_{rms}-dependence,

$$OPD_{rms} = C_w K_{GD} \rho_\infty \delta M^2 \sqrt{C_f} F(M) \tag{6.24}$$

where $F(M) = \left(1 + \frac{\gamma - 1}{2} M^2 \left[1 - r(U_c(M)/U_\infty)^2\right]\right)^{-3/2}$ for adiabatic walls.

Here $U_C(M)$ is the convective speed of aero-optical structures which, as will be shown later in this chapter, does depend on the Mach number; see Figure 6.13.

Note, that both the models, Equations(6.22) and (6.24), have the same functional form, $OPD_{rms} \sim K_{GD} \rho_\infty \delta M^2 \sqrt{C_f} F(M)$, but different Mach-number-dependent

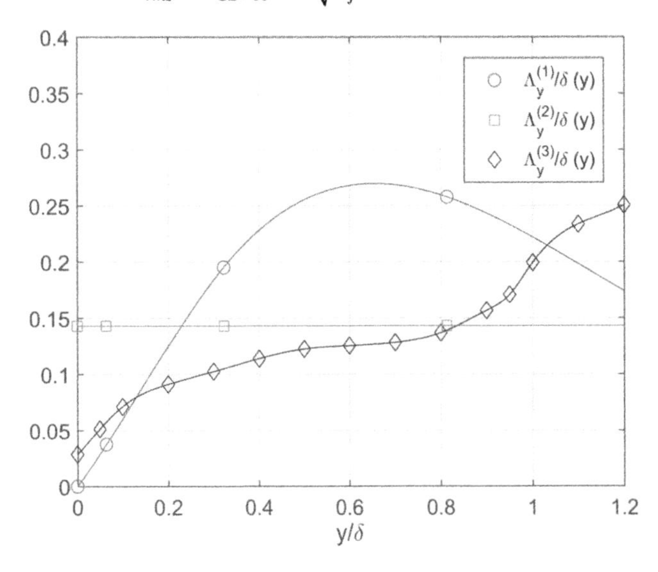

Figure 6.10 Different choices of the density correlation lengths: $\Lambda_y^{(1)}$ from .Wang and Wang (2012), $\Lambda_y^{(2)}$from Rose and Johnson (1982) and $\Lambda_y^{(3)}$ from Gilbert (1982). *Source:* Gordeyev et al. (2015), figure 10. Reproduced with the permission of AIP Publishing.

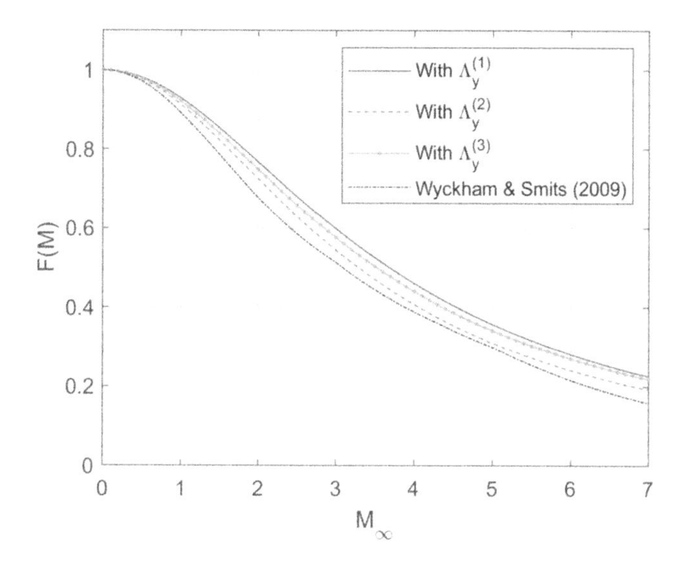

Figure 6.11 Mach dependent function, $F(M)$, defined in Equation (6.23), for different choices of $\Lambda_y(y/\delta)$, and $F(M)$ from Wyckham and Smits (2009), defined in Equation (6.24).

functions, $F(M)$. The F-function, defined in Equation (6.24), is also plotted in Figure 6.11 and generally agrees with other F-functions.

Finally, the B-constant in Equation (6.22) can be computed from Equation (6.23). Different choices of Λ give different values of the B-constant, $B = 0.17$ for $\Lambda_y^{(1)}$, $B = 0.15$ for $\Lambda_y^{(2)}$, and $B = 0.12$ for $\Lambda_y^{(3)}$. Since the shape of the F-functions in Figure 6.11 are approximately the same for all Λ's, we can determine the B-constant by comparing the model, Equation (6.22), with the experimental data. A comparison with experimental measurements in subsonic flow (Gordeyev et al. 2014), supersonic flow (Gordeyev et al. 2012, 2015b) and low-hypersonic flow (Gordeyev and Juliano 2016), are presented in Figure 6.12. The model, Equation (6.22), correctly predicts aero-optical distortions up to Mach 5, if $B = 0.19$. Although there is a small deviation at Mach 5.8, a deviation at some Mach number above 3 to 5 should be expected because some of the assumptions used to develop Equation (6.22), are not valid beyond Mach 3 to 5; in fact, it is somewhat surprising that it does as well as it does above Mach 3.

The Malley probe measures aero-optical distortions and their convective speeds. Figure 6.13 summarizes the results of the convective speed measurements at different facilities and Mach numbers. The convective speed monotonically increases with Mach number, from $0.82U_\infty$ at subsonic speeds to $0.95U_\infty$ at Mach 5.8. These convective speeds are similar to convective speeds reported by Speaker and Ailman (1966) and Spina et al. (1991). Numerical simulations by Maeder et al. (2001) for Mach numbers of 0.9 and 2.3, and Tromeur et al. (2003) for $M = 3, 4.5$,

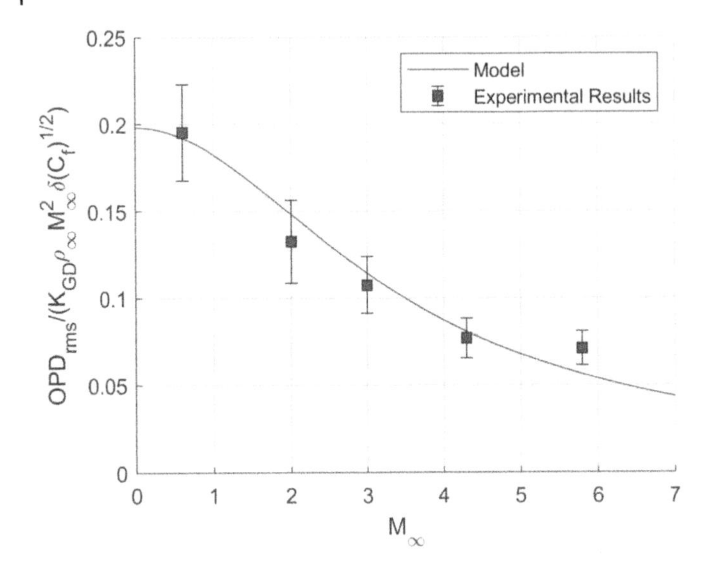

Figure 6.12 A comparison of the normalized OPD_{rms} scaling, Equation 6.22 with $\Lambda_y^{(2)}$ and $B = 0.19$ with experimental data for subsonic, supersonic and low-hypersonic Mach numbers. *Source:* Gordeyev S, Juliano TJ. 2016. Optical characterization of nozzle-wall Mach-6 boundary layers, AIAA Paper 2016-1586.

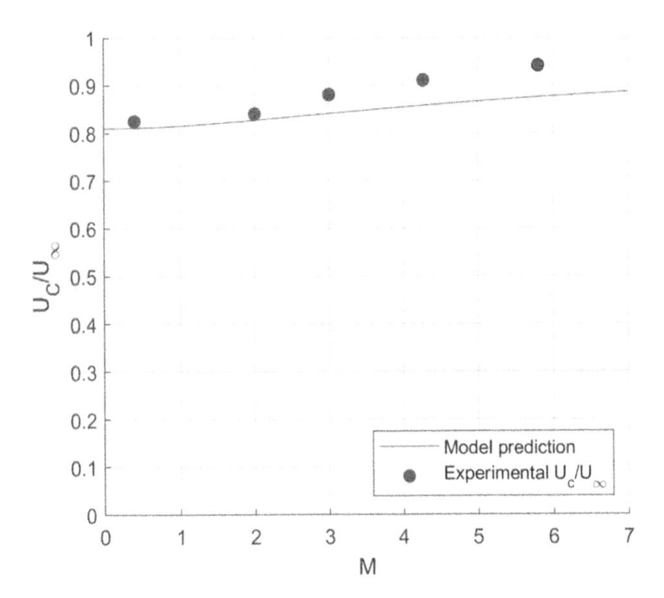

Figure 6.13 Experimentally measured normalized convective speeds of aero-optical structures and model prediction, Equation (6.25). *Source:* Gordeyev S, Juliano TJ. 2016. Optical characterization of nozzle-wall Mach-6 boundary layers, AIAA Paper 2016-1586.

and 6, showed that the density fluctuations for supersonic speeds are suppressed near the wall due to higher flow temperatures near the wall. At these speeds, density-related aero-optical structures away from the wall became relatively stronger, resulting in higher observed aero-optical convective speeds at supersonic speeds.

Since the optical aberrations are related to density fluctuations, it is possible to estimate the convective speed of aero-optical distortions. One plausible assumption is that the density structure convects at the local velocity, and the observed overall convective speed of the aero-optical structure is related to the velocity integral weighted by the density field. So, we can speculate that the convective speed of aero-optical structures can be presented as,

$$U_C = \frac{\int |\rho_{rms}(y)| U(y) dy}{\int |\rho_{rms}(y)| dy}$$

or

$$\frac{U_C}{U_\infty} = \frac{\int |\rho_{rms}(y)| f(y) dy}{\int |\rho_{rms}(y)| dy} \tag{6.25}$$

The result of this model is plotted in Figure 6.13. While the model slightly underpredicts the absolute value of the convective speed, it does correctly predict the experimentally observed increase of the convective speed with the Mach number.

To estimate aero-optical effects of the boundary layer, using Equation (6.22), one should know the boundary layer thickness. Sometimes it can be directly measured or guessed. In some cases, like for wind tunnels or aircraft fuselages, a streamwise Mach number distribution, $M(x)$, can be calculated knowing the nozzle geometry or the aircraft geometry. If the streamwise Mach number evolution along the surface is known, the boundary layer thickness can be estimated using a semi-empirical method presented by Stratford and Beaver (1961). For a given Mach number distribution, the method calculates the streamwise growth of the boundary layer by estimating an equivalent flat plate length, $X(x)$,

$$\begin{aligned} &X(x) = P(x)^{-1} \int_0^x P(x) dx, \quad \text{where} \\ &P(x) = \left[M(x) / (1 + 0.2 M(x)^2) \right]^4 \\ &\mathrm{Re}_X(x) = (a_0 / \nu_0) X(x) M(x) (1 + 0.2 M(x)^2)^{-(3-\omega)} \end{aligned} \tag{6.26}$$

Here $\mathrm{Re}_X(x)$ is the equivalent Reynolds number, corresponding to X, a_0 and ν_0 are the stagnation speed of sound and the dynamic viscosity, respectively, and ω

is the exponent in the viscosity–temperature relation. Typically, $\omega = 0.75$ is used. Using $X(x)$, $\mathrm{Re}_X(x)$, and the stagnation values, the boundary layer thickness can be computed for freestream Reynold numbers on the order of 10^7,

$$\delta(x) = 0.23X(x)\mathrm{Re}_X(x)^{-1/6}. \tag{6.27}$$

In order to estimate aero-optical distortions, using the model in Equation (6.22), the local skin friction, C_f, also should be known. If the boundary layer is approximately canonical, various empirical expressions can be used to estimate *incompressible* $C_{f,i}$, see Nagib et al. (2007), for instance. Here are several widely used expressions, which, in many cases, provide an estimate for $C_{f,i}$ within 5%

$$\begin{aligned} C_{f,i} &= \frac{1}{17.08\cdot[\log_{10}(Re_\theta)]^2 + 25.11\cdot\log_{10}(Re_\theta) + 6.012)} \\ C_{f,i} &= 2[2.604\cdot\ln(Re_{\delta*}) + 3.354]^{-2} \\ C_{f,i} &= 2[2.604\cdot\ln(Re_\theta) + 4.127]^{-2} \end{aligned} \tag{6.28}$$

Here Re_θ and $Re_{\delta*}$ are Reynolds numbers based on the local freestream conditions and the momentum thickness, θ, and the displacement thickness, δ^*. These thicknesses are related to the boundary layer thickness, δ, as $\theta = \frac{n}{(n+1)(n+2)}\delta$, $\delta^* = \frac{1}{(n+1)}\delta$, with $n = 7$ for $10^5 < \mathrm{Re}_x < 10^7$ and $n = 9$ for $10^6 < \mathrm{Re}_x < 10^7$ (Smits and Dussauge 1996).

Note that these methods estimate the *incompressible* skin friction coefficient only. For supersonic Mach numbers, the skin friction coefficient should be corrected for compressibility effects, $C_f = C_{f,i}\cdot(1 + 0.1M^2)^{-0.7}$ (Stratford and Beaver 1961).

Below is an example of how to estimate aero-optical effects from a boundary layer on a wall of a supersonic $M = 3$ tunnel 2.5 meters downstream of a stagnation chamber. The boundary layer is assumed to start developing immediately downstream of the stagnation chamber. The stagnation properties are, $T_0 = 300K, P_0 = 100\,psi$, the stagnation speed of sound $a_0 = 347\,m/\sec$, the stagnation viscosity $\mu_0 = 1.87\cdot10^{-5}\,Pa\cdot\sec$. The streamwise Mach number distribution is shown in Figure 6.14. Using Equation (6.26), the equivalent length, $X(x)$ and the estimated boundary layer thickness, $\delta(x)$, were calculated and plotted in Figure 6.14. The boundary layer, the displacement and the momentum thicknesses 2.5 meters downstream of the stagnation chamber were computed to be $\delta = 15.9\,mm$, $\delta^* = 1.6\,mm$ and $\theta = 1.3\,mm$, respectively. Equation (6.28) gives the estimate of $C_{f,i} = (1.88 \pm 0.04)\cdot10^{-3}$, and, after correcting for compressibility effects, $C_f = (1.20 \pm 0.02)\cdot10^{-3}$. Finally, For $M = 3$, from Figure 6.11 for $\Lambda_y^{(2)}$, $F(M) = 0.58$ and the overall level of aero-optical distortions is $OPD_{rms} = 0.08\,\mu m$.

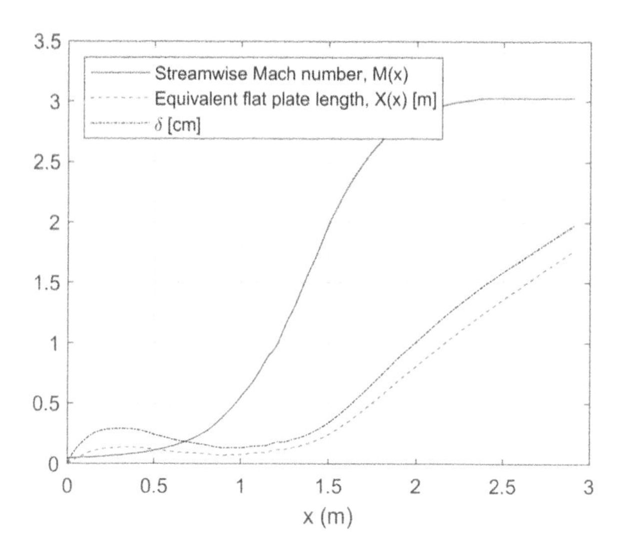

Figure 6.14 Streamwise evolution of various properties inside a representative $M = 3$ tunnel.

Both Malley probe and Shack-Hartmann wavefront sensors measure the local deflection angles, and the wavefronts can be reconstructed from these angles. If the deflection angles are measured with sufficient temporal resolution, the deflection angle spectra provide some additional useful information about the boundary layer. Figure 6.15 shows deflection-angle spectra for turbulent boundary layers at different Mach numbers up to $M = 5.8$, plotted as a function $St_{\delta} = f\delta / U_{\infty}$. All spectra approximately collapse into each other, with some deviations at high St_{δ}, indicating that the large-scale structure, which is responsible for most of the aero-optical distortions, does not change significantly with the Mach number. The peak in all spectra is approximately at $St_{\delta} = 0.9$, implying that the large-scale structure is about the boundary layer thickness. This spectra collapse was demonstrated at even higher Mach number of $M = 8$ (Lynch et al. 2021). The fact that the spectral peak location is the same over a wide range of Mach numbers, including the subsonic regime (Gordeyev et al. 2014), is a very useful result. It provides nonintrusive means of estimating the turbulent boundary layer thickness over a wide range of Mach numbers by simply sending a small-aperture laser beam normal to the boundary layer, measuring the resulting deflection angle with sufficient sampling rate. By finding the frequency of the peak f_{peak} in the deflection angle spectrum, and using the relationship $St_{\delta} = f_{peak}\delta / U_{\infty} = 0.9$, the boundary layer thickness can be estimated as $\delta = 0.9U_{\infty} / f_{peak}$, if the freestream speed, U_{∞}, is known.

From physical reasoning, the wavefront spectrum should approach a constant value at the limit of low frequencies. In this case, as it follows from Equation (4.3),

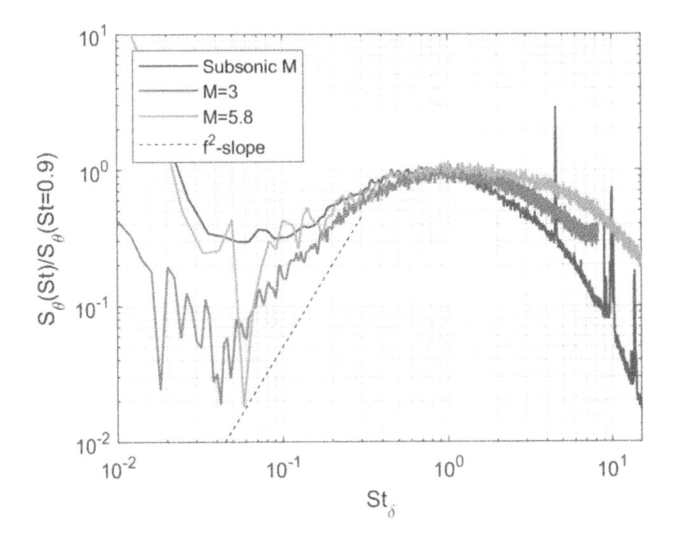

Figure 6.15 Normalized deflection angle spectra for the turbulent boundary layer as a function of normalized frequency, $St_\delta = f\delta / U_\infty$, at different supersonic Mach numbers. *Source:* Adapted from Gordeyev et. al. (2014), figure 10 and Gordeyev and Juliano (2016), figure 8.

the deflection angle spectrum should behave as $S(f) \sim f^2$ at low frequencies. This f^2-slope is also plotted in Figure 6.15. Clearly, most of the presented spectra deviate from the expected slope at low frequencies, as the deflection angle spectra are typically corrupted by the mechanical vibrations and other contaminating effects. If the deflection angle spectra are used to estimate overall value of OPD_{rms} via Equation (4.4), the low-frequency end spectrum should be properly cleaned; otherwise it will result in a higher value OPD_{rms}. One way to do it is to filter out the low-frequency end of the spectrum below $St_\delta = 0.1$ or to replace the low-frequency range with the f^2-fit.

6.3.2 Angular Dependence

When the laser beam travels through the boundary layer at an oblique angle, γ, it traverses a longer distance of $\delta / \sin(\gamma)$ inside the boundary layer, as schematically shown in Figure 6.16. By convention, when the beam traverses normal to the wall, $\gamma = 90$ degrees, forward-looking angles correspond to $\gamma < 90$ degrees, and back-looking angles have $\gamma > 90$ degrees. A simple correction to account for oblique propagation on OPD_{rms} would be to replace δ with $\delta/\sin(\gamma)$ in Equation (6.22). Alternatively, the angular dependence can be included in Equation (6.22) by replacing the B-constant with a function of the oblique angle, $B(\gamma) = 0.19 / \sin(\gamma)$. The numerical value of 0.19 was chosen to match the B-constant for the

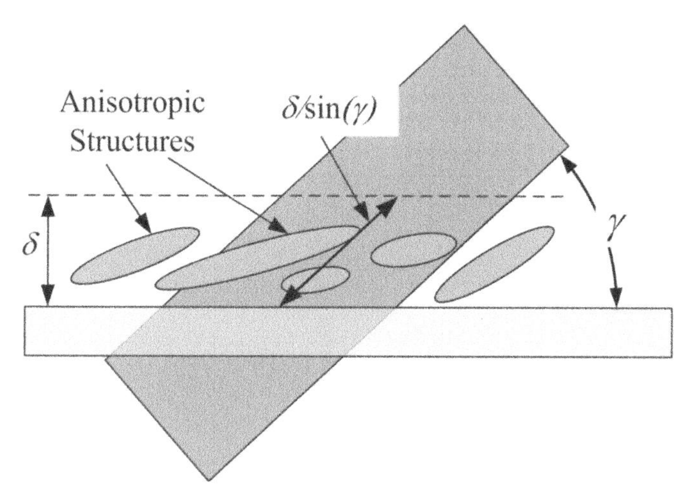

Figure 6.16 Beam traveling through the flow at an oblique angle, γ. Flow goes from left to right.

wall-normal angle of $\gamma = 90$ degrees. This correction assumes that forward-looking angle will result in the same increase in OPD_{rms}, as an equivalent back-looking angle; that is the angular dependence is isotropic.

Active research in dynamics of boundary layers in the last couple of decades undeniably showed that the turbulent boundary layer has packets of vortical structures with a preferred angular or *anisotropic* direction (Adrian et al. 2000; Adrian 2007; Hutchins et al. 2005; Robinson 1991; Wang and Wang 2012), which can be observed in Figure 6.8. As a consequence, one would expect that when the laser beam goes in the downstream direction, it travels along these elongated coherent structures and becomes more aberrated, as indicated in Figure 6.16, than when it travels in the symmetrical, but upstream direction. Naturally, aero-optical distortions in boundary layers, which are related to the large-scale structures, should also exhibit anisotropic behavior for different oblique angles. Various experimental (Gordeyev et al. 2014) and numerical (Truman and Lee 1990; Wang and Wang 2012; White and Visbal 2012) investigations provided some estimates of $B(\gamma)$. The results are shown in Figure 6.17. For all results, $B(\gamma)$ indeed shows anisotropic behavior, with back-looking angles above 90 degrees being more aero-optically aberrating than the forward-looking elevation angles below 90 degrees. The numerically obtained $B(\gamma)$ agree well with the experimental results, showing a very similar anisotropic dependence.

The isotropic oblique-propagation angular dependence, $B(\gamma) = 0.19/\sin(\gamma)$, is also plotted in Figure 6.17. While the isotropic approximation does a decent job for elevation angles between 70 and 120 degrees, it overestimates optical aberrations at forward-looking angles below 50 degrees and underestimates optical distortion above 130 degrees.

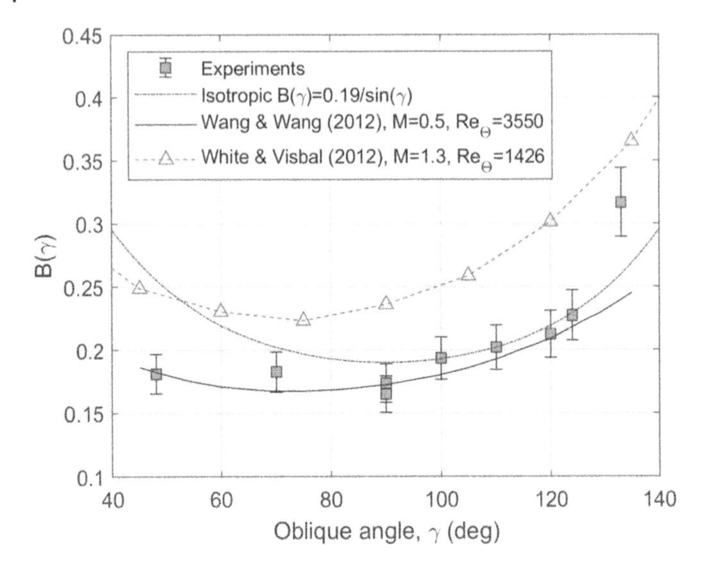

Figure 6.17 $B(\gamma)$ for single-boundary-layer data versus the oblique viewing angle, γ. *Source:* Gordeyev et. al. (2014), figure 17. Reproduced with permission of Cambridge University Press.

6.3.3 Finite Aperture Effects

All the presented scaling laws for turbulent boundary layers assume that the beam aperture size is much larger than the characteristic length scale of the turbulent boundary layer. In cases where the aperture is comparable or smaller than this scale, the aperture effects, discussed in Chapter 5, should be considered. One way to experimentally study these effects is to use the wavefront data over a sufficiently large aperture and apply progressively smaller apertures to the dataset. After removing the residual piston and tip/tilt components the resulting OPD_{rms} as a function of the applied aperture size can be computed. Experimental data of two-dimensional wavefronts, collected over an aperture size of 10 boundary layer thicknesses, were used to perform this analysis in De Lucca et al. (2014), and the results are presented in Figure 6.18. The $OPD_{rms}(Ap)$ results are normalized by the large-aperture OPD_{rms}, given by Equation (6.22). The level of aero-optical distortions is a monotonic function of Ap/δ and varies significantly for $Ap/\delta < 7$. For larger apertures, the ratio approaches the value of one, as expected.

In Chapter 5 it was shown that knowing the local deflection-angle temporal spectrum, we can compute OPD_{rms} for any aperture using Equation (5.8). Using the empirical fit for the streamwise deflection angle spectra, given in Gordeyev et al. (2014), the aperture effects on OPD_{rms} can be modeled for a range of different apertures. The prediction for the fit-based model is also presented in Figure 6.18 and shows a very good agreement with two-dimensional experimental data.

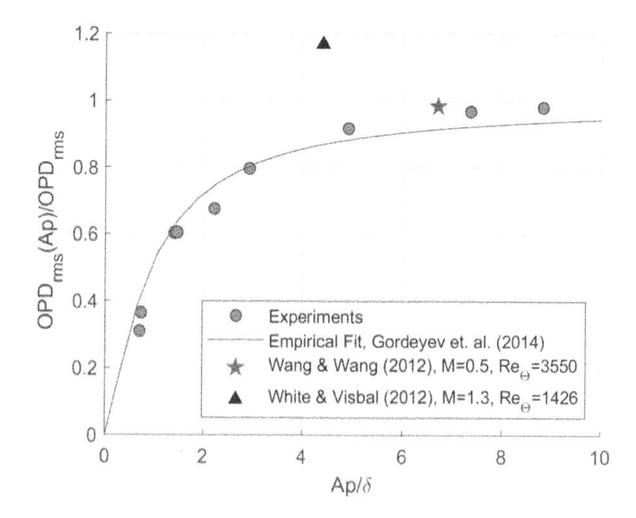

Figure 6.18 Predicted and measured OPD_{rms} for different aperture sizes. *Source:* Adapted from Gordeyev et al. (2014), figure 17. Reproduced with permission of Cambridge University Press.

Also shown in Figure 6.18 are results from numerical simulations performed by Wang and Wang (2012) and White and Visbal (2012). Overall, levels of aero-optical distortions agree well with experimental results, especially for a higher Re_θ of 3550. Difference could be contributed to the Reynolds number mismatch, as both numerical simulations were still performed for relatively low Reynolds number, so the numerically simulated boundary layers might have some low-Re transient features, while experimental values of Re_Θ were much higher, between 12,000 and 27,000.

As a final comment on the aperture effects, for very large apertures $Ap \gg 10\delta$ *Ap*, streamwise growth of the boundary layer cannot be ignored and, for an infinitely large aperture, aero-optical aberrations caused by boundary layers will be infinite. However, for most practical applications, aperture sizes are on the order of several boundary-layer thicknesses and, in this case, the boundary-layer can be assumed to be homogeneous in the streamwise direction.

6.3.4 Nonadiabatic Wall Boundary Layers

As mentioned before, the SRA, Equation (6.16), has been shown to be approximately valid for adiabatic turbulent boundary layers for predicting OPD_{rms} values and it was used to derive Equation (6.22). For nonadiabatic walls, where the wall temperature, T_W, is either higher or lower than the recovery temperature at the wall, T_r, additional convective heat transfer to or from the wall will affect the density field in the boundary layer and, consequently, the resulting aero-optical effects.

To account for the wall-temperature mismatch, $\Delta T = T_w - T_r$, an extended version of Strong Reynolds analogy, ESRA (also known is the modified Crocco relation and the Walz equation), can be used

$$\frac{\bar{T}(y)}{T_\infty} = \frac{T_w}{T_\infty} + \frac{T_r - T_w}{T_\infty}\left(\frac{\bar{U}(y)}{U_\infty}\right) - r\frac{(\gamma-1)}{2}M_\infty^2\left(\frac{U(y)}{U_\infty}\right)^2$$

$$\frac{T_{rms}(y)}{T_\infty} = \frac{\Delta T}{T_\infty}\left(\frac{u_{rms}(y)}{U_\infty}\right) - r\frac{U(y)u_{rms}(y)}{c_p} \tag{6.29}$$

The ESRA gives comparable results for wall-normal locations between the wall and half of the boundary layer thickness ($y < 0.5\delta$). Outside of this region, however, it introduces some errors, related to the constant total temperature assumption (Smits and Dussauge 1996). Nevertheless, as it will be shown later in this section, it is still instrumental to use ESRA to derive aero-optical distortions in the presence of the heat transfer to/from the wall.

To compute the density fluctuations, we can use the equation of state, assuming no correlation between the pressure and temperature

$$\left(\frac{\rho_{rms}}{\rho(y)}\right)^2 = \left(\frac{T_{rms}}{T(y)}\right)^2 + \left(\frac{p_{rms}}{P_\infty}\right)^2 \tag{6.30}$$

Here the local pressure is assumed to be constant across the boundary layer and equal to the freestream pressure. The local density can be found from the equation of state $\rho(y) = P_\infty / (RT(y)) = \rho_\infty (T_\infty / T(y))$, where the mean temperature profile is given in Equation (6.29).

Similar to the derivation of Equation (6.22), we will assume that the mean velocity profile is self-similar, $U(y)/U_\infty = f(y/\delta)$. In addition, we will use Equation (6.19) for the fluctuating velocity profile. To compute the fluctuating pressure profile, we will use the scaling proposed by Guarini et al. (2000).

$$\begin{aligned} p_{rms}/(\rho_w u_\tau^2) &= p_{rms}/(\rho_w U_\infty^2 (C_f/2)(\rho_\infty/\rho_w)) \\ &= p_{rms}/(\rho_\infty U_\infty^2 (C_f/2)) = h(y/\delta) \end{aligned} \tag{6.31}$$

These normalized functions *f*, g and *h* are presented in Figure 6.19.

Substituting Equations (6.29) into (6.30), and using Equations (6.19) and (6.31) gives the following relationship for ρ_{rms} in terms of the velocity and the temperature profiles in the wall normal direction for a given ΔT

$$\begin{aligned} \left(\frac{\rho_{rms}}{\rho_\infty}\right)^2 &= \left(\frac{T_\infty}{T(y)}\right)^2\left[\left(\frac{T_{rms}}{T(y)}\right)^2 + \left(\frac{p_{rms}}{P_\infty}\right)^2\right] = \left(\frac{T_{rms}}{T_\infty}\right)^2\left(\frac{T_\infty}{T(y)}\right)^4 + \left(\frac{T_\infty}{T(y)}\right)^2\left(\frac{p_{rms}}{P_\infty}\right)^2 = \\ &\left(\frac{T_\infty}{T(y)}\right)^3 (C_f/2)g^2(y)\cdot\left[\frac{\Delta T}{T_\infty} + (\gamma-1)M_\infty^2 f(y)\right]^2 + \left(\frac{T_\infty}{T(y)}\right)^2\left[\gamma M_\infty^2 (C_f/2)h(y)\right]^2 \end{aligned} \tag{6.32}$$

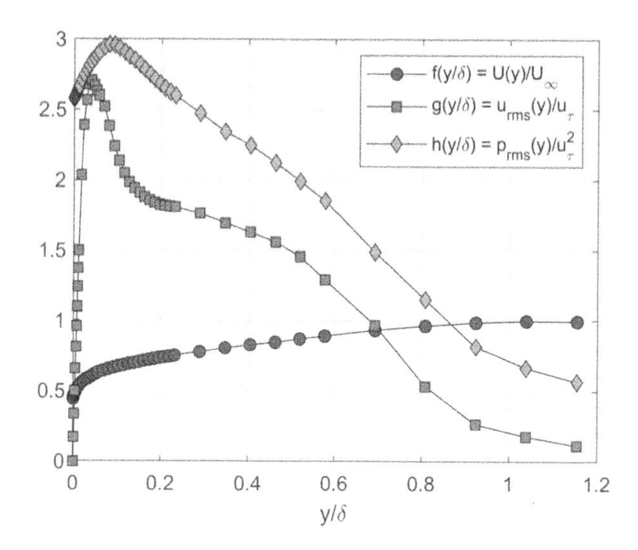

Figure 6.19 Normalized mean component, U, fluctuating velocity u_{rms}, and fluctuating pressure, p_{rms}, profiles. *Source:* Adapted from Gordeyev et al. (2015), figure 8. Reproduced with permission of AIP Publishing.

Finally, substituting Equation (6.32) into the linking equation, Equation (2.19), results in the following relationship,

$$OPD_{rms} = B_0 K_{GD} \rho_\infty \delta \sqrt{C_f} \left[M_\infty^4 + C_1 \frac{\Delta T}{T_\infty} M_\infty^2 + C_2 \left(\frac{\Delta T}{T_\infty} \right)^2 \right]^{1/2},$$

where,

$$\begin{aligned} B_0^2 &= \int_0^\infty [(\gamma-1) f(y) g(y)]^2 (T_\infty / T(y))^3 \Lambda(y) dy + \\ &\gamma^2 (C_f/2) \int_0^\infty g^2(y) h^2(y) (T_\infty / T(y))^2 \Lambda(y) dy, \\ C_1 &= 2(\gamma-1) \int_0^\infty f(y) g^2(y) (T_\infty / T(y))^3 \Lambda(y) dy / A_0^2 \\ C_2 &= \int_0^\infty g^2(y) (T_\infty / T(y))^3 \Lambda(y) dy / A_0^2. \end{aligned} \tag{6.33}$$

The model in Equation (6.33) provides a functional dependence on the temperature mismatch and Mach number. The constants B_0, C_1 and C_2 can be extracted from experiments or estimated using Equations (6.33).

Note that for the adiabatic wall boundary layer, that is, $\Delta T = 0$, Equation (6.33) reduces to the scaling relation for subsonic speeds, $OPD_{rms} \sim \rho_\infty \delta \sqrt{C_f} M_\infty^2$, Equation (6.22) with $F(M) = 1$.

Equation (6.33) shows that $\mathrm{OPD}_{\mathrm{rms}}^2$ is a quadratic function of ΔT and reaches a minimum at negative $\Delta T_{\min}$,

$$\frac{\Delta T_{\min}}{T_\infty M_\infty^2} = -\frac{C_1}{2C_2} = -\frac{(\gamma-1)\int_0^\infty f(y) g^2(y)(T_\infty / T(y))^3 \Lambda(y) dy}{\int_0^\infty g^2(y)(T_\infty / T(y))^3 \Lambda(y) dy} \tag{6.34}$$

with a value of

$$\frac{OPD_{rms}^2(\Delta T_{\min})}{OPD_{rms}^2(\Delta T = 0)} = \frac{\int_0^\infty \left(\begin{array}{l} g^2(y) \cdot (T_\infty / T(y)) \left[(\gamma - 1) M_\infty^2 f(y) - \dfrac{\Delta T_{\min}}{T_\infty} \right]^2 \\ + \gamma^2 (C_f / 2) h^2(y) \end{array} \right) (T_\infty / T(y))^2 \Lambda(y) dy}{A_0^2} \tag{6.35}$$

As the minimum happens at $\Delta T < 0$, OPD_{rms} is higher at $\Delta T = 0$ and still higher at $\Delta T > 0$. It implies that the cooling of the wall reduces aero-optical distortions from the boundary layer.

For positive $\Delta T > 0$, Equation (6.33) can be rearranged as,

$$OPD_{rms} = B_0 K_{GD} \rho_\infty \delta \sqrt{C_f} \left(M_\infty^2 + D_1 \frac{\Delta T}{T_\infty} \right) \left[\begin{array}{l} 1 + \dfrac{D_2}{2} \left(\dfrac{\Delta T / T_\infty}{M_\infty^2 + D_1 \Delta T / T_\infty} \right)^2 \\ + H.O.T. \end{array} \right] \tag{6.36}$$

where $D_1 = C_1 / 2$ and $D_2 = C_2 - (C_1 / C_2)^2$. As the last term, D_2, in the square brackets of Equation (6.36) will be shown later to be much less than one, and Equation (6.36) can be further simplified to

$$OPD_{rms} = B_0 K_{GD} \rho_\infty \delta \sqrt{C_f} \left(M_\infty^2 + D_1 \frac{\Delta T}{T_\infty} \right), \quad \Delta T > 0 \tag{6.37}$$

We can draw several important conclusions from this model:

1) *Cooling* the wall to the optimal $\Delta T_{\min}$ should *reduce* the aero-optical distortions.
2) Amount of optimal cooling, $\Delta T_{\min}$, is proportional to a square of the freestream Mach number.
3) *Heating* the wall will always *increase* the aero-optical distortions.

A series of experiments with a heated wall BL were conducted to validate the model and to find the empirical constant, D_1. To determine the value of the D_1 constant, Equation (6.37) can be re-arranged as,

$$\frac{OPD_{rms}(\Delta T) - OPD_{rms}(\Delta T = 0)}{OPD_{rms}(\Delta T = 0)} = D_1 \frac{\Delta T}{T_\infty M_\infty^2} \tag{6.38}$$

To heat the wall, a heated pad was added to the wall upstream of the measurement station and the overall aero-optical distortions were measured for various Mach numbers and positive ΔT's; details of the experiment and data reduction are provided in Gordeyev et al. (2015). The extracted values of D_1 are plotted versus Re_θ in Figure 6.20(a). The constant does not appear to be a function of the Reynolds number, at least in the studied range. The mean value of D_1 was found to be $D_1 = 2.13 \pm 0.07$, giving the experimental value of $C_1 = 2D_1 = 4.22 \pm 0.14$.

With the known constant D_1, Equation (6.37) can be used to find B_0-constant. Figure 6.20(b), shows the results, where OPD_{rms} data for the heated wall boundary layer, normalized by $K_{GD}\rho_\infty \delta \sqrt{C_f}$ are plotted versus $(M_\infty^2 + D_1(\Delta T / T_\infty))$. The linear scaling relationship successfully collapses OPD_{rms} values over a wide range of subsonic Mach numbers and positive temperature differences. The slope of the OPD_{rms} data was found to be $B_0 = 0.19$, which is consistent with the value of $B = 0.19$ for adiabatic boundary layers; see Figure 6.12.

It is apparent from these results that a positive mismatch between the wall temperature and the adiabatic wall temperature can greatly affect the OPD_{rms} value. Thus, the effect of positive temperature difference cannot be ignored in optical distortions of the turbulent boundary layer. Another important consequence is by introducing a positive temperature mismatch, it is possible to thermally "tag" and measure resulting aero-optical distortions using wavefront sensors in *incompressible* boundary layers. This approach was successfully used to study dynamics of the large-scale structures in incompressible turbulent boundary layers (Gordeyev and Smith 2016; Saxton-Fox et al. 2019).

For negative temperature differences, we need to use the full scaling relationship, Equation (6.33). Factoring out M_∞^2 from the right-hand side of Equation (6.33) gives the $OPD_{rms}(\Delta T)$, normalized by the OPD_{rms} for the adiabatic wall, $\Delta T = 0$,

$$\frac{OPD_{rms}(\Delta T)}{OPD_{rms}(\Delta T = 0)} = \left[1 + C_1 \frac{\Delta T}{\left(T_\infty M_\infty^2\right)} + C_2 \left(\frac{\Delta T}{\left(T_\infty M_\infty^2\right)}\right)^2\right]^{1/2} \tag{6.39}$$

As discussed earlier, for some negative temperature difference, given in Equation (6.33), aero-optical distortions should be at their minimum. Figure 6.21(a), shows the normalized OPD_{rms} data for three Mach numbers plotted versus $\Delta T / (T_\infty M_\infty^2)$

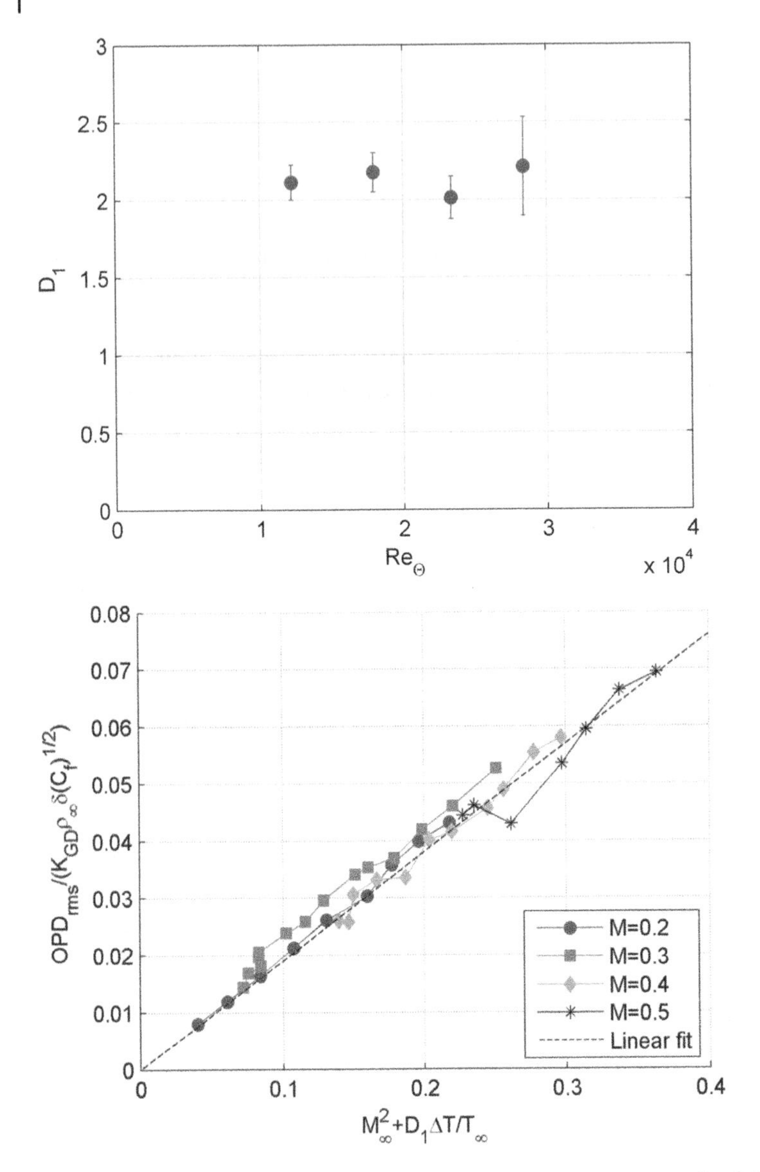

Figure 6.20 (a) D_1 versus Re_Θ for the heated boundary layer. (b) Normalized OPD_{rms} versus $M_\infty^2 + D_1(\Delta T / T_\infty)$ for subsonic Mach numbers and positive temperature differences. *Source:* Gordeyev et. al. (2015), figures 5 and 6. Reproduced with permission of AIP Publishing.

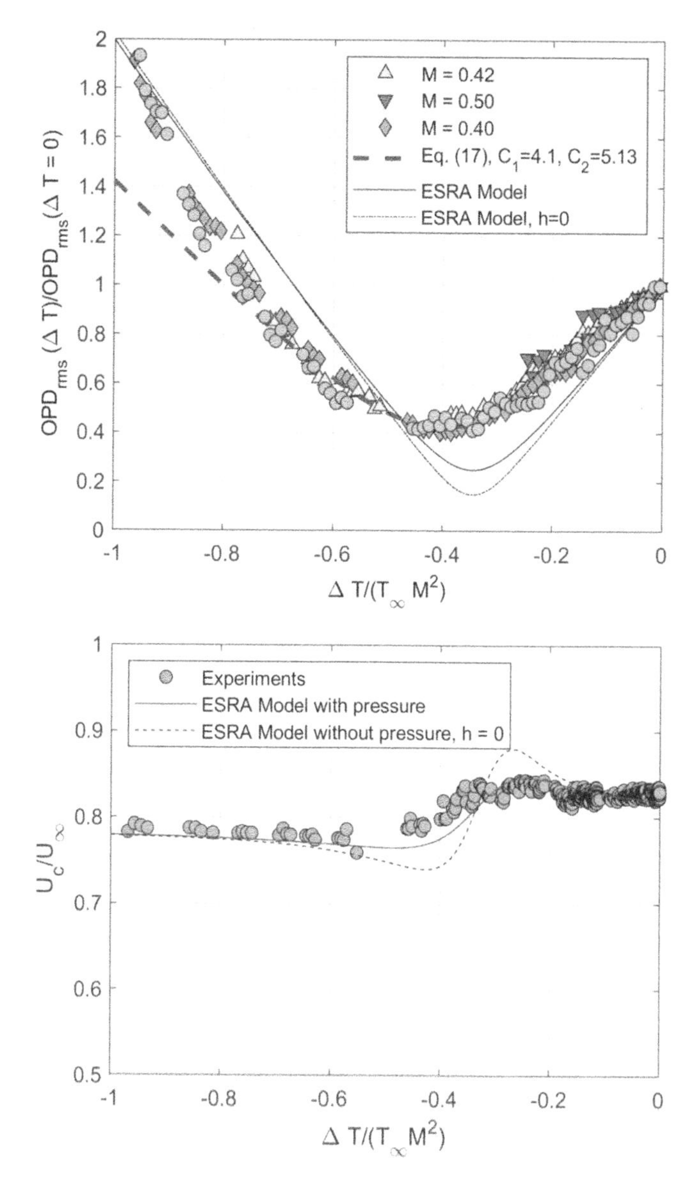

Figure 6.21 (a) OPD_{rms} normalized by the adiabatic-wall OPD_{rms} value versus $\Delta T / (T_\infty M_\infty^2)$, the best-fit and the model predictions with and without ($h = 0$) the pressure term. (b) Convective speed of aero-optical structure as function of the wall temperature and the model prediction with and without the pressure term. *Source:* Gordeyev et. al. (2015), figures 5 and 6. Reproduced with permission of AIP Publishing.

and the strong effect of the difference between the wall and adiabatic wall temperature is apparent in this figure. All experimental results successfully collapse onto one curve and the minimum of approximately 0.4 in the normalized OPD_{rms} value occurs for $\Delta T/(T_\infty M_\infty^2) = -0.4$. In other words, the optical aberrations in a turbulent boundary layer are decreased by as much as 60% at this temperature difference, compared to the adiabatic wall boundary layers. This is a dramatic decrease in the magnitude of optical aberrations, and it provides a promising passive way to significantly reduce aero-optical distortions caused by the turbulent boundary layers. Performing a least-square fit, the constants C_1 and C_2 were found to be 4.1 and 5.13, respectively. Figure 6.21(a) shows Equation (6.40) with these fitting constants, indicating a good prediction of the normalized OPD_{rms} values for the range of $\Delta T/(T_\infty M_\infty^2)$ between zero and -0.7. For $\Delta T/(T_\infty M_\infty^2) < -0.7$ experimental results are higher than Equation (6.40) predicts, indicating a possible departure from some of the assumptions used in deriving the model due to increased buoyancy effects, among others. The value of C_1 agrees within experimental error with the one obtained from the heated-wall experiment, mentioned earlier. In addition, the second term in Equation (6.36) is always less than $D_2/(2D_1^2) = 0.03$, verifying the assumption of neglecting this term in deriving Equation (6.37).

Using *f*, *g*, and *h* functions from Figure 6.19, constants C_1 and C_2 were also computed from Equation (6.33). The resulting model prediction is shown in Figure 6.21(a), as a thin solid line. The model predicts that the OPD_{rms} minimum happens at $\Delta T_{min}/(T_\infty M_\infty^2) = -0.34$, which is within 15% of the experimentally observed value 0f -0.4. In addition, the model predicts a larger decrease in the minima for aero-optical distortions, $OPD_{rms}(\Delta T_{min})/OPD_{rms}(\Delta T = 0) = 0.26$, underestimating the experimentally observed value by 30%. These deviations are expected, as the ESRA model is not exact and tends to underpredict the density fluctuations for nonadiabatic walls. Nevertheless, it correctly predicts the functional form, given in Equation (6.33) and, with experimentally obtained constants, still can be used to predict the aero-optical environment for turbulent boundary layers with moderately cooled walls, $-0.7 < \Delta T/(T_\infty M_\infty^2) < 0$.

As a last comment, if the pressure term, *h*, is not included in the model, the model, plotted as a dashed line in Figure 6.21(a), significantly underpredicts the decrease in $OPD_{rms}(\Delta T_{min})/OPD_{rms}(\Delta T = 0)$. Notice that the inclusion of the pressure term is important only in the temperature range of the lowest OPD_{rms} and does not significantly alter OPD_{rms} for the canonical adiabatic-wall boundary layer. It is generally accepted that in adiabatic-wall boundary layers the temperature fluctuations are several times larger than the pressure fluctuations, so the pressure fluctuations can be safely neglected in this case. But the presented model shows that in moderately cooled boundary layers the temperature fluctuations are suppressed, while the pressure fluctuations, which are related to velocity fluctuations, are not significantly modified. Thus, the cooled-wall boundary layer might be a more suitable flow to study pressure fluctuations inside the boundary layer.

The presented model can also be used to estimate the speed of the convective aero-optical structures for different wall temperature mismatches, using Equation (6.25). As a reminder, the similar approach was used to provide an explanation of the experimentally observed increase in the convective speed of aero-optical structures in a boundary layer at supersonic speeds, shown in Figure 6.13. Using density profiles from Equation (6.32), the convective speed prediction is plotted in Figure 6.21(b), as a function of the wall temperature difference, along with direct experimental measurements of the convective speeds. The model properly predicts all the experimentally observed trends, including the reduction of the convective speed for large negative $\Delta T/(T_\infty M_\infty^2) < -0.5$ and even absolute values of the convective speed for the adiabatic wall boundary layer. The model of the density fluctuations, given in Equation (6.32), shows that for large negative temperature differences, the density fluctuations near the wall are increased, while in the outer part of the boundary layer the density fluctuations are suppressed. Thus, the optical contribution from slower-moving structures near the wall start to increase. From Equation (6.25), it follows that the observed convective speed of aero-optical structure would decrease. It is important to notice that when the pressure fluctuations are ignored ($h = 0$ in Equation (6.32)), the model shows less agreement with the data, as demonstrated in Figure 6.21(b). Thus, the convective velocity data also indicate that the pressure term *must be accounted for* to study density fluctuations in moderately cooled boundary layers. This conclusion is consistent with the previous comments about the importance of considering pressure fluctuations to better predict overall aero-optical distortions in moderately cooled boundary layers.

To further study the effect of the pressure fluctuations, simultaneous measurements of the velocity field and the overall wavefronts in the subsonic boundary layer were performed by Gordeyev et al. (2015c) and Gordeyev and Smith (2016). By requiring pressure fluctuations to be zero, the density field and the resulting wavefronts can be computed from the velocity field, using an instantaneous version of SRA. By comparing the computed wavefronts using the velocity-field via SRA approach to the measured wavefronts, instances where the pressure fluctuations cannot be ignored were identified and studied. An identified instance, where pressure fluctuations appear, was found to be correlated to the presence of large-scale vortical structures. As discussed earlier, the pressure fluctuations inside shear layers with vortical structures significantly contribute to the overall aero-optical distortions; therefore, these simultaneous velocity/wavefront studies, as well as recent studies in adverse-pressure boundary layers (Schatzman and Thomas 2017), provide growing evidence that local shear-layer-type structures with associated lower pressure regions may play an important role in the boundary layer dynamics.

Another effort (Ranade et al. 2016) studied the effect of large-scale freestream pressure fluctuations imposed on turbulent boundary layers where the pressure fluctuations were found to modulate the amplitude of small-scale structures inside the boundary layer. The study was instigated by wavefront anomalies observed when measuring wavefronts for a large-aperture laser beam projected through a forced shear layer but also capturing the turbulence in the compressible boundary layer over the high-speed optical window (Duffin 2009).

Lastly, recently high-speed wavefront sensors were successfully used to study topology and dynamics of various transitional features, turbulent bursts and modal waves, in transitioning laminar hypersonic boundary layers (Gordeyev and Juliano 2016).

6.3.5 Instantaneous Far-Field Intensity Drop-Outs

Free-space communication systems use lasers to wirelessly transmit the data with high bandwidth, up to several Gigabit/sec, for telecommunication and computer networking (Kaushal et al. 2017). They use a laser transmitter to send the laser beam, encoded with the data, and a photosensitive receiver to process the signal. Significant drops in the laser intensity at the receiver due to various attenuations (fog, rain, clouds etc.) may lead to signal loss and disruption of wireless link. Turbulence will impose an additional effect of scintillation or rapid changes in intensity, which can also disrupt the link.

Because the time-averaged Strehl ratios for beams transmitted through most attached turbulent boundary layers are usually quite high, turbulent boundary layers have always been presumed to be an aero-optic nonissue. However, after performing high-speed measurements of aero-optical distortions in subsonic boundary layers (Gordeyev et al. 2003), clear evidence of intermittent intensity dropouts were found. Additional research conducted by Wittich et al. (2007) yielded a more complete *2-D* picture of the aberrating boundary-layer structures. Thus, while the general comments regarding time-averaged Strehl ratio remain unchanged, the warnings of possible dropout problems now seem more likely to cause a serious deterioration of transmitted laser communication signals.

Using experimentally measured time-resolved 2-D wavefronts over a range of aperture sizes, it is possible to compute the instantaneous far-field Strehl Ratio, using Equation (2.28). Examples of the time-resolved Strehl ratio as a function of time are plotted in Figure 6.22 for a communication laser wavelength of 1.5 μm for three aperture sizes, $Ap = 10$, 20, and 50 cm. The experimentally measured OPDs were rescaled in amplitude for a boundary-layer thickness of $\delta = 20$ cm which represent distances of approximately 12 m aft of the nose of an aircraft at $M = 0.8$. While the average intensity is still relatively high in all cases, it is clear that there are many dropouts, lasting a few milliseconds, especially for larger apertures. From a laser-based communication point of view, these energy drop-outs might translate into a loss of several Gigabytes of data during each drop-out, inevitably slowing the communication link, since the lost chunks of data must be constantly

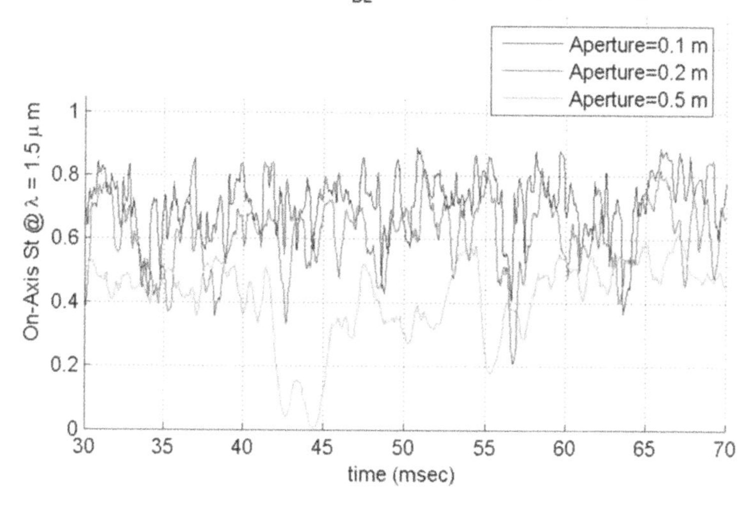

Figure 6.22 Instantaneous far-field Strehl ratio vs time for laser beams propagated through a subsonic boundary layer for different beam apertures of 0.1, 0.2, and 0.5 m for a boundary layer with $\delta = 20$ cm and a 1.5 μm laser.

retransmitted. We should point out that these predictions are based on the beam being projected normal to the boundary layer. As was discussed before, the *OPD* increases when the beam is projected through the boundary layer at oblique angles, thus making the intensity drop-out problem even worse. Also, it should be remembered that in all of these cases the tip/tilt was removed over the aperture. As discussed later in this book, beam jitter also adds to the intensity losses.

From Equation (2.30) it follows that if the optical wavefront has a normal distribution in space over the aperture, then the Maréchal formula, Equation (2.30), can be used to calculate the *instantaneous* Strehl Ratio for any $OPD_{rms}(t)$. Figure 6.23 presents spatial probability distributions of OPD for subsonic boundary layers. Indeed, *spatially*, wavefronts do have a normal distribution regardless of the aperture size. Thus, for turbulent boundary layers the instantaneous Strehl Ratio, $SR(t)$, is directly related to the instantaneous $OPD_{rms}(t)$ via Equation (2.31).

From a statistical point of view, the probability distribution of $OPD_{rms}(t)$ in time is more relevant than the time traces of $OPD_{rms}(t)$ themselves. Figure 6.24(a), shows a probability density function for the $OPD_{rms}(t)$ for the aperture of $Ap = 10\delta$ for $M = 0.4$ and 0.5. The shape of the PDF at each Mach number is well-approximated by a log-normal probability density function

$$PDF(OPD_{rms}) = \frac{1}{OPD_{rms} s\sqrt{2\pi}} \exp\left[-\frac{(\ln(OPD_{rms}) - m)^2}{2s^2}\right] \tag{6.40}$$

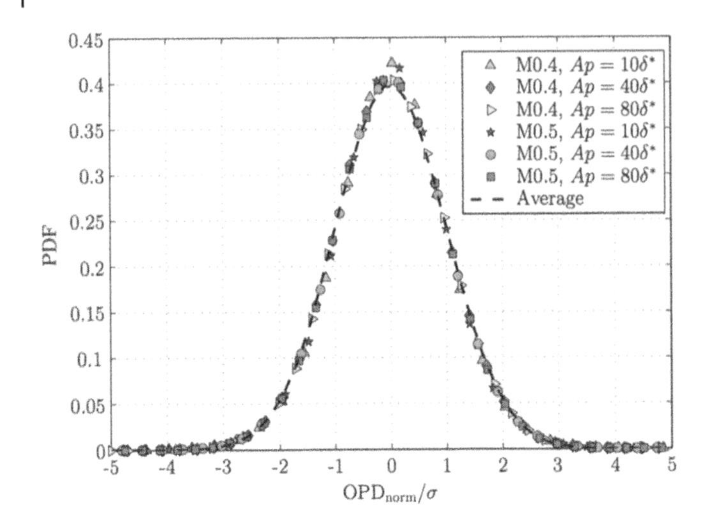

Figure 6.23 PDF for the $OPD(x)$ at $M = 0.4$ and 0.5 and three aperture sizes plotted as the probability distribution of OPD. *Source:* Gordeyev et al. (2013), figure 4. Reprinted with permission from DEPS. All rights reserved.

where m is the temporal mean and s is the temporal standard deviation of the natural log of $OPD_{rms}(t)$. The dashed lines in Figure 6.24(a), are log-normal distributions where the m and s parameters have been calculated from the experimental data at each Mach number. As these curves show, the log-normal distribution captures the general shape characteristics of the experimental data quite well.

Let us define a normalized, aperture-dependent wavefront as

$$OPD^{norm}(x,t;Ap) = \frac{OPD(x,t;Ap)}{OPD_{rms}(t;Ap=\infty)} \tag{6.41}$$

where the wavefront is normalized by the "infinite-aperture," time-averaged value of the OPD_{rms}. Probability distributions of the normalized spatial root-mean-square of $OPD^{norm}(x,t;Ap)$, labeled $OPD_{rms}^{norm}(t)$, for $M = 0.4$ and 0.5 for $Ap = 10\delta$, are shown in Figure 6.24(b). The probability density functions for different Mach numbers are now collapsed onto a single curve, but the shape of the curve is a function of the aperture. Changing the size of the aperture results in different values of the mean and standard deviation for the normalized $OPD_{rms}^{norm}(t)$. The values of the temporal mean, μ, $\mu(Ap) = \dfrac{\overline{OPD_{rms}(Ap)}}{OPD_{rms}(Ap=\infty)}$, and the temporal standard deviation or the spread, $\Sigma(Ap)$, of the $OPD_{rms}^{norm}(t)$ versus the aperture size were calculated from experimental data and are shown in Figure 6.25 for the $M = 0.4$ and 0.5. The slight variation between the different data sets is primarily

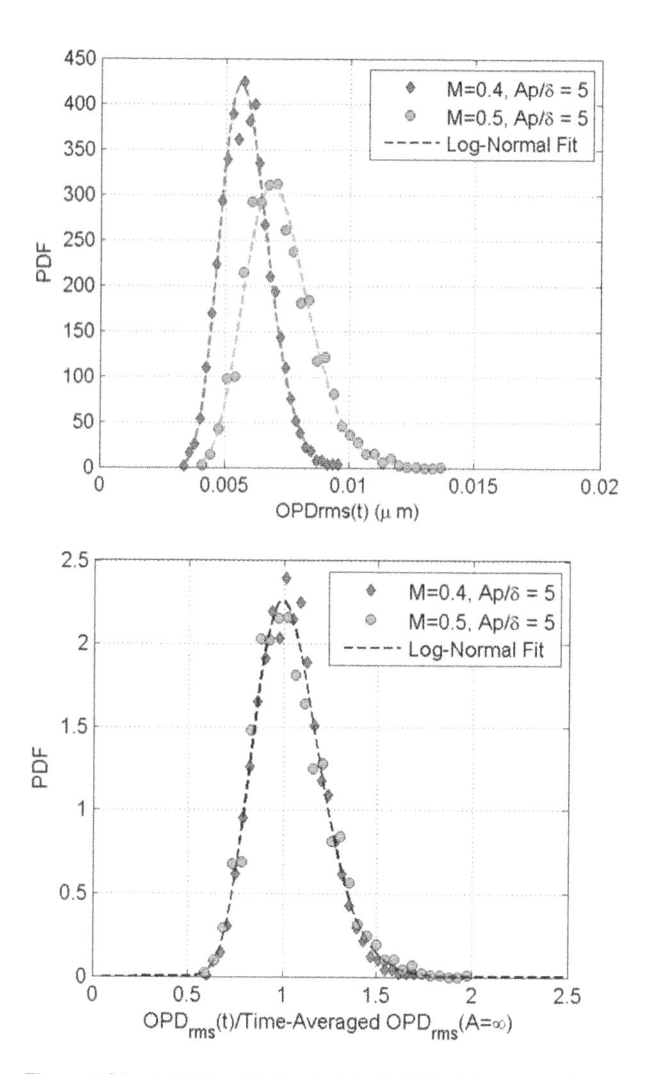

Figure 6.24 (a) PDF of $OPD_{rms}(t; Ap)$. (b) PDF of the normalized $OPD_{rms}(t; Ap) / \overline{OPD_{rms}(t; Ap = \infty)}$. M = 0.4 and 0.5. *Source:* Gordeyev et al. (2013), figure 6. Reproduced with permission of DEPS. All rights reserved.

from experimental errors in the estimation of the boundary-layer thickness. The time-averaged value of $OPD_{rms}(t; Ap)$ monotonically increases with aperture size and approaches the "infinite" aperture value of $\overline{OPD_{rms}(t; Ap = \infty)}$; therefore, μ approaches unity when the size of the aperture is increased. From the experimental data it can be seen that this unity value is achieved when the size of the aperture is larger than 8δ. From the plot of the spread, Σ of the$OPD_{rms}^{norm}(t; Ap)$ in Figure 6.25, the spread initially increases as the aperture size increases, but at approximately $Ap = 4\delta$ the value of Σ begins to decrease. Keeping in mind that tip/tilt is removed from the wavefronts, the initial increase is the result of the aperture being smaller than the characteristic size of the optically active structures in the boundary layer. The spread continues to increase with increasing aperture size until several complete optically active structures are within the aperture at a given instance (which occurs at approximately 4δ). However, once the aperture is larger than the characteristic size of several optically active structures, the spread of $OPD_{rms}(t; Ap)$ decreases. If the aperture were allowed to continue to increase in size until it was infinitely large, the value of the spread, Σ, would go to zero while the mean value, μ, would become one; thus, for an "infinite" aperture, the PDF of $OPD_{rms}(t; Ap = \infty)$ would become a delta-function centered at unity. However, as mentioned before, for very large apertures the streamwise variation of the boundary layer should be taken into account and the presented simplified analysis will no longer be valid.

The PDF of the $OPD_{rms}^{norm}(t)$, Equation (6.40), can be defined in terms of the mean value, μ, and the spread, Σ, which are in turn functions of the aperture size, as shown in Figure 6.25. These parameters are related to the m and s parameters in Equation (6.40) as,

$$m = \log\left(\frac{\mu}{\sqrt{1 + (\Sigma/\mu)^2}}\right), \quad s^2 = \log(1 + (\Sigma/\mu)^2) \tag{6.42}$$

Knowing the PDF of nondimensional $OPD_{rms}^{norm}(t)$, it is possible to reconstruct the actual PDF of *dimensional* $OPD_{rms}(t; Ap)$, using Equations (6.40) and (6.42) for any aperture size, using the data from Figure 6.25 and the scaling law for $\overline{OPD_{rms}(t; Ap = \infty)}$, Equation (6.22).

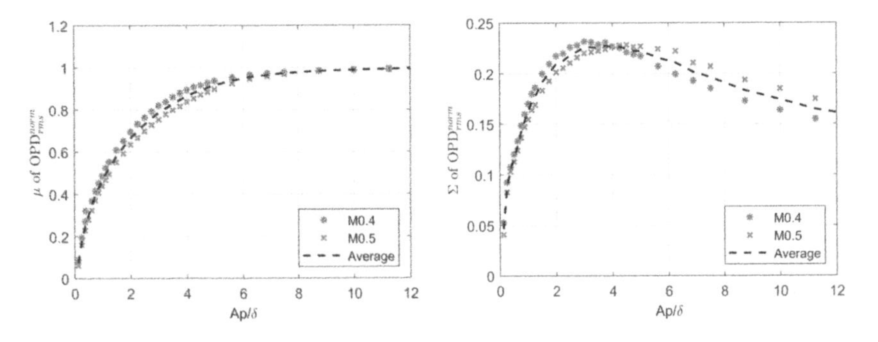

Figure 6.25 The temporal mean, μ, and the spread, Σ, of $OPD_{rms}^{norm}(t)$ for different aperture sizes at Mach numbers of 0.4 and 0.5. *Source:* Gordeyev et al. (2013), figure 7. Reproduced with permission of DEPS. All rights reserved.

It is often of interest to find the statistical properties of the instantaneous far-field Strehl Ratio, such as the percentage of time below a certain threshold value, which is directly related to potential data loss for laser-based communication systems (Majumdar and Ricklin 2008). Note that losing the signal for short periods of time does not necessarily mean losing data, as different encoding schemes, such as interleaving or Forward-Error-Correction codes can be employed to send a redundant signal and tolerate a certain amount of data losses; see Majumdar and Ricklin (2008), for instance. Thus, the signal still can be transmitted through a noisy channel, but it will require decoding to make a redundant signal, inevitably increasing the amount of data to be transmitted to send the original signal. The knowledge of relative amount of time of intensity drop-outs, drop-out durations and frequencies are helpful in choosing a proper encoding scheme to maximize the *original data* transmission rate.

If a system operates on the absolute value of $SR(t)$, the link is presumed to be lost if the absolute value of $SR(t)$ drops below a prescribed value. Other systems depend on a relative intensity variation, $SR(t)/\overline{SR(t)}$, and the communication link is considered to be lost if the relative intensity drops below a certain value. We will consider both cases.

6.3.5.1 Absolute SR Threshold

If optical communication systems require that the laser signal strength at the far-field receiving station remains above a minimum value, the communication link can only reliably operate when the Strehl Ratio is above a certain system-defined threshold value, TH_{SR}. Below this threshold value, the link is considered to be broken.

The Maréchal formula, equation (2.32), can be rearranged to solve for OPD_{rms} as a function of SR as $OPD_{rms}(t) = \frac{\lambda}{2\pi}\sqrt{-\ln[SR(t)]}$, or it can be rewritten in terms of the $OPD_{rms}^{norm}(t)$ as,

$$OPD_{rms}^{norm}(t) = \frac{\lambda}{2\pi\overline{OPD_{rms}(Ap=\infty)}}\sqrt{-\ln[SR(t)]} \tag{6.43}$$

Using Equation (6.43), the threshold value, TH, can be found as a function of TH_{SR}, the laser wavelength, λ, and $\overline{OPD_{rms}(Ap=\infty)}$ as,

$$TH = \frac{\lambda}{2\pi\overline{OPD_{rms}(Ap=\infty)}}\sqrt{-\ln[TH_{SR}]} \tag{6.44}$$

If the instantaneous value of the normalized OPD_{rms}^{norm} goes *above* the threshold value, TH, then the Strehl ratio goes *below* TH_{SR}, and the optical communication system is considered inoperable, and the data are lost. To determine the amount of data lost at the far-field, or, equivalently, the total percentage of time that the normalized $OPD_{rms}^{norm}(t)$ is above the given threshold value, TH, the complementary cumulative distribution function (*CCDF*) can be used. For the log-normal distribution given by equation 6–40, the log-normal complementary cumulative distribution function is defined as

$$CCDF(OPD_{rms}^{norm} > TH) = 1 - \frac{1}{2} erfc\left(-\frac{\ln(TH) - m}{s\sqrt{2}} \right) \tag{6.45}$$

where *erfc* is the complementary error function. Figure 6.26 shows the CCDF, or the percentage of the $OPD_{rms}^{norm}(t)$ signal above the threshold value, TH, for different aperture sizes. For example, for the aperture of $Ap = 10\delta$, when the threshold value, TH, is less than 0.5, 100% of the optical aberrations are larger than the threshold value, meaning that in the far-field the entire signal will be below the required operational Strehl Ratio threshold and no signal will be registered at the receiver. Increasing the threshold value allows durations of the $OPD_{rms}^{norm}(t)$ to begin dropping below the threshold, TH, permitting portions of the signal bit stream to reach the far-field with an acceptable Strehl Ratio. For threshold values, $TH > 1.8$, none of the normalized $OPD_{rms}^{norm}(t)$ is above the threshold and the entire signal reaches the far-field above the threshold Strehl Ratio. It is important to note that this limitation on TH is stricter than for energy-deposition systems operating only on the time-averaged intensity on the target.

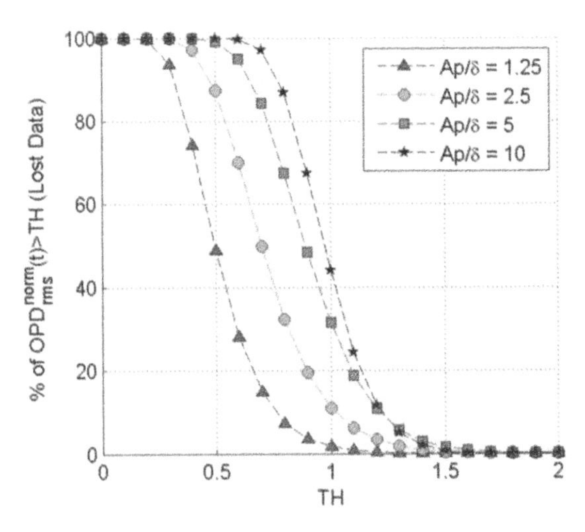

Figure 6.26 CCDF of a log-normal PDF showing the percentage of $OPD_{rms}^{norm}(t)$ that is above the threshold value versus TH for various aperture sizes. *Source:* Gordeyev et al. (2013), figure 8. Reproduced with permission of DEPS. All rights reserved.

Summarizing, the percentage of data lost due to boundary-layer aero-optical aberrations for given flight conditions and the aperture size can be estimated as follows:

1) Calculate $\overline{OPD_{rms}(t; Ap=\infty)}$ for the anticipated boundary layer parameters using Equation 6.22. The boundary layer thickness can be measured experimentally using a hot-wire, or a Pitot-probe rake. It can be also estimated by measuring the deflection angle spectrum. As it was demonstrated in Figure 6.15, the frequency peak, f_{peak}, in the deflection angle spectrum is approximately at $St_\delta = f_{peak}\delta / U_\infty = 0.9$. Knowing the freestream speed, the boundary layer thickness can be estimated as $\delta = 0.9U_\infty / f_{peak}$. Finally, the boundary layer thickness can be obtained using Equation (6.27), or from numerical estimations.
2) For a given threshold of Strehl Ratio, TH_{SR} determine the threshold value, TH, for $\overline{OPD_{rms}(t; Ap=\infty)}$ and the laser wavelength, λ, using Equation 6.44.
3) For the given Ap/δ value, find the mean, μ, and the spread, Σ, values from Figure 6.25.
4) Using Equation 6.42, calculate the m and s parameters defining the log-normal distribution of the normalized $OPD_{rms}^{norm}(t)$.
5) Calculate the amount of data lost for the given m, s, and TH parameters using the CCDF(TH) function, Equation 6.45.

To illustrate the procedure, let us compute the amount data loss for the subsonic boundary layer with the following parameters: the boundary layer thickness of $\delta = 10$ cm, $M = 0.8$, an altitude of 5,000 ft, the viewing angle normal to the wall and the aperture of $Ap = 5\delta = 0.5$ m. Using Equation (6.22), the level of aero-optical distortions would be $\overline{OPD_{rms}(t; Ap=\infty)} = 0.11$ μm. For a laser wavelength of $\lambda = 1$ μm and the Strehl Ratio threshold of $TH_{SR} = 0.5$, from Equation (6.44), TH can be calculated as 1.2. Finally, Figure 6.26 gives that the relative amount of time, when the intensity is below the threshold, is equal to 10%.

For reliable operation, most free-space communication receivers require the received intensity to be above a particular threshold value. If the absolute threshold is given, one can also calculate drop-out durations and time intervals between consecutive drop-outs (relative occurrence of drop-outs). Probability distributions for drop-out durations and times in-between drop-outs for the aperture of $Ap/\delta = 2$ for different values of thresholds, TH, are presented in Figure 6.27. For the large threshold of $TH = 1.3$, the relative amount of the "lost" data is small, about 1%, and the most probable drop-out duration is about $0.5\delta / U_\infty$; the probability distribution for the time interval between drop-outs is wide, indicating intermittent nature of drop-out events, with the average time interval between drop-outs of $20\delta/U_\infty$. When the threshold is decreased to $TH = 0.9$, the amount of "lost" data becomes about 13%. For this threshold, the most probable

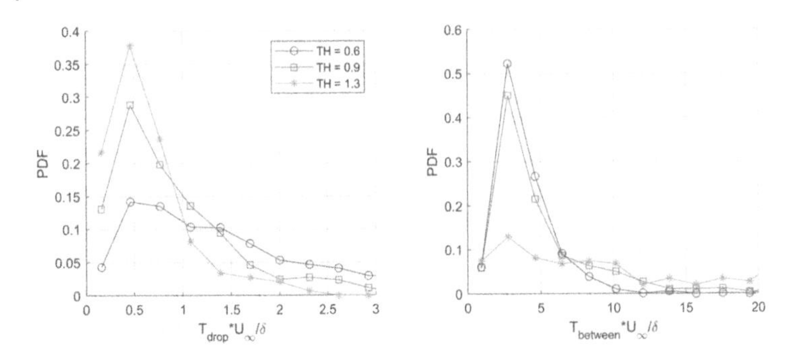

Figure 6.27 Probability of the drop-out durations, T_{drop}, and time interval between successful drop-outs, $T_{between}$, for different threshold values, *TH*. *Source:* Gordeyev et al. (2013), figure 9. Reproduced with permission of DEPS.

drop-out duration is still about $0.5\delta / U_{\infty}$, but the probability tail becomes thicker for this lower threshold, indicating a wider range of drop-out durations. The frequency of drop-out events is increased, with the averaged time interval between drop-outs becoming $4\delta / U_{\infty}$. When the threshold is decreased even further, to $TH = 0.6$, it results in "losing" almost 60% of the data, with drop-outs becoming even longer, as the averaged drop-out duration becomes about $2\delta / U_{\infty}$; the average time between drop-outs is decreased to $3\delta / U_{\infty}$. The typical drop-out duration due to the boundary layer is of the order of the δ / U_{∞}, or, for a typical transonic boundary layer, on the order of a millisecond. This seemingly short drop-out might potentially result in a loss of several Gigabytes of data during the drop-out, thus definitely requiring some sort of interleaving coding scheme to reliably send data through free-space, laser-based communication channels.

Although the presented analysis is based on the experimental data collected at subsonic speeds of $M = 0.4$ and 0.5, intensity drop-out properties for the supersonic boundary layers can be computed in a similar fashion.

6.3.5.2 Relative Intensity Variation

Optical distortions caused by the beam propagation through atmosphere over long distances result in intensity fluctuations on the target, which are characterized by a relative intensity variation on the target, $Z = I(t) / \overline{I(t)} = SR(t) / \overline{SR(t)}$. For Kolmogorov-type (statistically homogeneous and isotropic) atmospheric turbulence, these fluctuations have a log-normal distribution and are usually described by the log-intensity variance, $\sigma^2_{\ln Z} = \overline{(\log Z)^2} - (\overline{\log Z})^2$ (Tatarski 1961). For weak atmospheric fluctuations and a planar wave, by the log-intensity variance approximately becomes the well-known Rytov variance, $\sigma^2_{\ln Z} \approx 1.23 C_n^2 (2\pi / \lambda)^{7/6} L^{11/6}$ (Andrews et al. 2001), where L is the traveling distance. C_n^2 is the index-of-refraction structure constant. It is defined as a constant in the optical structure function,

$D(r) \equiv \langle (W(r_0,t) - W(r_0 + r,t))^2 \rangle_{All\, r_0,t} = C_n^2 r^{2/3}$, if the separation values are smaller than the Kolmogorov microscale (Tatarski 1961). For aero-optical distortions, though, the distribution of the relative intensity variation, Z, is clearly not a log-normal one, as it follows from Equation (6.40). Nevertheless, we can still compute the log-intensity variance as a function of the overall level of aero-optical distortions caused by boundary layers, $\overline{OPD_{rms}(t; Ap = \infty)}$, for different apertures as

$$\sigma_{\ln Z}^2 = (2\pi \overline{OPD_{rms}(Ap = \infty)} / \lambda)^4 \cdot G_A(Ap / \delta) \tag{6.46}$$

where $G_A(Ap / \delta) = \exp(4\mu + 4\Sigma^2) \cdot (\exp(4\Sigma^2) - 1)$ accounts for finite-aperture effects. The log-intensity variance as a function of the relative aperture size, Ap / δ, and $\overline{OPD_{rms}(t; Ap = \infty)} / \lambda$ is presented in Figure 6.28(a). The log-intensity variance increases with the increasing OPD_{rms} as the fourth power of OPD_{rms}, or, recalling Equation (6.22), as the fourth power of the boundary-layer thickness, δ. Also, it is inversely proportional to the fourth power of the laser wavelength. Clearly, these functional dependencies for boundary-layer aero-optical-related effects are quite different from the atmospheric optical effects, expressed in the Rytov variance. In Figure 6.28(b), $G_A(Ap / \delta)$ is plotted versus the aperture size. $G_A(Ap / \delta)$ and, therefore, the log-intensity variance initially increases with the aperture size, reaches the maximum around $Ap / \delta = 5$ and then starts decreasing for larger apertures. Again, this behavior is different from atmospheric optical effects, where the log-normal

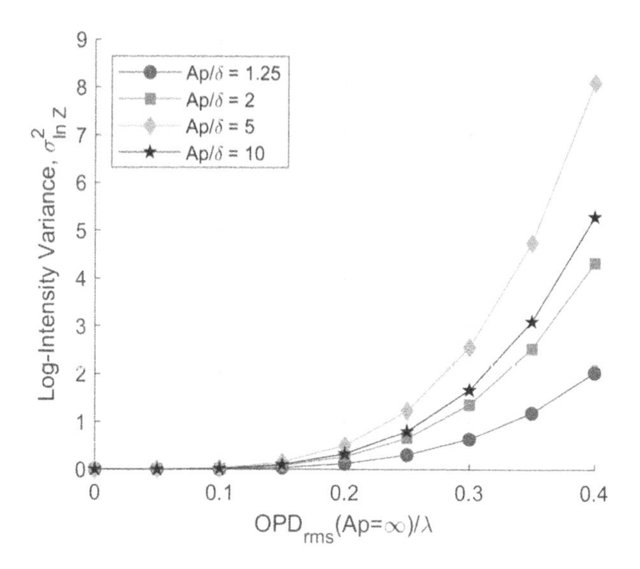

Figure 6.28 (a) $\sigma_{\ln Z}^2$ as a function of $\overline{OPD_{rms}(t; Ap = \infty)} / \lambda$ for different apertures. (b) G_A as a function of Ap / δ. *Source:* Gordeyev et al. (2013), figure 10. Reproduced with permission of DEPS.

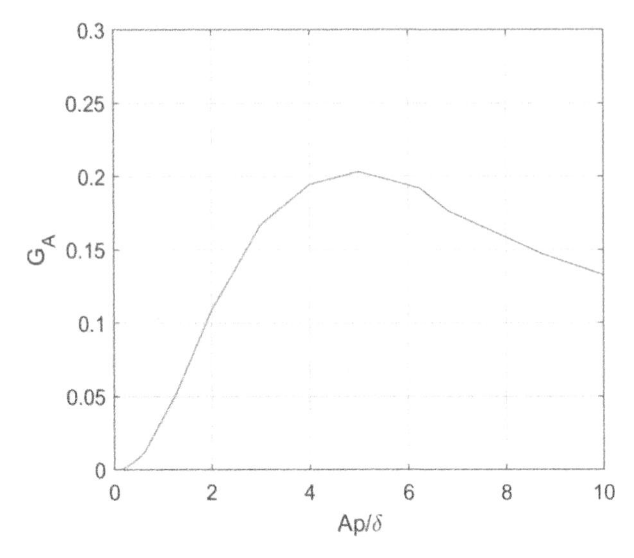

Figure 6.28 (Continued)

variance monotonically decreases with the aperture size, so-called aperture-averaging effects; see Tatarski (1961) or Andrews et al. (2001), for instance.

Knowing the log-intensity variance caused by turbulent boundary layers, we can compare it to the Rytov variance and find the "equivalent additional" distance the laser beam needs to propagate through the atmosphere to have similar intensity scintillations. Using boundary layer parameters from the example in the previous sub-section, the log-intensity variance can be calculated using Equation (6.46) as $\sigma^2_{\ln Z} = 0.05$, and, for a strong-turbulent atmosphere with $C_n^2 = 10^{-14} m^{-2/3}$, the "equivalent additional" distance is approximately 330 m. The same boundary-layer parameters, but at a higher Mach number of $M = 2$, $\sigma^2_{\ln Z} = 7.1$, with the "equivalent additional" distance of more than 5 km.

6.4 Turrets

Up to this point, we have considered aero-optical effects from some fundamental flows, specifically shear and boundary layers. But for practical applications, hemisphere-on-cylinder turrets, with a few examples shown in Figure 6.32, are optimal platforms with large fields-of-regard to project or receive laser beams to or from a target. They provide a convenient way to point and keep the beam in a desired direction for a transmitting station or to keep a lock on an incoming beam for a receiving station. Land-based observatories for telescopes are perfect examples of such turrets. Naturally, most airborne laser-based systems, both past and present, use some form of the turret geometry.

Airborne optical turrets were extensively studied in the 1970s and early 1980s. For long-wavelength, around 10 microns, lasers being considered for airborne lasers, these studies showed that, at low speeds, the turrets produce only steady-lensing aberrations and unsteady optical aberrations were found to be a contributing factor only at transonic and supersonic speeds, when unsteady density fluctuations become significant. Good summaries of extensive experimental and modeling studies of optical turrets at transonic and supersonic speeds prior to the mid-1980s can be found in Gilbert and Otten (1982) and Sutton (1985).

As mentioned in the Introduction, airborne lasers under consideration during the 1970s and 1980s had wavelengths around 10 μm. Toward the end of 1980s, advancements in laser technology made near-IR lasers (with wavelengths ~ 1 μm) good candidates to be used for airborne lasers. So, while absolute optical distortions are relatively small around turrets at moderate subsonic speeds (OPD_{rms} ~ 0.1μm), relative phase distortions, $2\pi OPD_{rms} / \lambda$, imposed on much-shorter-wavelength laser beams were increased ten-fold or so, thus making unsteady optical distortions, caused by a separated flow behind a turret large enough to significantly reduce the far-field intensity. Combined with a significant progress in wavefront measurement instrumentation, it spurred a renewed interest in studying and mitigating optical aberrations caused by turrets.

Before we start discussing the flow topology and the aero-optical distortions around turrets in the later sections, we would like to briefly discuss different angular frames of references, used to describe the beam direction, also called viewing direction. The turret viewing direction is commonly described with two angles, the azimuthal angle (Az), and the elevation angle (El), as indicated in Figure 6.29. From a flow perspective, however, a different, flow-direction-based coordinate system is preferred, α, called the viewing or line-of-sight (LOS) angle

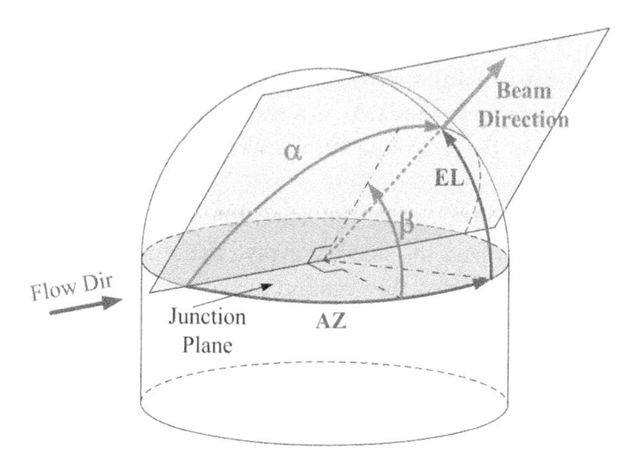

Figure 6.29 Definitions of different angular frames of references to describe the beam direction (viewing angle). *Source:* Flight Measurements of Aero-Optical Distortions from a Flat-Windowed Turret on the Airborne Aero-Optics Laboratory (AAOL), AIAA Paper 2011-3280.

and β, the modified elevation angle. For pure spheres, the flow field would only be a function of how far upstream or downstream the turret is looking, which is represented by only the viewing angle α. The modified elevation angle accounts for changes in the flow that are derived from the symmetry-breaking presence of the cylinder under the hemisphere. The transformation between these two frames of references is

$$\alpha = \cos^{-1}[\cos(Az)\cos(El)], \quad \beta = \tan^{-1}\left[\frac{\tan(El)}{\sin(Az)}\right] \tag{6.47}$$

and

$$El = \sin^{-1}[\sin(\alpha)\sin(\beta)], \quad Az = \tan^{-1}[\tan(\alpha)\cos(\beta)].$$

6.4.1 AAOL

Between 2000 and 2010, several government-sponsored programs were focused on developing airborne turret laser systems. At the same time, the fundamental research into aero-optical environments around turrets was still very limited due to several limiting factors. As discussed later, turret sizes should be large enough to create realistic turbulent flow environments, so the aero-optical studies should be performed in large (1 m x 1m or even larger) aerodynamic tunnels. These tunnels should have good optical access to transmit large-diameter beams at different angles. The vibrational environment of these tunnels should be sufficiently small to prevent the optical data from being corrupted by these vibrations. As a result, the experiments using turrets were performed only in a few locations, like Wright-Patterson AFB, AEDC, and US Air Force Academy. Wavefront sensors available at that time were capable of collecting the wavefronts at sampling rates of only few Hertz, so the data were limited sequences of uncorrelated wavefronts. The low sampling rate also prevented the use of any spectral-based data analysis and corruption-removal techniques that are now standard. As a result, only a handful of meaningful aero-optical data of the flows around turrets were collected during this period; see Figure 6.30(a).

A real game changer came in 2007, when the High-Energy Laser Joint Technology Office (HEL JTO) recognized the need to evolve the study of aero-optics to in-flight research and thus funded the creation of the Airborne Aero-Optics Laboratory (AAOL) Program. Managed by Dr. Eric Jumper from the University of Notre Dame, the AAOL program began collecting wavefront data in realistic flight environments for different turret geometries. The in-flight data have now proliferated through government, industry, and university communities, and has become the mainstay of research in understanding and mitigating aero-optic effects. The AAOL program coincided with emergence of high-speed

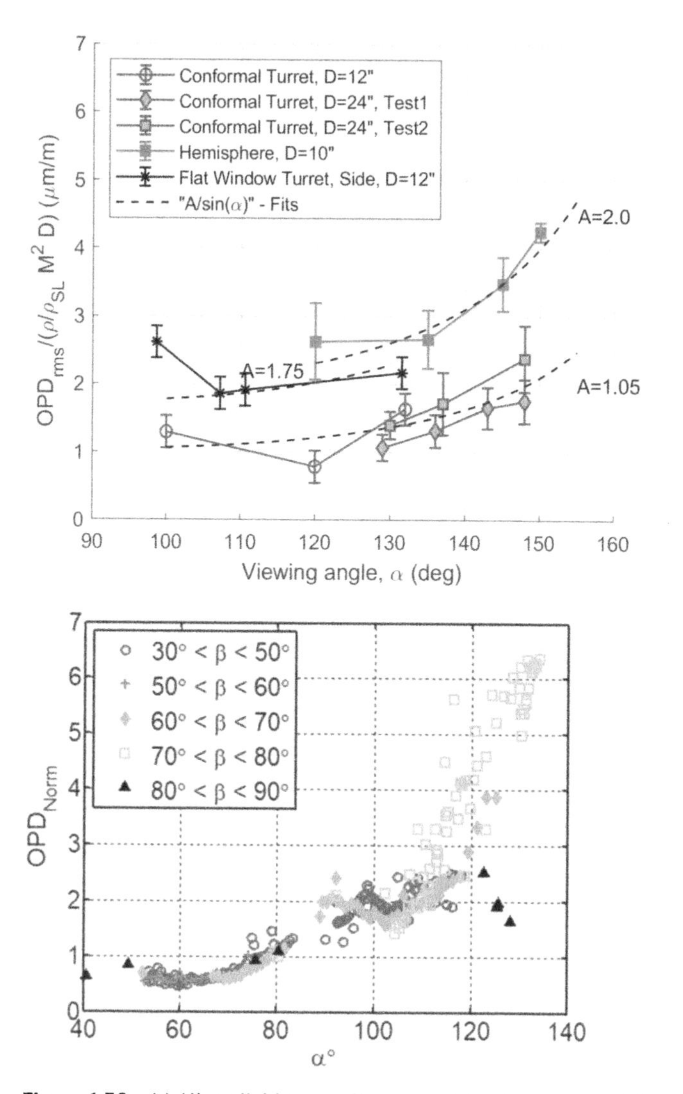

Figure 6.30 (a) All available open-literature data of aero-optical distortions for turrets prior to 2010. *Source:* Adapted from Gordeyev and Jumper (2010), figure 7. Reproduced with the permission of Elsevier. (b) Aero-optical data from a single flight campaign. *Source:* Flight Measurements of Aero-Optical Distortions from a Flat-Windowed Turret on the Airborne Aero-Optics Laboratory (AAOL), AIAA Paper 2011-3280.

wavefront sensors, thus allowing a collection of large amounts of spatiotemporally resolved aero-optical data over short times; see an example of data collected in a single one-week flight campaign in Figure 6.30(b).

The ultimate objective of the AAOL program was to advance the understanding of aero-induced effects in realistic flight environments and investigate possible means of their mitigation. The AAOL program developed an affordable path to obtain in-flight data about the effect that the various types of turbulent flow over and around a turret has on wavefronts. Due to reciprocity, aero-optical distortions of a laser beam propagated from the turret will be the same for the incoming beam. If the source of the incoming beam is from a distant target or guide star, the incoming beam is already imprinted with aberrations due to its traverse through the atmosphere. In order to avoid this, and thus to simplify the interpretation of wavefronts obtained through the turret on the laboratory aircraft, the AAOL program proposed using a beam from a source aircraft flying in relatively close formation to the laboratory aircraft. Based on the analysis, described in Jumper et al. (2013), the nominal distance between the aircraft should be approximately 50 m to minimize various sources of optical error. But this posed a further dilemma of generating a "pristine" beam that arrives at the laboratory aircraft's turret pupil without having been corrupted by aero-optical effects from the flow around the source aircraft. The final proposed concept was to have the source beam leave as a small diverging beam, originating from the source aircraft with a beam diameter of only a few millimeters and then diverging to overfill the pupil aperture on the laboratory aircraft turret by a factor of two. The rationale for the use of the small beam at the source was that the beam would be small compared to the coherence length of the optically relevant turbulent structures inside a thin turbulent boundary layer present on the skin of the laser aircraft. The aperture effects for the beam's small diameter would then only allow the boundary-layer turbulence on the source aircraft to impose only a slight tip/tilt on the beam. By the time the beam reaches the turret, installed on the laboratory aircraft, the wavefront on the beam would nominally be spherical so that any tip/tilt on the beam at the source would not affect the spherical figure on the arriving beam at the laboratory aircraft. It is important to separate the diverging-beam curvature from the aero-optical distortions around the turret measurement. In order to do so, the concept included removing the incoming, diverging-beam curvature by mechanically adjusting the turret's telescope prescription to remove the curvature imposed by a nominal 50 m radius.

To limit the cost, the AAOL program made use of commercially available business jet aircraft so that the overall cost of flying the aircraft would be shared with other uses of the aircraft by switching the aircraft in and out of experimental status. After seeking quotations from several business jet companies that were willing to take their aircraft in and out of passenger status, Cessna Citations, with a top speed of $M = 0.62$, were chosen as the airborne platforms for the AAOL program.

The AAOL flight program consists of two aircraft flying in formation at a nominal separation distance of 50 m. A diverging, small-diameter, continuous Yag:Nd laser beam sent from a chase plane to an airborne laboratory; see Figure 6.31. The turret on the laboratory aircraft consists of a one-foot-diameter (30.5 cm) turret with a 10.16 cm (4 inches) clear-aperture; the window can be either flat or conforming to the spherical figure of the turret (i.e., conformal). Pictures of various test turrets are shown in Figure 6.32. The turret itself presents a mold line that is a hemisphere on a cylindrical base, which when installed in the aircraft protrudes out the side of the aircraft through a modified crew escape hatch. The turret can be extended so that the cylindrical base protrudes into the airstream by different amounts. Once the laser and turret systems are tracking each other, a 20 mm stabilized beam emerges from the turret mounting and optical "box" onto the optical bench in the laboratory aircraft, as schematically

Figure 6.31 Airborne Aero-optics Laboratory. (a) Schematic of the two aircraft flying in formation. (b) A picture taken from the source aircraft during a flight test, with green laser illuminating the turret on the laboratory aircraft. Courtesy of University of Notre Dame.

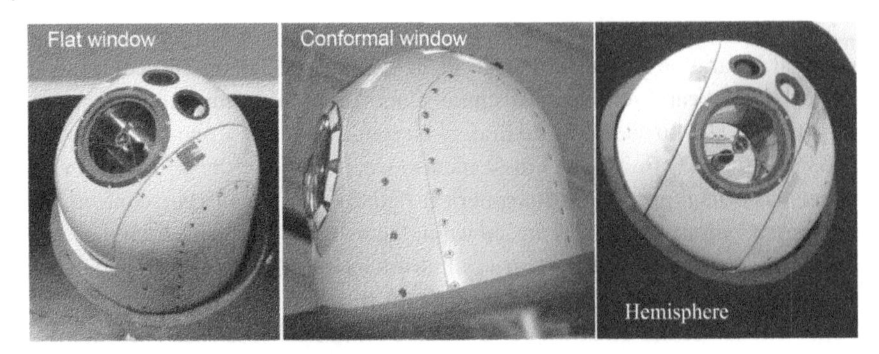

Figure 6.32 Various turret geometries tested in the AAOL programs. *Source:* Courtesy of Stanislav Gordeyev.

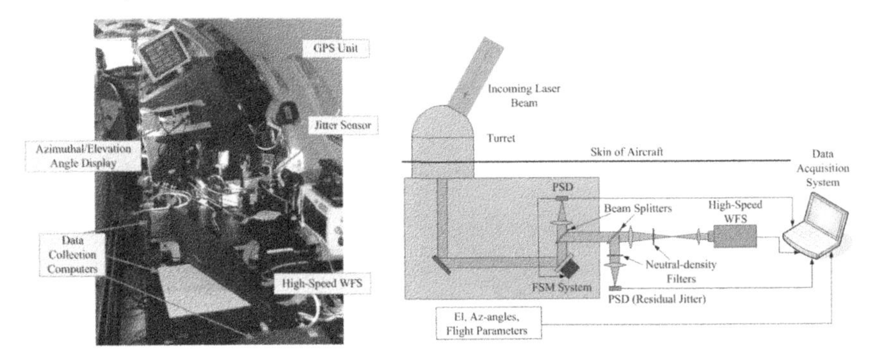

Figure 6.33 A picture and schematic of the optical setup on the laboratory aircraft. *Source:* Jumper et. al. (2013), Reproduced with the permission of SPIE.

shown in Figure 6.33. The "stabilization" of the beam is performed by a closed-loop fast-steering mirror (FSM) system that reimages the turret pupil and is able to reduce the beam's overall jitter to a cutoff frequency of approximately 200 Hz, thus acting as a high-pass jitter filter. The "stabilized beam" is then split between the various sensors on the optical bench onboard the laboratory aircraft.

The AAOL program ran between 2007 and 2012. In 2012, Cessna Citation aircraft were replaced with faster Falcon 10 aircraft, capable of reaching $M = 0.82$ in level flight. These higher speeds opened a door to extensive studies of unsteady shock effects appearing on the turret for transonic speeds $M > 0.6$. For this reason, the new program was called the Airborne Aero-Optics Laboratory-Transonic (AAOL-T). It ran between 2012 and 2017, and was responsible for fully mapping out the aero-optical environment for different turret geometries for Mach numbers between 0.5 and 0.8. At the point of this writing, the third installment of the program, the Airborne Aero-Optics Laboratory–Beam Control (AAOL-BC), to be completed by January of 2023, continues investigating different aspects in aero-optical effects and various means to mitigate them. In total, more than 30 one-week flight campaigns were performed. Combined with numerous tunnel tests of the same flight turret assembly in the Mach 0.6 White Field Tunnel at the University of Notre Dame. In all, The AAOL, AAOL-T and AAOL-BS have led to fully quantifying the aero-optical environment for different turret geometries over a wide range of flight speeds and altitudes, and to test various aero-optical mitigation schemes.

6.4.2 Flow Topology and Dynamics

While turrets provide a convenient means of pointing and tracking and then projecting the laser beam, their less-than-ideal aerodynamic shape creates complex flow fields consisting of all major fundamental turbulent flows: boundary layers, separated shear layers, wakes, necklace vortices, and other large-scale vortical structures. A schematic and surface-flow visualization of the flow topology around a nominal turret are presented in Figure 6.34. The incoming boundary layer experiences an adverse pressure gradient at the front of the turret, causing the boundary layer to separate. The separated boundary layer forms a vortical tube, with both ends (or legs) of the tube extending downstream of the turret on both sides. This vortical structure is called a necklace (or horseshoe) vortex (Baker 1979). Above the necklace vortex, the flow is attached at the front part of the turret, while the adverse pressure gradient at the aft part of the turret forces the flow to separate. The separation region interacts with the necklace vortex legs and creates a complex three-dimensional flow field behind the turret, with reverse flow downstream and at the bottom of the turret, as well as secondary vortices, the so-called horn vortices, on both sides of the turret. When transmitted through the separated region, a laser beam experiences significant unsteady aero-optical aberrations, even at relatively low subsonic speeds. Aero-optical distortions in this

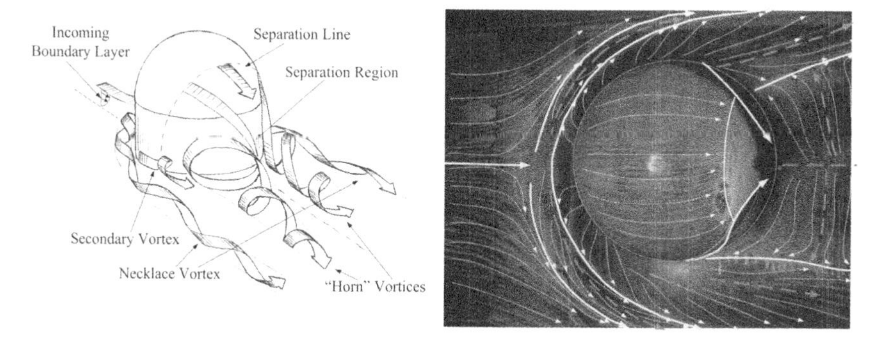

Figure 6.34 Schematic of the subsonic flow around the turret and surface flow topology on and around turret. *Source:* Gordeyev and Jumper (2010).

region are predominantly caused by shear-layer vortical structures and the separation bubble formed downstream of the turret. At transonic and supersonic speeds, shock-induced aero-optical effects with large density gradients are added to this already complicated picture.

When the hemispherical part of the turret is well above the necklace vortex, the flow around the hemispherical section can be approximated by the flow around a sphere. Subsonic flow around spheres at high Reynolds numbers has been extensively studied by Achenbach (1972). He found that when the Reynold number, based on the turret diameter, Re_D, is less than 200,000, the boundary layer is laminar before separation and the separation occurs around a look-back angle of 80–85 degrees. When the Reynolds number is above 300,000, the boundary layer at the front portion of the sphere experiences a laminar-to-turbulent boundary-layer transition, so the boundary layer is turbulent at the apex of the sphere. As a result, the separation point over a sphere is delayed until approximately 120 degrees.

Cp-distribution along the centerline for various turret geometries, including a hemisphere only, for a wide range of Re_D are presented in Figure 6.35(a). Pressure distributions at high Reynolds numbers, $Re_D > 10^6$, are nearly identical and primarily independent of the Reynolds number. At these Reynolds numbers, the boundary layer is turbulent on top of the turret, so the flow separates between 115 and 120 degrees, consistent with separation locations over spheres at similar Reynolds numbers. For a relatively small $Re_D = 190{,}000$, the separation point moves upstream and occurs around 100 degrees. The separation location is further downstream of the separation point at 82 degrees observed over spheres at low Reynolds numbers. This discrepancy is probably due to the boundary layer on top of the turret starting transitioning from the laminar to the turbulent state. In the separated region, the pressure is nearly constant at $C_p = -0.3$, regardless of the Reynolds number. This value is also quite similar to the static pressure inside the separation region reported behind a sphere.

When only the hemisphere is placed on the surface, the presence of the necklace vortex at the bottom of the hemisphere affects the C_p-distribution in this region. Side-views of the flow topology around the hemisphere-on-cylinder turret and the hemisphere-only are schematically demonstrated in Figure 6.35(b). The necklace vortex pushes the stagnation point in front of the hemisphere to approximately $\alpha = 15$ degrees; see Figure 6.35(a). The C_p-value on top of the hemisphere ($\alpha = 90$ degrees) is slightly lower than on top of the turret. The flow also separates around 120 degrees, but the C_p-values are not constant inside the separated region, although Cp eventually reaches the same value as for the hemisphere-cylinder turret of $C_p = -0.3$ for $\alpha > 140$ degrees.

The Cp-distribution for a potential flow around the sphere,

$$C_p(\alpha) = 1 - \frac{9}{4}\sin^2(\alpha),$$

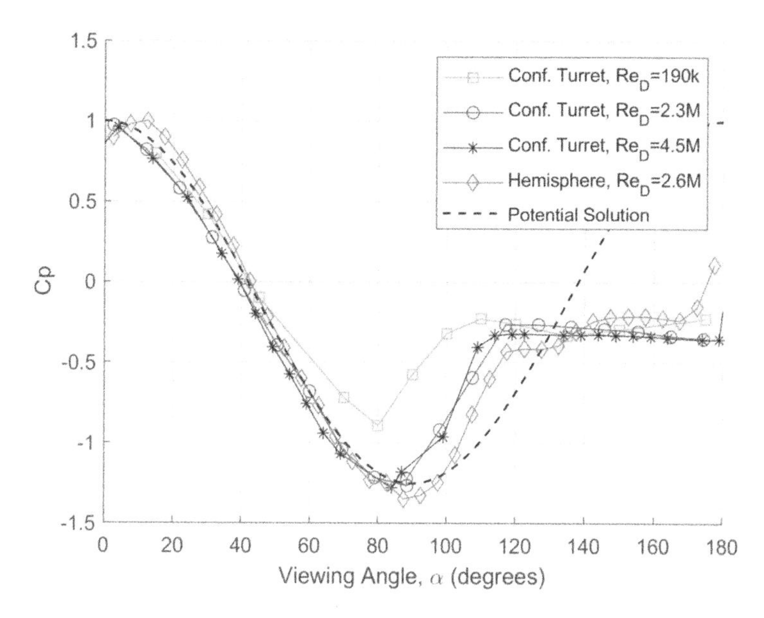

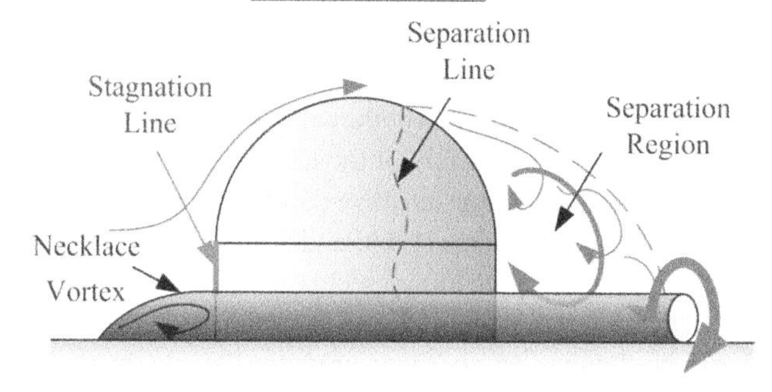

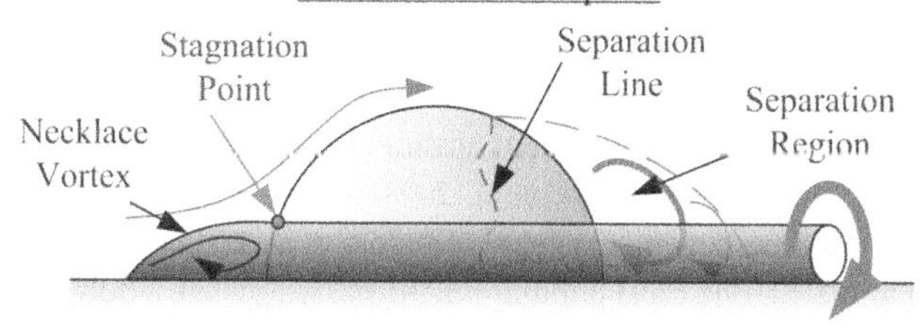

Figure 6.35 (a) Pressure coefficients along the turret's center-plane, for turrets at different Re_D. The potential solution for the flow around a sphere, Equation (3), is given as a dashed line. (b) The schematic of the flow topology around the hemisphere-on-cylinder turret and the hemisphere. *Source:* Gordeyev and Jumper (2010), figure 5. Reproduced with permission of Elsevier.

is also plotted in Figure 6.35(a). For the turret configuration, the necklace vortex is well below the hemispherical part of the turret and the potential C_p-solution describes quite well the pressure distribution on the front part of the turret. For the hemisphere, as mentioned before, the necklace vortex is present at the bottom of the hemisphere, thus slightly changing the C_p-distribution at the front portion from that of the potential solution.

As the turret is not rigid, its elastic motion can introduce a jitter onto optical components used to project a laser beam from the turret. From a practical perspective, jitter of only a few microradians can force a beam to miss a distant target or at least to average down the intensity on the target by "painting" the aim point. One source of turret vibrations arises from unsteady surface pressure fluctuations and resultant local forces that are due to the turbulent flow features around the turret. The beam jitter related to these flow-induced vibrations is termed the aero-mechanical jitter of the turret. While the turret vibrational response depends on its internal structure and mounting arrangement, the unsteady pressure field depends only on the turret geometry and incoming Mach number. In addition, the unsteady pressure fluctuations on the surface of the turret are directly related to the turbulent structures, causing aero-optical global jitter and higher-order aberrations (De Lucca et al. 2012).

One way to investigate the time-changing pressure field on the surface of the turret is to use an array of unsteady pressure sensors, although in practice this usually gives a fairly coarse spatial resolution. Another alternative is to use a pressure-sensitive paint (PSP), where, by the nature of the technique, a very detailed spatial resolution can be achieved. Pressure-sensitive paint is an optical method for measuring surface pressures (Bell et al. 2001; Liu and Sullivan 2005), which is based on oxygen quenching of excited-state luminescence of a painted model. The intensity of the emitted light from the model is inversely proportional to the local partial pressure of oxygen, and due to Henry's law, pressure. The distinct advantages of PSP are that it offers very high spatial resolution at relatively low cost. While the PSP technique has been known since the 1980s, its temporal response was very low, so results were limited to mostly steady-state pressure distributions. The response time is primarily governed by the paint thickness and the diffusivity of the paint binder (Gregory et al. 2014). Recent advances in the development of fast-response porous PSP coatings (Gregory et al. 2008) have allowed increased frequency response of PSP up to several kilohertz (Gregory et al. 2008, 2014; Peng et al. 2013; Crafton et al. 2015). The most popular PSP formulations are a polymer-ceramic binder with either platinum tetra(pentafluorophenyl) porphyrin (PtTFPP) (Crafton et al. 2015) or bathophen ruthenium (Hayashi and Sakaue 2017) as a luminophore.

Due to its excellent spatial resolution, sufficient frequency response and ability to implement it on complex surfaces, the PSP technique has been used to extract and study spatially and temporally resolved pressure fields for on the surface of various turrets. However, the PSP response also depends on the temperature variations, potentially affecting the accuracy of the pressure measurements, with the time-averaged component of the pressure field being the most affected (Hayashi and Sakaue 2020). Consequently, it is common practice that, after extracting the surface pressure fields, the time-averaged fields are removed from the instantaneous pressure fields to produce the fluctuating pressure fields. Various techniques, like statistical analysis, modal, and conditional techniques, then can be used to analyze these fields.

A typical spatial distribution of the fluctuating pressure field, expressed as root-mean-squares of the fluctuation pressure coefficient, $C_{p,rms}(s) = \sqrt{\overline{C_p^2(s,t)}}$, is shown in Figure 6.36. The unsteady pressure field is approximately symmetric in the spanwise direction. The unsteady necklace vortex, formed in front of the turret, is responsible for the increase in $C_{p,\ rms}$ in front of the hemisphere. The unsteady separation over the hemisphere is responsible for the increase in the pressure fluctuations, visible as a narrow region aligned in the spanwise direction on top of the hemisphere. Finally, an inherently unsteady reattachment line downstream of the hemisphere creates a crescent-shaped region of large pressure fluctuations. Very similar spatial pressure distributions were observed for hemisphere-on-turret configurations (Gordeyev et al. 2018).

The location of the separation line defines the size of the separated region downstream of the turret. As a reminder, the pressure on the surface changes

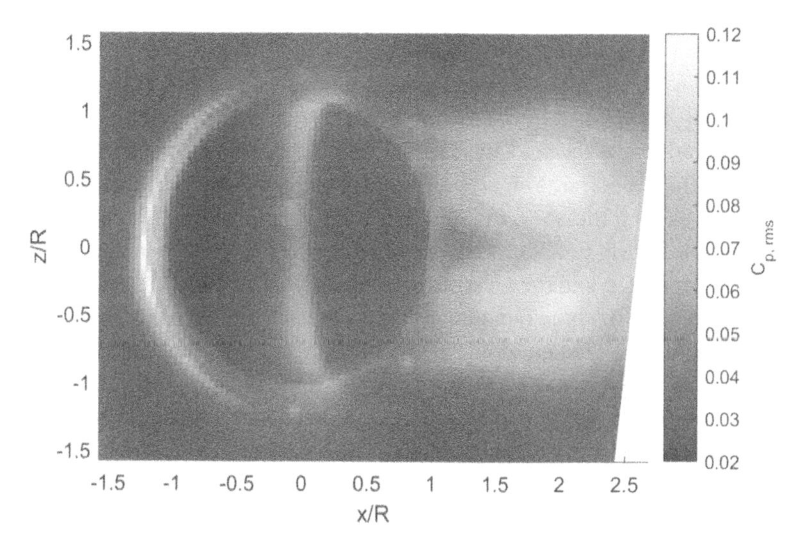

Figure 6.36 Spatial distributions of $C_{p,rms}(s)$ on the surface of the hemisphere and in the wake region. Flow foes from left to right. *Source:* Roeder et al., (2022), reproduced with permission of Elsevier.

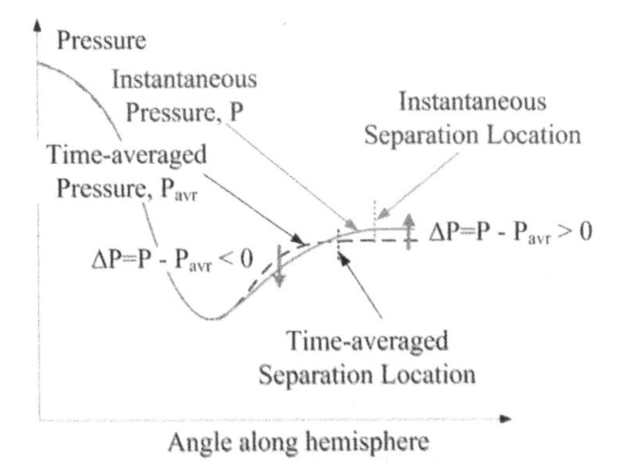

Figure 6.37 Conceptual description of the pressure spatial distribution near the unsteady separation point over a hemisphere. *Source:* Gordeyev et. al. (2018) / American Institute of Aeronautics and Astronautics" and Wittich" with "Gordeyev et. al. (2018).

upstream of the separation point, then becomes approximately constant inside the separated region. If the separation line is unsteady, its unsteadiness will affect the dynamics of the separated region. The separation line can be tracked as the location where the constant pressure region begins. But as mentioned before, in case of fast-response PSP, the mean surface pressure is corrupted by temperature changes in the flow, and it is often removed from the total pressure, leaving only the fluctuating component of the pressure field for the analysis. In this case, it is useful to establish the relation between the unsteady separation locations over the turret and related changes in the unsteady pressure field, as schematically illustrated in Figure 6.37. Here, the time-averaged pressure distribution over a hemisphere is shown as a dashed line. Pressure is highest at the stagnation point in front of the turret, where the velocity is zero. From the stagnation point, the flow accelerates at the front portion of the hemisphere, causing the pressure to decrease. After passing the apex, the flow starts slowing down and the pressure begins to raise, resulting in an adverse pressure gradient environment. At some point, the flow separates, and the pressure becomes approximately constant inside the separation region. At some moment, a change in the wake topology results in decreasing the pressure gradient downstream of the apex. Because of the smaller pressure gradient, the flow stays attached longer, and the separation point moves downstream, relative to the time-averaged separation location. Also, the pressure after the separation is slightly higher than the time-averaged pressure. Recall that the mean pressure is removed from the pressure field, so here only the unsteady pressure is analyzed. Looking at the difference between the instantaneous and the time-averaged pressure distributions, the downstream shift in the separation point would result in a negative change in the pressure field (a negative fluctuating pressure) just upstream of the instantaneous separation location, and possibly in a small positive pressure change (a positive fluctuating pressure) near and after the separation location. Similarly, if the separation point moves upstream of the time-averaged location, it would result in the positive pressure change, followed by the negative pressure change. In other words, the fluctuating pressure field near the separation location should be negatively correlated with the fluctuating pressure inside the separated region.

A very useful way to analyze spatiotemporal pressure fields is to apply Proper Orthogonal Decomposition or POD techniques, as outlined in Chapter 4. It decomposes the pressure field, $p(s,t)$, into independent orthogonal spatial modes, $p_n(s)$, multiplied by the corresponding temporal coefficients, $a_n(t)$, $p(s,t) = \sum_n a_n(t) p_n(s)$. An example of the first six POD spatial modes on the surface of a hemisphere are presented in Figure 6.38. The first dominant mode, as well as the fourth and the sixth modes are anti-symmetric in the spanwise direction, while modes #2, #3, and #5 are symmetric in the spanwise direction. The first six POD modes are responsible for 80% of the fluctuating pressure "energy." The dominant mode #1, which holds more than 40% of the energy, is anti-symmetric in the spanwise direction and primarily nonzero near the separation line. Remember, that POD modes are multiplied by the temporal coefficients, and the coefficients can be either positive or negative. Thus, using the argument outlined above, the positive value of the temporal coefficient would correspond to

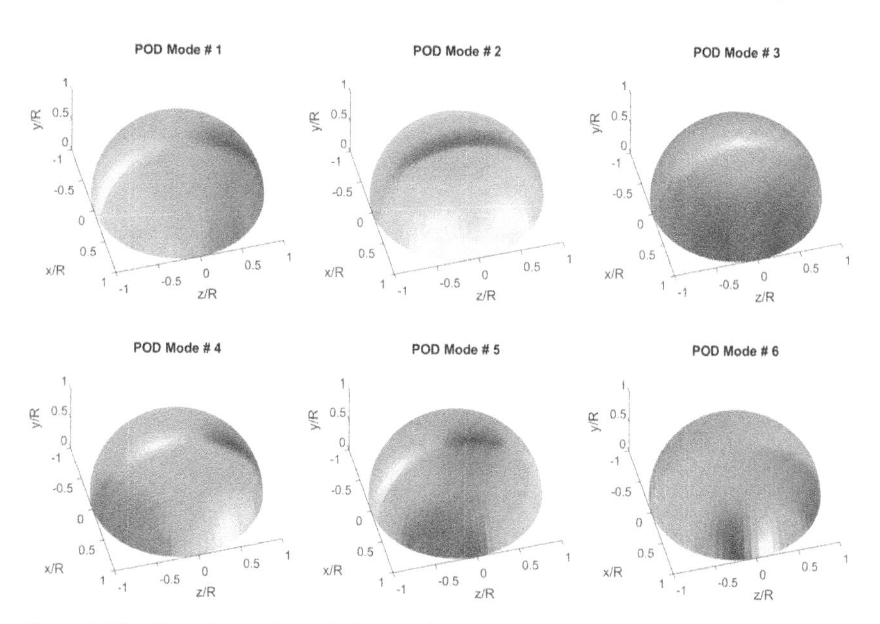

Figure 6.38 First six dominant POD spatial modes on the hemisphere. Flow goes along x-direction from negative to positive values.

the separation line located downstream of the time-averaged location, and the negative coefficients would indicate times where the separation line is upstream of the time-averaged location. It represents the situation when the separation line on one side of the hemisphere moves upstream, while the separation line on the opposite side shifts downstream. This anti-symmetric separation motion is associated with the anti-symmetric vortex shedding and was shown to be related to the global shifting motion of the wake, shown in Figure 6.39. Higher-order modes #4 and #6 can be viewed as perturbations of the separation line, as well as the pressure signature of vortical structures present in the separation region. The most dominant (~20% of the energy) symmetric mode #2 reveals a region of negative pressure change near the separation line, followed by a global positive pressure change in the separated region. Thus, mode #2 and similarly, mode #3 describe the symmetric motion of the separation line near the hemisphere apex in the streamwise direction, with the related changes in the separated region. These modes are associated with the symmetric vortex shedding off the hemisphere. The corresponding wake dynamics are associated with a wake global breathing mode, schematically shown in Figure 6.39.

The spectra of the temporal POD coefficients for the first 6 POD modes are shown in Figure 6.40. The main shifting mode, Mode #1, has distinct peaks in its spectra at approximately $St_D = fD/U_\infty = 0.1-0.2$, where D is the hemisphere diameter and U_∞ is the incoming freestream speed. This peak was also observed in studies of hemisphere-on-cylinder turrets (Gordeyev et al. 2014; De Lucca et al. 2018) and also was found to be approximately constant for a range of Mach numbers between 0.35 and 0.66. The spanwise-symmetric modes #1 and #3 have most of their energy contained in low frequencies.

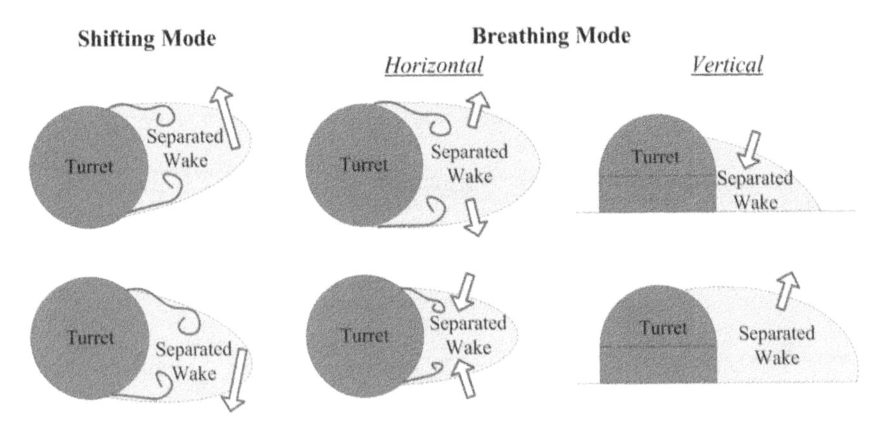

Figure 6.39 Schematics of shifting and breathing wake modes of the turret wake. *Source:* Adapted from De Lucca et al. (2018b), figure 8. © 2018 by De Lucca, Gordeyev, Morrida, Jumper, and Wittich.

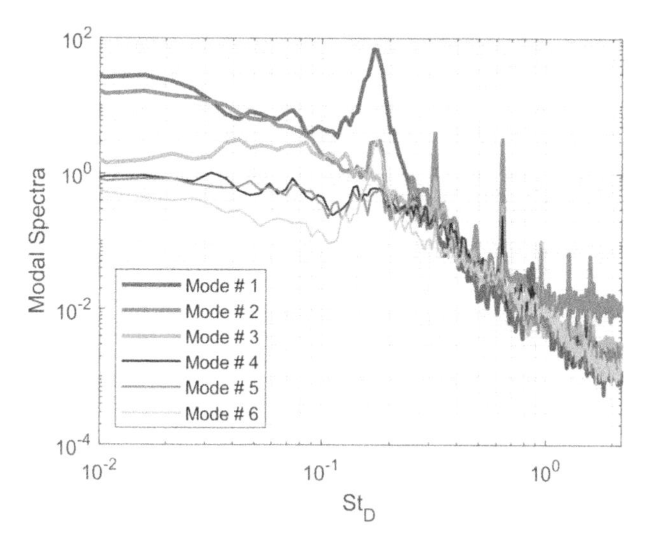

Figure 6.40 Spectra of POD temporal coefficient for the first six dominant POD modes for the hemisphere, presented in Figure 6.38.

Detailed analysis of unsteady pressure fields of a hemisphere-on-cylinder turret with realistic surface features such as gaps and "smile" cutouts for both the conformal- and the flat-window apertures is discussed in Gordeyev et al. (2014). They found that the slope discontinuity caused by the flat-window aperture introduced additional localized vortical structures and related pressure fluctuations when the window was faced either forward or sideways. Except facing forward, when the flat window tripped the incoming flow and introduced strong vortical structures on top of the turret, the presence of the flat window introduced local structures only and did not significantly change the overall unsteady pressure field. The significant contribution to the pressure variations was found to be from the "smiles." These large, cavity-like cutouts were found to significantly affect the instantaneous pressure field, depending on the "smiles" positions relative to the incoming flow. They concluded that in order to reduce the pressure fluctuations and the resulting unsteady forces, acting on the turret, the optimal turret geometry should be a conformal turret without any significant slope discontinuities.

6.4.3 Steady-lensing Effects at Forward-looking Angles

The turret changes both the steady and the unsteady components of the surrounding density, which in turn imposes steady and unsteady wavefront components on the outgoing laser beam. As discussed in the previous section, the flow around the forward half of the turret (or nose-mounted turrets) is attached and it can be fairly accurately described by a potential inviscid flow solution around a sphere,

$$u_r(r,\alpha)=U_\infty \sin(\alpha)\left[1-\left(\frac{R}{r}\right)^3\right],\quad u_\alpha(r,\alpha)=-U_\infty \cos(\alpha)\left[1+\frac{1}{2}\left(\frac{R}{r}\right)^3\right]$$

where u_r and u_α are the radial and the window-angle velocity components, r is the distance from a point in the flow from the sphere center, U_∞ is the freestream velocity and $R=D/2$ is the sphere radius. Knowing the velocity distribution in space, the density variations can be estimated assuming weakly compressible flow; the velocity field creates pressure variations through Bernoulli's equation, $p'+1/2\rho_0(u_r^2+u_\gamma^2)=p_0=const$ and the density and pressure are related via an isentropic equation, $\gamma\rho'/\rho_0=p'/p_0=p'/(\rho_0 a^2)$, where a is the speed of sound and γ is the ratio of specific heats. Finally, the density field can be integrated along the lines parallel to the beam-propagation direction starting at the aperture to get the pseudo-steady-lensing, that is the optical aberrations due to a steady density field around the turret

$$OPD_{steady} \sim K_{GD}\int_a^\infty \rho'(\vec{x})dl \sim -K_{GD}\left(\frac{1}{2c^2}\rho_0\right)\int_a^\infty u^2(\vec{x})dl \sim -K_{GD}\left(\frac{U_\infty^2}{2c^2}\rho_0 R\right)\int_a^\infty f(r/R,\alpha)d(l/R)$$

$$OPD_{steady}=(\rho_0/\rho_{SL})M^2 D f(Ap/D,\alpha)$$

where $f(Ap/D,\alpha)$ is a function of the relative aperture size and the window angle. It follows from this equation, the pseudo-steady-lensing optical aberrations also scale as $OPD_{steady}\sim\rho_0 M^2 D$. Examples of "normalized" wavefronts, $f(Ap/D,\alpha)$, for several forward-looking window angles in the center-plane for $Ap/D=0.33$ are shown in Figure 6.41. For forward-looking angles, the steady aero-optical distortions are dominated by the tilt-component due to the spatially varying density field over the front portion of the turret. At the apex of the turret, the flow reaches the maximum speed and starts decelerating at the aft portion of the turret. The resulting steady wavefront at the viewing angle of 90 degrees has no residual tip/tilt and only higher-order distortions, which is mostly astigmatism (Zernike mode $Z_2^2(r,\varphi)$). For simple turret geometries, including the hemisphere-only geometry,

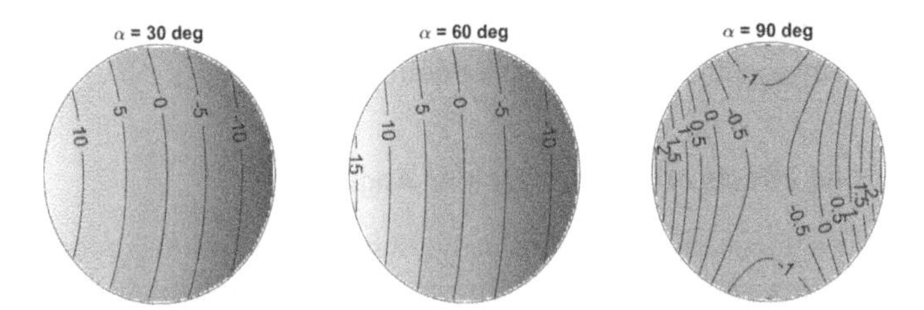

Figure 6.41 Normalized steady-lending wavefronts for several forward-looking window angles around a sphere. *Source:* Gordeyev and Jumper (2010), figure 6. Reproduced with permission of Elsevier.

the potential approach can be extended to transonic and supersonic flows to calculate the pseudo-steady-lensing aberration environment (Fuhs and Fuhs 1982).

The proposed potential approach obviously fails to take into account optically aberrating effects of sharp density gradients caused by shocks, which, as will be discussed later, are present around the turret at high transonic ($M > 0.6$), and supersonic speeds.

6.4.4 Aero-optical Environment at Back-looking Angles

In the beginning of this chapter, we discussed a general scaling law, Equation (6.3), for aero-optical effects, presented again here for convenience

$$\frac{OPD_{rms}}{D} = \frac{\theta}{D}\frac{\rho_0}{\rho_{SL}} G(M) g\left(\frac{\delta}{D}, \alpha, \beta, \mathrm{Re}; geometry\right).$$

For a large class of turrets, we can simplify this equation. Often, the turret size, D, is much larger than the incoming boundary layer, δ For a sufficiently small incoming boundary layer, $\delta / D \ll 1$, most of the boundary layer on the turret-mounting surface upstream of the turret gets wrapped into the necklace vortex and does not *directly* affect optical distortions over the turret. A new boundary layer on the hemispherical portion of the turret starts regrowing from the stagnation point on the front of the turret. Assuming that the Reynolds number is above the critical value, the boundary layer growth is almost linear with distance before separation. Therefore, the boundary-layer thickness before it separates from the turret will be proportional to the turret diameter only. Consequently, the large-scale structures inside the separated shear layer, θ will also be proportional to the turret diameter, $\theta / D \simeq 1$. When the diameter-based Reynolds number, Re_D, is larger than ~500,000, the boundary layer becomes turbulent before it separates. As discussed before, the main mechanism for creating the large-scale structures inside the separated shear layer is the inviscid inflectional mechanism, which is mostly independent of Re. The Reynolds number affects only the small structures in the shear layer. Due to the small correlation size and weak associated density fluctuations, these small structures typically do not add any significant optical distortions when compared to the distortions caused by the large separated-flow structures. Therefore, as long as the Reynolds number is greater than the critical value, the optical distortions can be assumed to be independent of Re. Finally, for subsonic and low transonic speeds, $G(M) \sim M^2$. Therefore, the aero-optical scaling for large turrets can be re-written as,

$$OPD_{rms}^{Norm} \equiv \frac{OPD_{rms}}{(\rho_0 / \rho_{SL}) M^2 D} = B(\alpha, \beta, geometry) \tag{6.48}$$

This scaling was confirmed for both tunnel (Gordeyev and Jumper 2010) and flight (Porter et al. 2013b) data. This similarity law is often called the "ρM^2"-scaling.

Values of normalized OPD_{rms} for a flat-window turret for different (α,β) - angles are shown in Figure 6.42. At forward-looking angles, $\alpha < 80$ degrees, the normalized values remained around one, indicating that the aero-optical environment at these angles was not largely affected by higher-order aero-optic aberrations. This is an expected result, as the flow is attached at the front portion of the turret. Beyond a viewing angle of 80°, the normalized values begin to increase, reaching a local maximum between 2 and 2.5 at approximately $\alpha = 90$ degrees, but only for the flat-window configurations. This peak, also visible in Figure 6.30 at different β-angles, is attributed to the presence of the unsteady separation bubble, forming over the flat-window aperture at this range of viewing angles (Gordeyev et al. 2011). The flow separates at about 115 degrees, so at large viewing angles the beam propagates through a shear-layer-dominant separated region formed downstream of the turret. As the viewing angle continues to increase toward 120 degrees, the beam propagates through a larger portion of the wake. Simultaneously, the vortical structures in the shear-layer grew larger in this region of the wake and the normalized OPD_{rms} increased rapidly, more than doubling over a few degrees, and reaching normalized levels of 5–6 at large back-looking angles $\alpha > 130$ degrees.

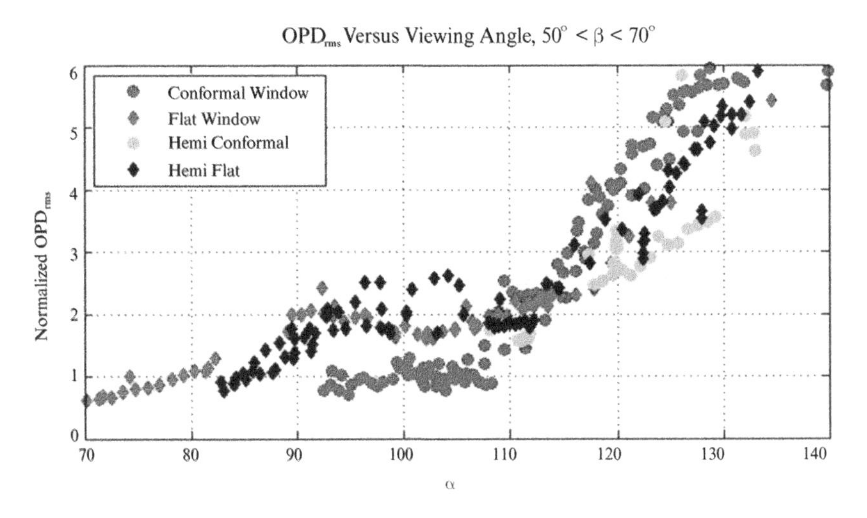

Figure 6.42 Normalized OPD_{rms} as a function of the viewing angles for different turret configurations, collected in-flight. *Source:* De Lucca et al. (2013), figure 9. Reproduced with permission of SPIE.

To demonstrate the damaging far-field effects due to aero-optical distortions in the turret wake, let us use the "ρM^2"-scaling law, shown in Equation (6.47), and compute the actual levels of aero-optical distortions for a flight at sea-level at $M = 0.5$and a turret diameter of $D = 0.2$ meters. For the back-looking angle of $\alpha = 100$ degrees, $OPD_{rms} = 0.1$ microns. Using the Large-Aperture Approximation, Equation (2.32), with the laser wavelength of $\lambda = 1\mu m$, the expected reduction in the far-field intensity will be 0.67. However, for $\alpha = 130$ degrees, OPD_{rms} becomes 0.3 microns, resulting in only 2% of the diffraction-limited intensity at the target. The aero-optical effects will be even more devastating at faster speeds and larger turrets, as OPD_{rms} increases with the Mach number and turret size.

As discussed in Chapter 4, the modal analysis, particularly the POD analysis, is very useful in identifying coherent structures. To investigate the presence of the organized vortical structures in the wavefronts, the POD analysis of the wavefront data at different viewing angles was performed. Figure 6.43(a), shows the normalized OPD_{rms} for different viewing angles. Several viewing angles, indicated by solid circles were selected and the POD analysis of the wavefront data at these angles was performed. The first four dominant POD modes are presented in Figure 6.43(b). For the viewing angles below 90 degrees, there is no visible spatial structure in the POD modes. But when the laser beam is sent through the separated wake region, the first four POD modes for $\alpha > 90$ degrees clearly show the

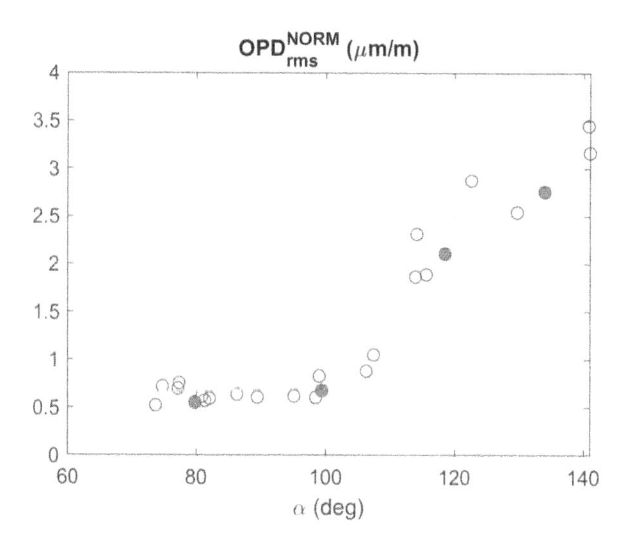

Figure 6.43 (a) Normalized OPD_{rms} for different viewing angles collected in-flight for the conformal turret at $M = 0.6$. POD analysis was performed for selected viewing angles, indicated by solid circles. (b) First 4 POD modes of wavefronts data at selected viewing angles. Mode numbering goes from left to right. Note the dominance of nearly spatially periodic shear-layer structures at the viewing angles above 100 degrees.

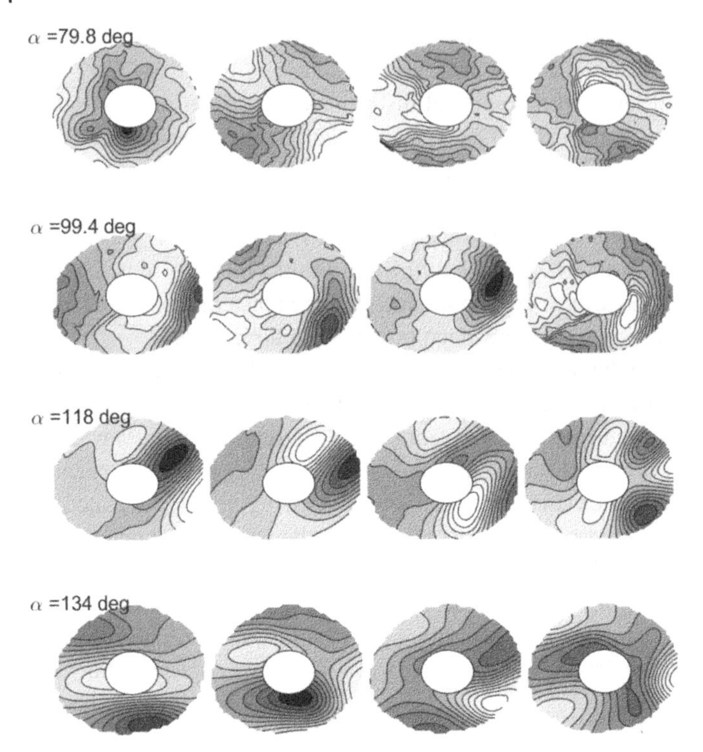

Figure 6.43 (Continued)

presence of the organized structures. The structure size increases with the viewing angle, as the shear layer structure grows in streamwise size farther downstream from the separation point. This modal analysis demonstrates that the complex region in the separated wake downstream of the turret is dominated by the organized vortical structures, present in the shear-layer region of the wake. It should be noted that for the same turret geometry, the POD modes are essentially the same for all flight tests at the same azimuth/elevation angles.

6.4.5 Shock-effects at Transonic Speeds

The previous discussion of the flow topology and optical distortions caused by turrets is valid only for the fully subsonic regime, where the flow around the turret is subsonic everywhere. Since the flow accelerates over the turret, the local Mach number will reach a sonic value at some incoming Mach number, which is called the critical Mach number. For incoming Mach numbers above the critical Mach number, the flow will become locally supersonic. Assuming isentropic flow, C_p as a function of the local Mach number can be derived from the compressible Bernoulli's equation as

$$C_p(M) = \frac{2}{\gamma M_\infty^2}\left[\left(\frac{1+\frac{\gamma-1}{2}M_\infty^2}{1+\frac{\gamma-1}{2}M^2}\right)^{\gamma/(\gamma-1)} - 1\right]$$

If the incompressible pressure distribution C_{P0} is known, it is possible to estimate the local compressible C_p using a compressible Karman-Tsien correction (Shapiro 1953),

$$C_P = \frac{C_{P0}}{\sqrt{1-M_\infty^2} + \left(\frac{M_\infty^2}{1+\sqrt{1-M_\infty^2}}\right)\frac{C_{P,0}}{2}}$$

Combining these two equations, the critical incoming Mach number, M_{cr} M_{cr}, can be computed for the flow to reach the sonic speed at a given location knowing the incompressible $C_{p,0}$ value. From Figure 6.35, left, the lowest $C_{p,0}$ on the turret is approximately −1.25, giving the critical incoming Mach number of 0.55. At this incoming Mach number, the flow reaches the sonic speed on top of the turret. If the incoming Mach number is above this critical Mach number, the flow, after passing the sonic speed, will continue to accelerate on top of the turret, forming a local supersonic region on top of the turret. To match the subsonic flow downstream of the turret, a normal shock will form at the end of the supersonic region, as schematically indicated in Figure 6.44. The boundary-layer-shock interaction will cause the flow to prematurely separate on top of the turret and form a larger separation region. When the incoming Mach number increases further, the supersonic region on the turret will grow in size and, at some point, extend to the base of the turret. This simple analysis does not take into account the turbulent

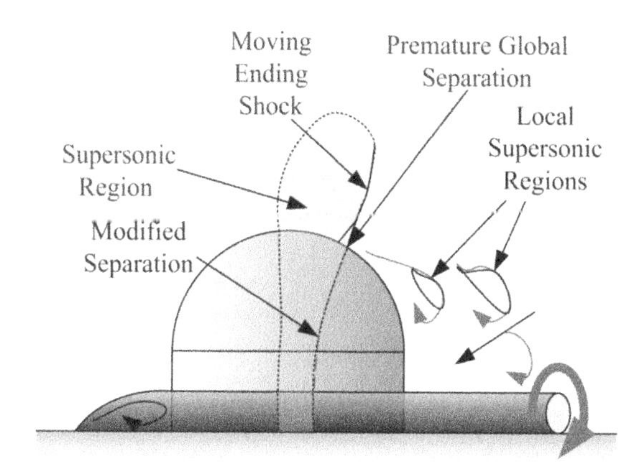

Figure 6.44 Transonic flow features on the turret. *Source:* Gordeyev and Jumper (2010), figure 8. Reproduced with permission of Elsevier.

wake behind the turret, so the real flow around the turret at transonic speeds is even more complex, with possible unsteady pockets of supersonic flows and weak shocks appearing in the wake, as shown in Figure 6.44.

The premature shock-induced separation will result in a larger wake for the transonic regime. Combined with the increased density variations across the oblique shock and local supersonic regions in the turret wake, it will undoubtedly impose additional aero-optical distortions. Therefore, the overall aero-optical distortions at transonic speeds are expected to be higher than predicted by the "ρM^2"-scaling. This effect was observed experimentally for the first time in 2008 (Vikasinovic et al. 2008). At $M = 0.65$, the level of optical aberrations around the hemisphere was found to be almost two times higher than the subsonic "ρM^2"-law predicts. The AAOL-T program provided very detailed information of the shock related aero-optical effects. Figure 6.45 shows normalized OPD_{rms} values for the flat- and conformal-window turrets for a range of Mach numbers between 0.5 and 0.8. While technically the shock appears above the critical Mach number of 0.55, the shock was found to be weak and intermittent up to $M = 0.65$ and did not impose significant aero-optical effects (De Lucca et al 2013). But at higher transonic Mach numbers of 0.7 and 0.8, the unsteady shock intensifies and additional optical distortions related to the shock motion clearly appear in the range of viewing angles of between 70 and 95 degrees, seen as a local increase of the normalized OPD_{rms} in Figure 6.45 for both turret configurations.

Although not perfect, the approximate collapse of aero-optical data in the wake region for $\alpha = 100 - 120$ degrees indicates that the "ρM^2"-scaling, used to normalize aero-optical data, still can be used to rescale data to different turrets at subsonic and transonic speeds. In addition, it indicates that after the flow is separated, the presence of the shock does not affect the dynamics of the shear-layer structures in the separation region.

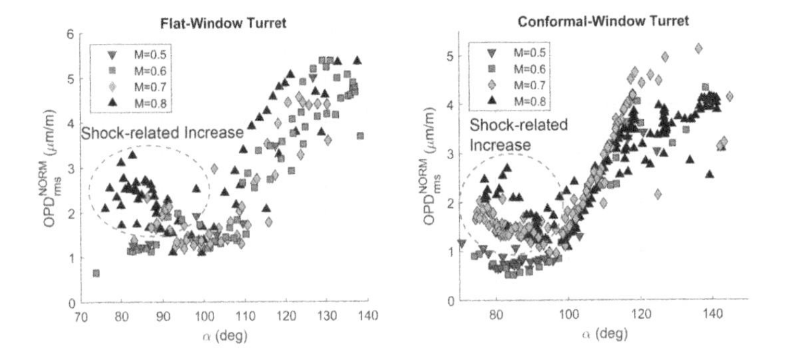

Figure 6.45 Normalized OPD_{rms} for different viewing angles for flat- and conformal-window configuration at different Mach numbers, collected in-flight. *Source:* Adapted from Morrida et. al. (2017), figure 5.

As mentioned in Chapter 4, another way to quantify the aero-optical distortions is to compute the spatial map of the temporal variance of the wavefronts at different points over the turret,

$$S(\alpha,\beta) = (\overline{OPD^{Norm}(\alpha,\beta;t)^2})^{1/2}.$$

$S(\alpha,\beta)$ describes the temporal variation of aero-optical wavefronts at a given point over the turret, also characterized by α and β, and it is directly related to the strength of aero-optical flow features. As the conformal window does not change the turret shape, the temporal variance is independent of the aperture location and is only related to optical features in the separated wake. The spatial map of the temporal variances of wavefronts for different aperture positions were projected onto the turret and were averaged in overlapping regions. The spatial map of the wavefront distortions over the turret surface for $M = 0.6$ is presented in Figure 6.46(a). Aero-optical distortions are small at the forward portion of the turret where the flow is attached. A region of increased distortion, related to the separation or wake region, which is dominated by "horn" vortices, is clearly visible. As the vortices are present on both sides of the turret, they create an additional velocity downwash in between them along the centerline, see Figure 6.34(a). This increased downwash results in a smaller wake size along the center plane, with correspondingly smaller levels of aero-optical distortions for an

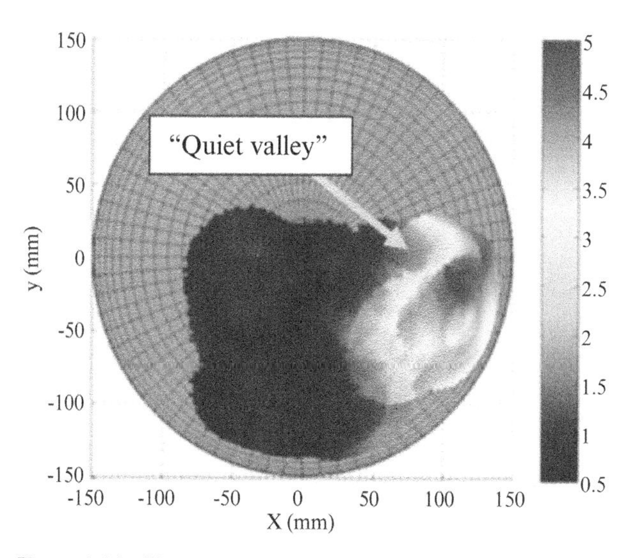

Figure 6.46 The top view of temporal variance of the normalized wavefronts, S(α,β), (in μm/m) for the conformal-window turret at (a) $M = 0.6$, *Source:* Adapted from Morrida et. al. (2017), figure 5 (b, top) $M = 0.7$ and (b, bottom) $M = 0.8$. Flow goes from left to right. *Source:* Morrida et. al. (2017), Reproduced with permission from Optical Society.

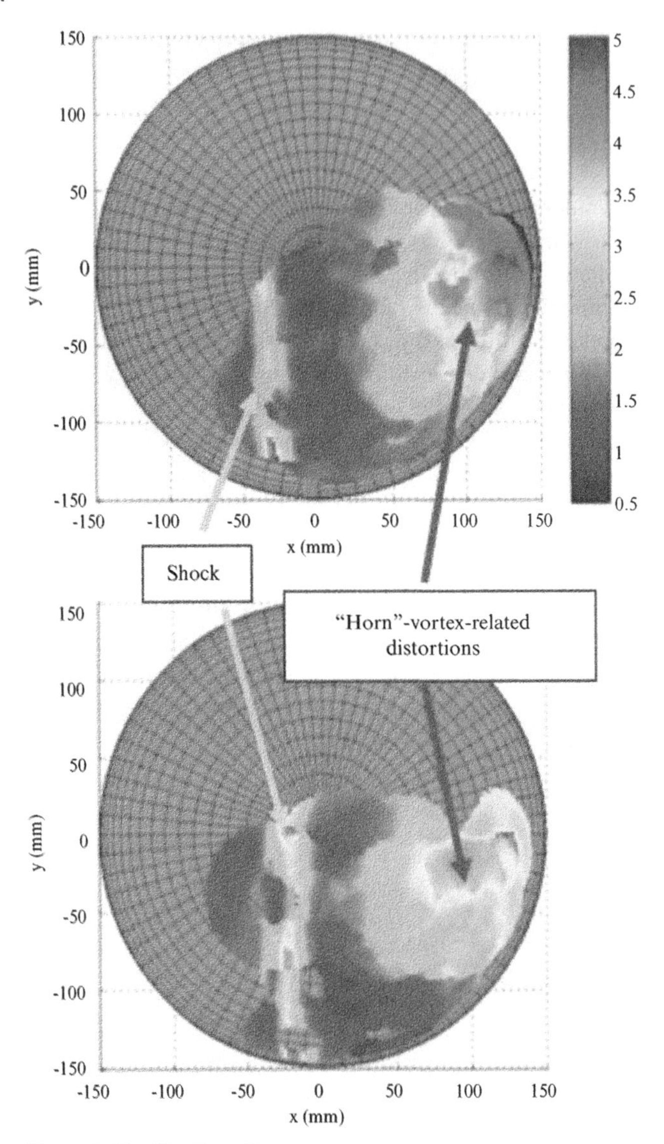

Figure 6.46 (Continued)

increased range of α up to 125–130 degrees. This region is labeled as a "quiet valley" in Figure 6.46 and was also observed around flat-window turrets (Porter et al. 2013b), and in numerical simulations of the flow around the conformal-window turret by Mathews et al. (2016).

Similarly, the spatial maps of the temporal variances of the mapped wavefronts were made for transonic Mach numbers of $M = 0.7$ and 0.8 and the results are presented in Figure 6.46(b). The shock creates additional localized distortions and they are clearly visible as a line of the increased distortions around $\alpha = 80$ degrees. The shock-related optical intensity increases with Mach number, as expected. The average shock location is mostly independent of β. The "quiet valley," while still present at these transonic speeds, is weakened and does not extend as far downstream as for $M = 0.6$case. The "horn"-vortex-related region weakens as well and moves closer to the center plane at $M = 0.8$, compared to $M = 0.7$ case; it indicates that the changes in the separated wake dynamics are due to the unsteady shock. The same trends were also observed around hemisphere-only turrets.

The shock on top of the conformal turret has no fixed point to "anchor" itself, so originally it was expected to oscillate rapidly due to the shock-boundary-layer interaction (Smits and Dussauge 1996). However, the conditional and spectral analysis of aero-optical data ((De Lucca 2013; Morrida et al. 2017) revealed that the shock oscillates at a much lower frequency, comparable with the dominant frequency of the wake. This indicates the existence of a dynamic locking between the separated wake and the shock. Figure 6.47(a), shows aperture-averaged wavefront spectra, normalized by $(\rho_\infty M^2)^2$, for a side-looking angle, $\alpha = 80$ degrees, for several Mach numbers. At $\alpha = 80$ degrees, the flow is attached over the aperture for $M = 0.5$ and the wavefront spectrum does not have any significant optical features. For $M = 0.6$, a very weak shock appears over the aperture, which results in

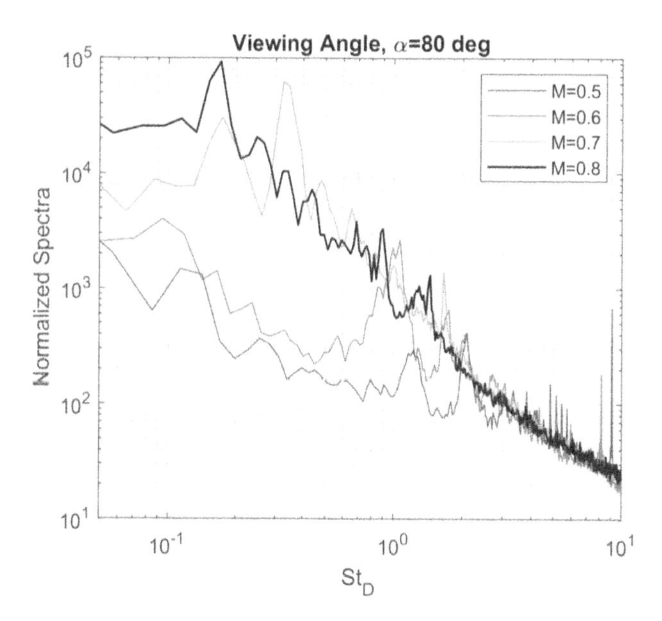

Figure 6.47 Normalized aperture-averaged wavefront spectra for several Mach numbers for (a) $\alpha = 80$ degrees and (b) $\alpha = 120$ degrees.

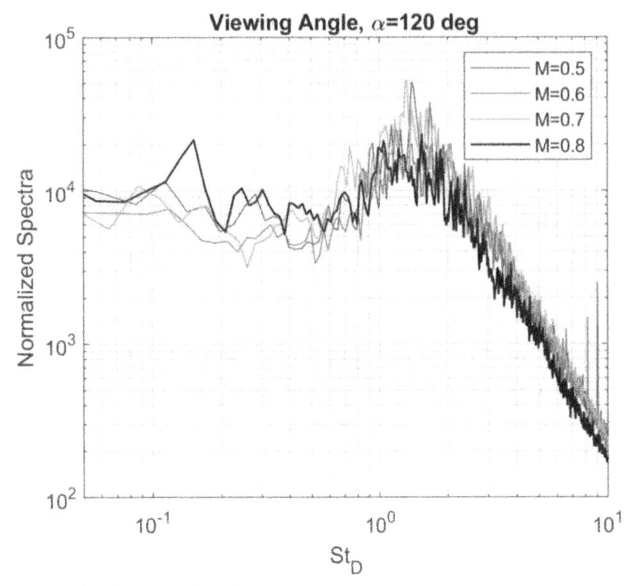

Figure 6.47 (Continued)

the appearance of the small peak around $St_D = 1$; otherwise, the spectrum stays unchanged. For $M = 0.7$, the shock becomes stronger, and the spectrum is significantly increased at low $St_D < 1$. At this Mach number, the spectrum has several peaks, with the dominant ones around $St_D = 0.3$ and 0.18. For $M = 0.8$, the spectrum has a single peak around $St_D = 0.18$. So, these spectra indicate that while at low transonic speeds the shock oscillates at a higher frequency range of St_D between 0.3 and 1, while at high transonic speeds, the shock dynamics is locked in with the separation dynamics with a lower frequency.

For $\alpha = 120$ degree, shown in Figure 6.47(b), the beam traverses through the separated region, dominated by shear-layer structures. Normalized spectra for all Mach numbers, including transonic ones, show a very good collapse. The peak in spectra is around $St_D = 1$, which corresponds to a typical frequency for shear layer structures at this viewing angle. Once the flow is separated, the presence of the shock does not significantly affect the structures in the separation region. Only at high subsonic speed of $M = 0.8$, the secondary, separation-line-related peak appears around $St_D = 0.18$, indicating that the shock motion starts modulating or affecting the temporal evolution of the shear-layer structures in the wake at high transonic speeds. This is also consistent with the lock-in mechanism between the shock and the separation region, discussed earlier.

7

Aero-Optical Jitter

Up to this point, we were primarily focused on the statistics of the high-order component of the wavefront, since it directly affects the instantaneous intensity distribution of the laser beam on a target. However, as we showed in Equation (2.36), the tip/tilt component will also affect the far-field pattern to shift or point away from the target. If the tip/tilt is time-dependent, also called jitter, it will also degrade the beam intensity of the target by "painting" the initial pattern over the aim point on the target. In a time-averaged sense, jitter enlarges the effective beam spot size on the target, and therefore significantly impacts target irradiance. Thus, understanding and ultimately mitigating various sources of jitter are important to maintain a proper pointing of the beam on the target.

One component of jitter is caused by the turbulent flow structures on the order of the aperture size. The component of jitter is called the aero-optical jitter. Another component of the jitter is caused by either the unsteady motion of the platform itself or by the vibrations in the beam director induced by the platform motion. This component is called the base-motion-related jitter. Finally, the turbulent flow will create an unsteady pressure field around the beam director, like a turret. These pressure-related forces might cause either the whole beam director or some parts of it to oscillate due to aero-elastic effects. The resulting jitter is called the aero-mechanical jitter (Kalensky et al. 2021). Obviously, all three components of jitter will affect the beam irradiance on the target.

Both the aero-optical and aero-mechanical jitter stem from the presence of the turbulent flow around the aircraft. This fact makes it very difficult to separate and study these components. Therefore, the overall instantaneous jitter, which includes both the aero-optical and mechanical jitter, is typically removed from the measured wavefronts. When this necessary step is taken, information pertaining to large-scale optical structures is lost.

Various adaptive-optics systems have been developed that are able to compensate for a significant portion of the mechanical jitter. For instance, an optical

Aero-Optical Effects: Physics, Analysis and Mitigation, First Edition. Stanislav Gordeyev, Eric J. Jumper, and Matthew R. Whiteley.

inertial reference unit in conjunction with a fast-steering mirror and an appropriate controller can reject mechanical jitter disturbances up to 1 kHz (Burns et al. 2015; Burns 2016). A further discussion of this is provided in Chapter 8. Even if the jitter rejection system were to operate such that all mechanical jitter could be compensated for, the aero-optical jitter would remain. This aero-optical jitter will affect the time-averaged far-field irradiance pattern. Therefore, it is desirable to not only quantify the overall jitter of the system but to develop a method that can decouple aero-optical and mechanical jitter for additional studies.

7.1 Local and Global Jitter

7.1.1 Local Jitter

Recall that from Hyugens principle, a wavefront travels normally to itself. The local direction of the propagation of the beam, relative to the undistorted direction, z, known as the local deflection angles (θ_x, θ_y). For collimated beams, these local deflection angles are gradients of the wavefronts, as shown Equation (2.6). The local deflection angles are often called local jitter. These quantities are directly measured by a Shack-Hartmann wavefront sensor or a Malley probe, and used to reconstruct the underlying wavefronts.

From Equation (2.6) it follows that, if the aero-optical flow has a typical structure of size, L, $\left|\vec{\theta}(x,y,t)\right| \sim \dfrac{W(x,y,t)}{L}$. Taking the spatial root-mean-square and time-averaging gives a relationship between θ_{rms} and $W_{rms} = OPD_{rms}$ as,

$$\theta_{rms} \sim \frac{OPD_{rms}}{L} \tag{7.1}$$

If the scaling for OPD_{rms} is known, Equation (7.1) provides a useful scaling law for the deflection angles. For instance, for the turbulent boundary layer, Equation (6.22) provides the following scaling, $OPD_{rms} \sim \rho_\infty M_\infty^2 \delta \sqrt{C_f} F(M)$, giving the scaling for local jitter as $\theta_{rms} \sim \rho_\infty M_\infty^2 \sqrt{C_f} F(M)$. For a shear layer, Equation (6.15), $OPD_{rms} \sim \rho_\infty M_c^2 L$, resulting in $\theta_{rms} \sim \rho_\infty M_c^2$. It is interesting to note that in both cases, the levels of the local jitter *do not* depend on the thicknesses of the turbulent flows, but only on the freestream density and the freestream Mach number.

7.1.2 Global Jitter

As shown in Chapter 6, local unsteady deflection angle, or local jitter, measured by the Malley probe or over a single sub-aperture in the SHWFS, carries a lot of useful information about the underlying flow physics of optically aberrating structures, like the temporal spectra and typical scales of large-scale structures. But, as demonstrated in Equation (2.36), only an overall unsteady tip/tilt component over the entire aperture, called a global jitter, directly affects the beam pointing. In Chapter 5 it was demonstrated that the global tip/tilt depends on the aperture size and shape, and explicit equations for a single harmonic were calculated for the tip/tilt component for one-dimensional and circular apertures. Below we will perform a similar harmonic analysis of the global tip/tilt. For simplicity, we will consider the one-dimensional aperture only as it is straightforward to repeat it for the circular aperture.

As before, let us consider a simple one-dimensional harmonic wavefront, $W(x;\Lambda,\phi) = \sin(2\pi x/\Lambda + \phi)$, where Λ is the spatial wavelength of the optical wavefront, and ϕ is some arbitrary phase offset. The local jitter in the middle of the aperture is $\theta_x(x=0;\Lambda,\phi) = -\left.\frac{dW(x;\Lambda,\phi)}{dx}\right|_{x=0} = -\frac{2\pi}{\Lambda}\cos(\phi)$. We can define the global jitter transfer function, G_{GJ}, as a ratio between the global jitter in the streamwise direction, denoted in this chapter as $\theta_G(Aperture)$, (which is mathematically equal to B_1, given in Equation (5.5)), and the local jitter, θ_x, in the middle of the aperture,

$$G_{GJ}(z) = \frac{\theta_G(Aperture)}{\theta_x} = \frac{3\left[\sin z - z\cdot\cos z\right]}{z^3} \tag{7.2}$$

where, as before, $z = \pi Ap/\Lambda$. The plot of this global jitter transfer function is presented in Figure 7.1(a). For large wavelengths, z is small and the wavefront can be faithfully represented by a linear component only, as seen in Figure 5.1(b). As a result, the global jitter transfer function is close to one for small z. However, for wavelengths comparable with or smaller than the aperture size, the wavefront varies significantly over the aperture and the linear fit does a poor job representing the wavefront; see Figure 5.1(a), leading to small values of G_{GJ}-function. Moreover, for some wavelengths, for instance $Ap/\Lambda = 2$, the G_{GJ}-function becomes negative, meaning that the global jitter is out-of-phase with the local jitter. Overall, the global jitter transfer function behaves as a low-pass filter.

For a single frequency harmonic, G_{GJ}-function relates the amplitude of local jitter to the amplitude of global tilt over a given aperture. As a consequence, the autocorrelation spectral densities for local jitter and global tilt are related as well as,

$$S_G(k_x, Ap) = \left[G_{GJ}(z = \pi Ap/\Lambda)\right]^2 S_\theta(k_x = 2\pi/\Lambda) \tag{7.3}$$

(a)

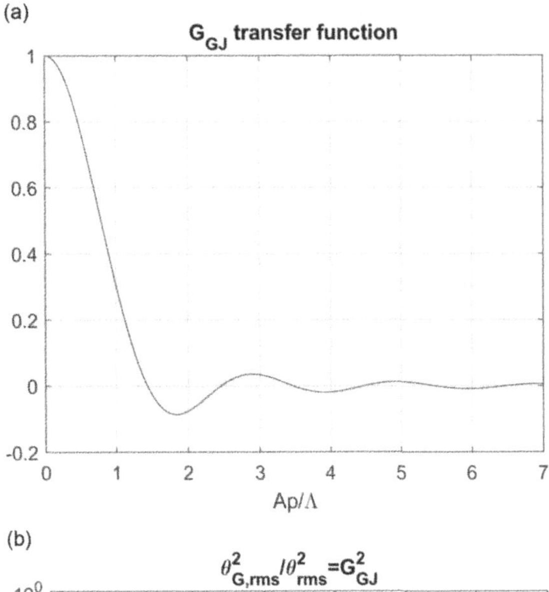

(b)

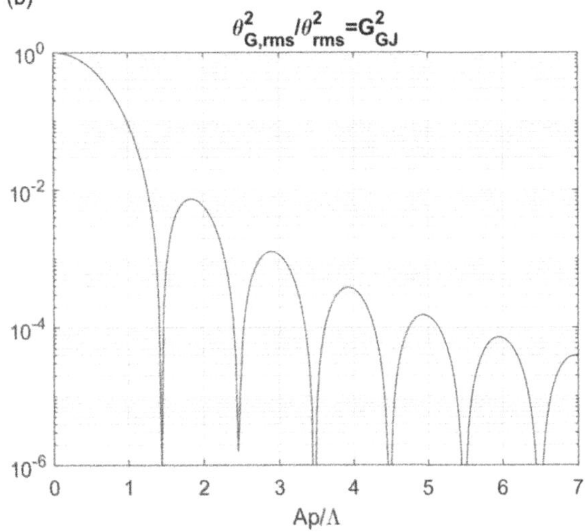

Figure 7.1 (a) Global jitter transfer function, G_{GJ}. (b) The energy attenuation between the global and the local jitter, given as G_{GJ}^2.

The plot of $(G_{GJ}(z))^2$-function is presented in Figure 7.1(b). Thus, by measuring only the spectrum of the local jitter, S_θ, we can compute the spectrum of the global jitter and determine the global jitter rms-values, $\theta_{G,rms}$, for different apertures

$$\theta_{G,rms}^2(Ap) = \frac{1}{\pi}\int_0^\infty [G_{GJ}(z)]^2 S_\theta(k_x)dk_x$$

For the case of the purely traveling homogeneous flows, the above equation can be re-written in terms of the temporal spectrum of the local jitter as

$$\theta_{G,rms}^2(Ap) = 2\int_0^\infty \left[G_{GJ}(z = \pi f Ap / U_C)\right]^2 S_\theta(f) df \qquad (7.4)$$

Thus, by measuring the local jitter *only* in the middle of the aperture, using the Shack-Hartmann sensor or a Malley probe, we can approximate the global tilt for *any aperture*, as long as the wavefronts can be treated as purely convective and homogeneous in the streamwise direction. Because local jitter is a point quantity, and therefore is not influenced by aperture effects, this is a very useful result.

7.2 Subaperture Effects

As discussed in Chapter 3, the Shack-Hartmann sensor measures the wavefronts by breaking the beam into smaller sub-aperture beams using a lenslet array and measuring the resultant deflection angles at many points. The Malley probe also uses a small, but still a finite diameter laser beam to measure the time-series of the deflection angles at a single point. Thus, due to the finite small-aperture beam size, denoted as A_{sub}, both sensors will technically measure the global jitter, θ_G, over the small aperture, rather than the true local jitter, θ. These effects are called subaperture effects and should be accounted for, if someone needs to properly measure the local jitter.

Both sensors measure the time series of the deflection angles averaged over a small sub-aperture (or beam size) of size A_{sub}. If the convective speed is known, Equation (7.3) can be rewritten in the frequency domain as $S_G(f, A_{Sub}) = \left[G_{GJ}(z = \pi f A_{Sub} / U_C)\right]^2 S_\theta(f)$. If the sub-aperture size is given, this equation can be used to correct the measured deflection angle spectrum, $S_G(f, A_{Sub})$, in order to extract the true deflection angle spectrum, $S_\theta(f)$. From Figure 7.1, the global jitter significantly deviates from the local jitter for z-values of $z > 1$. It gives an estimate the range of frequencies, affected by the sub-aperture effects as, $f > 0.3 U_C / A_{Sub}$.

7.3 Techniques to Remove the Mechanically Induced Jitter

Remember, that the local jitter still can be corrupted by mechanical vibrations, so additional data processing should be performed to recover either the local or the global jitter spectra due to aero-optical effects only. Below we discuss several approaches on how to remove the aero-mechanical and base motion induced jitter from the overall global jitter and recover the turbulence induced component of global jitter. This is a post-processing technique and should not be taken as a method of cleaning jitter on an outgoing beam.

7.3.1 Cross-correlation Techniques

One way to remove the mechanically induced jitter is to measure or to estimate the instantaneous mechanical jitter separately and to subtract it from the overall jitter. It could be done, for instance, by independently recording mechanical motion of various optical components, such as mirrors, using accelerometers or laser vibrometers, simultaneously with the measurements of the overall jitter. Alternatively, various correlation-based techniques can be used to estimate the aero-mechanical component of the jitter. Below we will present a spectral Least-Square Estimation (LSE) technique, which was demonstrated in De Lucca et al. (2012) to be effective in estimating the spectrum of the aero-mechanical jitter and removing it from the total jitter. A similar procedure, but applied to remove higher-order aero-optical distortions due to the mechanical vibrations of large windows in a wind tunnel from the wavefront measurements is presented in De Lucca et al. (2014).

In order to implement the LSE technique, simultaneous measurements of the total jitter, denoted $\theta_T(t)$, and mechanical motion of the various optical components via several accelerometers, denoted $y_i(t)$, are required. In general, the LSE technique generates an estimate of a measured quantity using a set of other measured quantities. It does this by using the correlation between a measured jitter quantity and the additional measured quantities. The LSE technique was developed by Adrian (1977), and it can be written as

$$\tilde{\theta}_M(t) = \sum_{i=1}^{N} L_i y_i(t) \tag{7.5}$$

Here, the estimate of the mechanically induced jitter, labeled $\tilde{\theta}_M$, is approximated using a linear sum of the accelerometer measurements in N locations, $y_i, i = 1.N$, multiplied by the influence coefficients L_i. The influence coefficients can be computed using the total jitter signal, θ_T, as $L_i = \overline{\theta_T(t) y_i(t)} \cdot \left[\overline{y_j(t) y_k(t)}\right]^{-1}$. The term in the square brackets is the $[N \times N]$ matrix of all possible cross-correlations between the accelerometer signals. After estimating the mechanically induced jitter, it can be removed from the total jitter to recover the aero-optical-only jitter, $\theta_{Aero}(t) = \theta_T(t) - \tilde{\theta}_M(t)$. This method has several significant advantages. It makes no assumption about the relationship between the measured quantities; it only utilizes the statistical correlation between them. The influence coefficients automatically account for any dimensional differences between the measured quantities (to the point of accounting for calibration coefficients automatically). Finally, the influence coefficients only need to be solved once, making the LSE technique computationally efficient.

If only the spectrum of the aero-optical jitter is needed, LSE technique can be applied in the Fourier domain. In this case, the influence coefficients become a function of frequency, $L_i(f) = \langle \hat{\theta}_T(f)\hat{y}_i^*(f)\rangle \cdot \left[\langle y_i(f)\hat{y}_k^*(f)\rangle\right]^{-1}$, where the hat symbol denotes a Fourier transform and the asterisk denoted a complex conjugate. The estimate of the Fourier transform of the mechanically induced jitter becomes, $\hat{\theta}_M(f) = \sum_{i=1}^{N} L_i(f)\hat{y}_i(f)$. Finally, the Fourier transform aero-optical jitter is found by subtracting this estimate from the total jitter, $\hat{\theta}_{Aero}(f) = \hat{\theta}_T(f) - \hat{\theta}_M(f)$. Figure 7.2 presents an example of various components of the jitter spectra, computed using the spectral version of the LSE technique. The measured total jitter is plotted as a black line, the estimated mechanically induced jitter is plotted as a red line, and the recovered aero-optical jitter is shown as a blue line. The low-end of the total jitter spectrum below 1 kHz is almost entirely corrupted by mechanical vibrations, evidenced by complete overlap between the total and the mechanical jitter spectra. Because in this frequency range the total and the mechanical components of the jitter are almost the same, the resulting difference, which is the aero-optical component, is very sensitive to experimental errors. As a result, the LSE technique might not be effective in this low-frequency range. Still, the LSE technique was capable of recovering the smooth aero-optical jitter in the

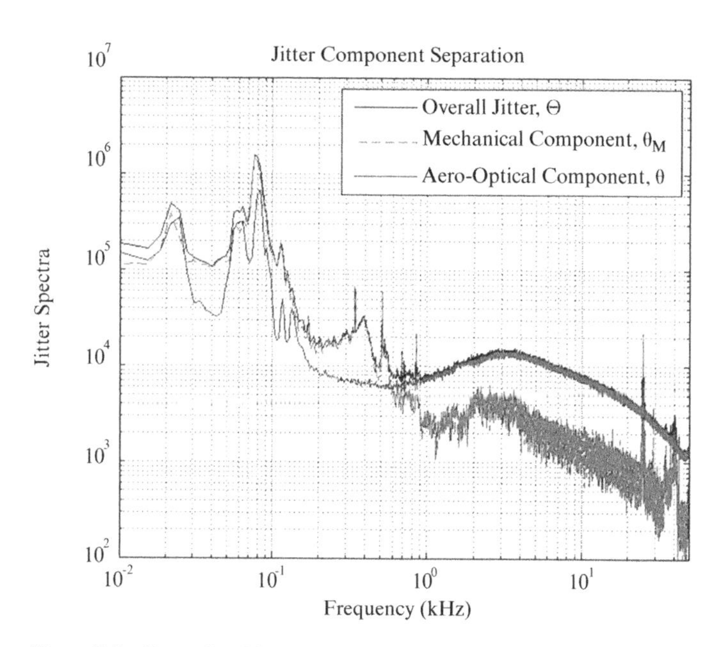

Figure 7.2 Example of jitter spectra separation using spectral LSE technique. *Source:* De Lucca et. al. (2012) / American Institute of Aeronautics and Astronautics.

range between 0.2 and 1 kHz. Above 1 kHz, the analysis revealed that the mechanically induced jitter is small, compared to the aero-optical jitter, so the total and the aero-optical jitter are virtually the same.

7.3.2 Large-Aperture Experiments

Another way to remove the mechanically induced jitter from the wavefront data is to recognize that if the aperture is sufficiently large, the aero-optical component of global jitter is small. Thus, any global tip/tilt components, present in the large-aperture wavefronts, are from mechanically induced jitter or other corruptions. Therefore, removing all global jitter from the wavefront data will significantly reduce the mechanical contamination. After *all* global tip/tilt is removed from every wavefront frame in the time series, the local jitter can be recovered by computing wavefront gradients. If the aperture is sufficiently large, the large aperture wavefronts can be re-apertured to form new datasets with varying aperture sizes, less than the original aperture size, and the aperture effects on the global jitter for these smaller apertures can be studied.

To demonstrate this type of approach, wavefront measurements were conducted in a subsonic wind tunnel, see Kemnetz and Gordeyev (2017) for details. For this experiment, the aperture size in the streamwise direction was chosen to be $Ap = 6.37\delta$, where δ is the local boundary layer thickness. As a typical structure size in the wavefronts due to the turbulent boundary layer approximately equal to δ, these structures will not significantly contribute to the aero-optical global jitter at this large aperture size. Therefore, any tip/tilt, present in the wavefront data, are due to mechanical vibrations. As discussed before, the global tip/tilt component was removed from each time instance. The resulting large wavefront can be re-apertured for different smaller apertures, as schematically shown in Figure 7.3. For each aperture size, the small aperture was moved along the large wavefront at a given convective speed, and values of global jitter were

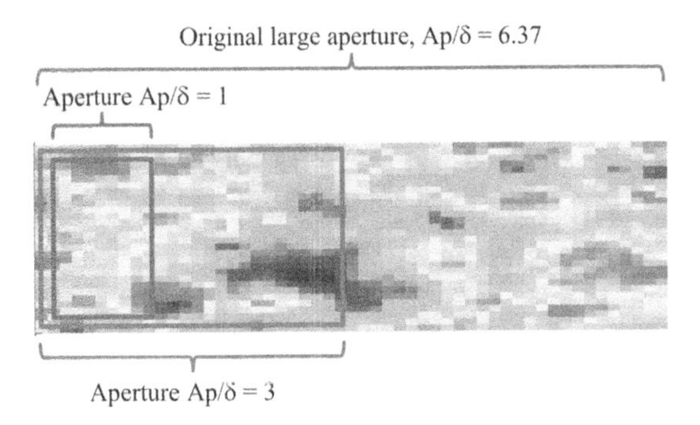

Figure 7.3 Variable apertures, applied to the large aperture wavefront data. *Source:* Kemnetz et al., (2017), American Institute of Aeronautics and Astronautics.

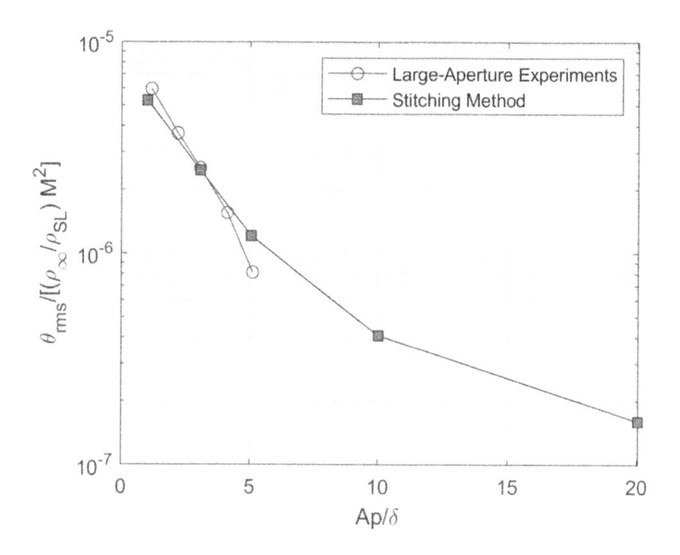

Figure 7.4 Normalized global jitter, $\theta_{G,rms}$, for different aperture sizes, extracted from the turbulent boundary layer experiments with a large streamwise aperture size, $Ap = 6.37\delta$, and estimated using the stitching method. *Source:* Adapted from Kemnetz (2019), Figure 7.9. Courtesy of Matthew Kemnetz.

extracted for each aperture streamwise location. Knowing the convective speed, the streamwise locations of the aperture inside the large wavefront can be traded for time, resulting in time series of global jitter. Values of $\theta_{G,rms}$, were then computed and the normalized values for different smaller apertures, $Ap/\delta < 5$ are presented in Figure 7.4. As expected, the global jitter is maximum at the very small aperture sizes and is significantly reduced for larger apertures. Also presented in Figure 7.4 are the estimates of the global jitter, computed using the stitching method, which will be discussed in the next section. The agreement between the direct measurements using the large aperture and estimates using the stitching method is good.

These types of results can be used to estimate the levels of aero-optical jitter due to the turbulent boundary layer present over the aperture, if a laser system is mounted inside an aircraft. For instance, for an aircraft flying at $M = 0.5$ at sea-level, with the boundary layer thickness of $\delta = 5$ cm, and the aperture size of $Ap = 5$ cm, the estimated aero-optical jitter will be $\theta_{G,rms}(Ap) = 0.75\mu rad$.

7.3.3 Stitching Method

So far, we have discussed two approaches to recover the aero-optical jitter, which would require either large-aperture measurements, or additional measurements of vibration-related quantities, using sensors such as accelerometers. Both of these approaches can be expensive and/or harder to set-up. Secondly, if the wavefront data are already collected, these approaches do not provide a clear path to reconstruct the global jitter statistics from the existing data. In Kemnetz and Gordeyev (2022), a new algorithm was presented, that takes advantage of the advective nature of aero-optical aberrations to recover the time-dependent tip/tilt and piston modes initially removed from experimentally measured wavefronts by again removing global tip/tilt and the mean (piston) components from every frame in the time series. It was demonstrated that the restoration method enables the aero-optical component of jitter to be quantified, and information pertaining to the global jitter due to large-scale structures of the flow can be regained. The restoration method is referred to as the "stitching method," and it necessitates making two assumptions in order to implement it. First, the flow must be primarily convective; that is, the wavefronts can be approximated as $W(x-U_C t,y)$ at least over an overlap region, as it will be described below. Second, the wavefront must be continuous in both space and time. The convective and continuity requirements prevent the stitching method from being used to analyze flows that contain stationary or non-convective features, such as unsteady shocks at transonic and supersonic speeds.

The stitching method requires gathering wavefronts at a sufficient frame rate so that each wavefront contains overlapping portions of both the preceding and past wavefronts. An example of two consecutive wavefronts is shown in Figure 7.5. The two wavefronts have the same structure, outlined by a dotted line, located at different spatial regions for each wavefront. If the wavefront, collected at a later time, $t_0+\Delta t$, is shifted properly, the outlined structure in the shifted wavefront will match the structure in the overlapping region in the preceding time, t_0. Recall that both wavefronts are a part of the same wavefront, but measured at different times over the fixed aperture. Thus, the portions of the wavefronts in the overlapping region should have the same piston and tip/tilt components. By

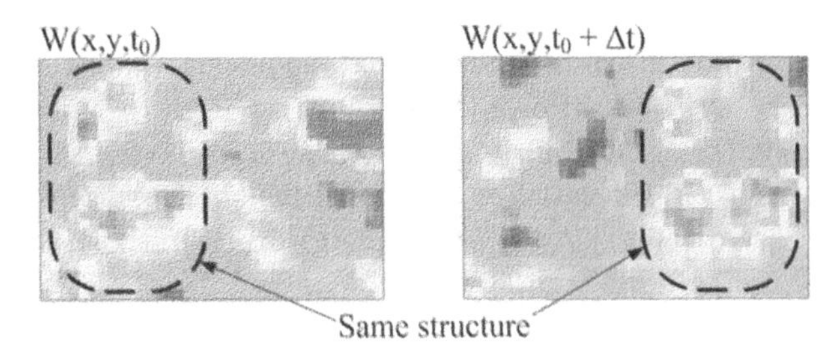

Figure 7.5 Two successive wavefronts, which have a common aero-optical structure. The flow moves from right to left. *Source:* Kemnetz et al., (2016), American Institute of Aeronautics and Astronautics.

reintroducing proper piston and tip/tilt components, we can make the difference between the wavefronts in the overlapping region to be zero to satisfy the continuity property of the wavefront

$$\left[W(x-\Delta x, y, t_0+\Delta t) - W(x, y, t_0) - A - Bx - Cy\right]_{Overlap} = 0$$

Remember, that all jitter, both mechanical and aero-optical, was removed from the original data. So, the reintroduced piston and tip/tilt are only due to optical turbulence induced distortions, both aero-optical and atmosphere-related ones. We can also blend the consecutive wavefronts to create a longer, smoothly blended, or "stitched" wavefront that is now longer than the original aperture size. A graphical depiction of this process is presented in Figure 7.6. The entire process can be repeated for each wavefront given in the time sequence. At every step, another properly shifted and linearly corrected wavefront is added to the previously stitched wavefronts, forming an ever-increasing continuous wavefront in the direction of the flow. Essentially, this process trades time for space to create a long strip of wavefront data.

At the end of the procedure, the overall long stitched wavefront might have a nonzero global tip/tilt and piston component, because the tilt-removed wavefront at the initial moment was used as an anchor to start the stitching method, with the unknown amount of piston and tip/tilt removed from it. By design of the stitching method, it recovers only a differential piston and tip/tilt between the adjacent wavefronts in order to recover the wavefront continuity. Thus, the unknown tip/tilt, removed from the first wavefront, was essentially removed from all

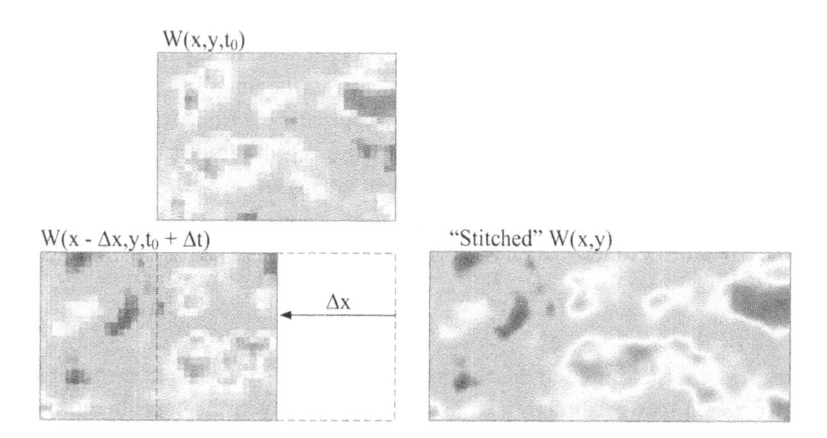

Figure 7.6 Two successive wavefronts stitched together by adjusting the piston and tip/tilt of the wavefront at time $t_0 + \Delta t$ to the wavefront at time t_0, shifted by $\Delta x = U_C \Delta t$, and the resulting blended "stitched" wavefront. The flow moving from right to left. *Source:* Kemnetz et al., (2016), American Institute of Aeronautics and Astronautics.

reconstructed wavefronts, resulting in a global nonzero tip/tilt in the reconstructed stitched wavefront. As discussed before, the overall global tip/tilt in the reconstructed stitched wave front is very close to zero for a sufficiently large reconstructed wavefront strip. As a last step, we can simply remove the global piston and tip/tilt components from the final reconstructed stitched wavefront, $W_{stitch}(x,y)$. By design, only the aero-optical piston and tip/tilt was reintroduced to this stitched wavefront. Details about implementing the stitching method algorithm are provided in Kemnetz and Gordeyev (2022).

Once the wavefronts have been stitched and corrected for the tip/tilt removal, we can then begin the process of recovering the aero-optical component of the jitter for different apertures. We can use the reconstructed stitched wavefront and re-aperture them to many individual frames with an arbitrary fixed aperture size. Schematically it is demonstrated. Piston and tip/tilt information from each frame can be extracted and various statistics of the global jitter, like probability density functions or the overall levels of the global jitter, expressed as $\theta_{G,rms}$, can be extracted as a function of the aperture size.

If needed, time can also be reintroduced into the stitched wavefront, since we treat the wavefront as purely convective in the streamwise direction, $W_{stitch}(x - U_C t, y)$. In other words, a fixed aperture region can be moved along the long-stitched wavefront with the given convective speed, and at any particular time, the tip/tilt over the imposed aperture can be determined. In doing so, a time series of the aero-optic jitter can be determined, and spectra can be calculated, for instance. Also, the resulting time series of the global jitter can then be used to determine the bandwidth of a FSM system needed to correct for the aero-optically imposed jitter.

In reality, the consecutive wavefronts will not perfectly match in the overlapping region due to the not pure convective nature of the wavefronts and the experimental noise present in the wavefronts. In this case, the amount of piston and tip/tilt needed to be reintroduced to the consecutive wavefront to ensure the wavefront continuity can be found by minimizing the difference between the wavefronts in the overlapping region in the least-square sense,

$$\int_{Overlap} \left[W(x-\Delta x, y, t_0+\Delta t) - W(x,y,t_0) - A - Bx - Cy\right]^2 dxdy \to \min.$$

To execute the stitching method, the convective speed should be known. This convective speed can be directly extracted from the wavefront data, by either implementing the direct or the spectral cross-correlation methods, as discussed in Chapter 4.

The accuracy of the stitching method depends on the size of the overlap region, which can be defined as *Overlap* ratio, $Overlap = (Ap - U_C \Delta t)/Ap = 1 - U_C/(f_{samp} Ap)$.

Here we denote the sampling frequency as f_{samp}. It also depends on the amount of the experimental noise present in the data. A comprehensive analysis of the uncertainty of the stitching method can be found in Kemnetz and Gordeyev (2022). Based on this analysis, they recommended keeping the Overlap parameter to at least $Overlap = 50\%$ to keep the relative error in the reconstructed global jitter below 20%. In practical terms, the sampling frequency should be $f_{samp} > 2U_C / Ap$. For better results, the sampling frequency should be even larger.

The stitching technique was successfully demonstrated on two fundamental turbulent flows, a turbulent boundary layer (Kemnetz and Gordeyev 2016, 2017) and a planar forced shear layer (Kemnetz and Gordeyev 2022). Figure 7.4 shows the extracted global jitter for the turbulent boundary layer, using the stitching method and direct measurements, using the large aperture. As one can see, the stitching method does a good job in predicting the global aero-optical jitter for various apertures. Probability density functions were also computed and are shown in Figure 7.7. The probability density functions for a wide range of apertures collapse on a normal distribution of zero mean and a standard deviation of 1.

A planar forced shear layer presents a particularly interesting case. Because of the single-frequency forcing, the dominant shear layer structure is nearly periodic in the streamwise direction, see Figure 7.8. Remember that for a purely periodic wavefront, we already derived the relationship between the local and the global jitter, presented in Equation (7.2) and plotted in Figure 7.1. Thus, we can compare

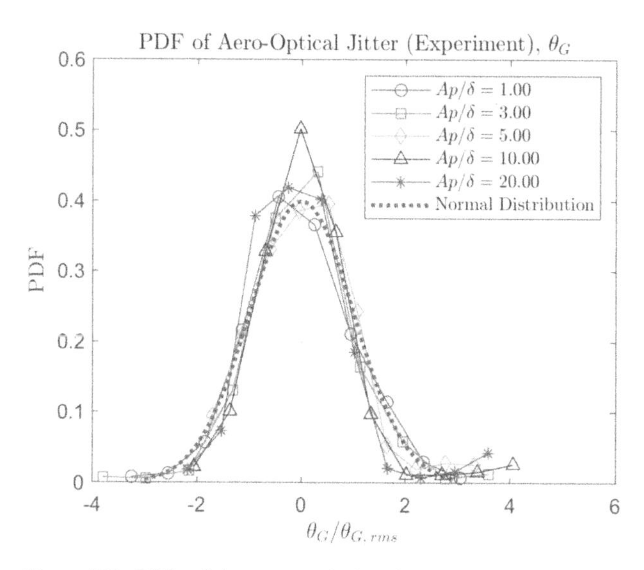

Figure 7.7 PDFs of the aero-optical global jitter for the turbulent boundary layer. *Source:* Adapted from Kemnetz and Gordeyev (2017), figure 14.

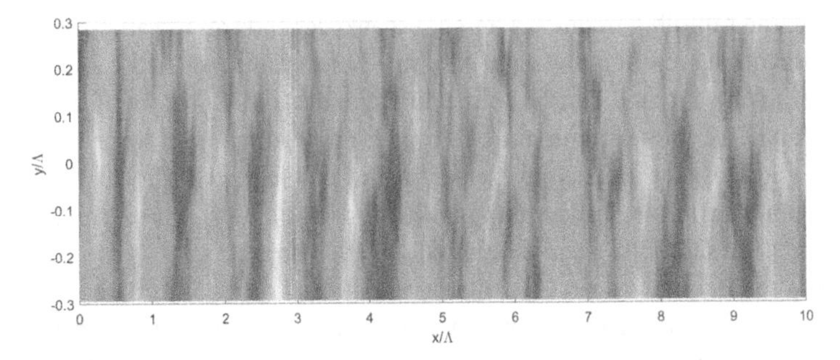

Figure 7.8 A portion of the reconstructed stitched wavefront for the forced shear layer. *Source:* Kemnetz et al., (2022), American Institute of Aeronautics and Astronautics.

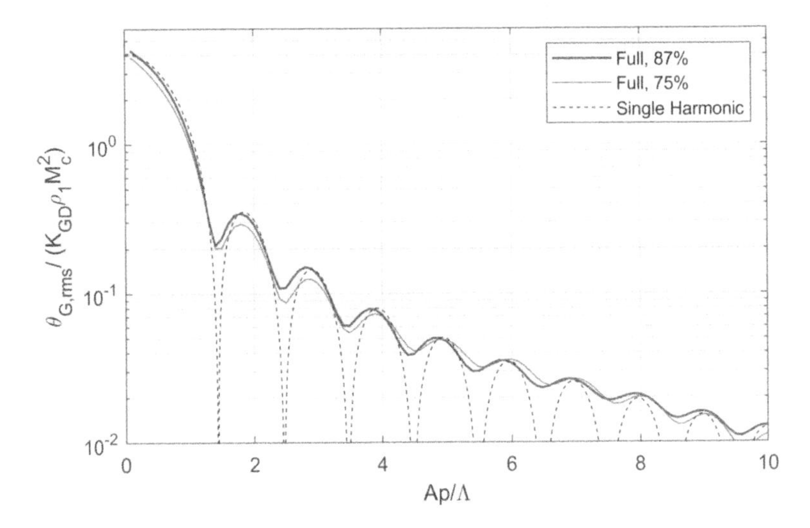

Figure 7.9 Normalized values of the global jitter, $\theta_{G,rms}$, reconstructed using the stitching method for two different Overlap-values. Analytical prediction using the single harmonic model, Equation (7.2), is also shown for comparison. *Source:* Kemnetz & Gordeyev (2022), figure 18. Reproduced with permission of the American Institute of Aeronautics and Astronautics, Inc.

the levels of the global jitter, extracted from applying different apertures to the stitched wavefronts in Figure 7.8 and the prediction from the single harmonic analysis. The values of the rms of global aero-optical tilt, $\theta_{G,rms}$, for a range of the apertures are shown in Figure 7.9. Similar to the global jitter for the boundary layer, the global tilt reaches the maximum at $Ap = 0$, where it equals to the local jitter and generally decreases with the increasing simulated aperture, with dropouts around $Ap / \Lambda \approx 1.4, 2.5, 3.5, 4.5$, and so on. These dropouts occur for the aperture sizes where the global jitter transfer function, G_{GJ}, in Equation (7.2), is equal to zero and the global tilt values are expected to significantly decrease. In reality, other harmonics are present in the wavefronts, resulting in non-zero values of the global jitter at these apertures. The results for the *Overlap* value of 87% and 75% are nearly identical, showing the robustness of the stitching method to the *Overlap* parameter.

In summary, the developed stitching method provides a robust algorithm to restore information pertaining to the aero-optical component of the jitter, if the wavefronts can be treated as primarily convective and continuous (no shocks etc.). If wavefronts are sampled at sufficiently high frequency, there will be an overlap of aero-optical structures between consecutive wavefronts. By adjusting tip/tilt and the piston in the overlapped region of the consecutive wavefronts to ensure the wavefront continuity in space, the removed tip/tilt and piston components can be determined. These components can then be added back into the wavefronts to restore information pertaining to aero-optical jitter only, and various statistical properties of the global jitter, such as the overall levels, temporal spectra and probability distribution functions, can be computed and analyzed as a function of the aperture size. If a closed-adaptive optic system with a fast-steering mirror is implemented to mitigate the aero-optical jitter, this information could help in the selection of the parameters of a fast steering mirror, such as frequency bandwidth and the maximum angular deflection.

8

Applications to Adaptive Optics

To this point, we have covered everything from the cause of aero-optic disturbances to their measurement and characterization; however, we have yet to address their mitigation. In this chapter we will briefly address mitigation, but it would take another book to properly approach this subject. To start this discussion, we will discuss the various objectives and effects that must be addressed and how they might be mitigated. From the first chapter we have inferred that the object of any mitigation is to be able to deliver the highest on-average intensity of a laser projected from an airborne platform onto a distant aim point. Therefore, we need to make the far-field pattern as compact as possible and second, we need to stabilize that pattern on the aim point. These two objectives involve different functional, electro-optical devices. Taken together, the overall system is referred to as the beam-control system, each part of which has its own set of complexities.

8.1 Beam-Control Components

A typical beam-control adaptive optics system is broken into a number of components. The first is the laser-beam originator, which includes the creation of a high-energy beam that often needs beam-quality cleanup devices. For our purpose here, we will leave that to others, but we will take for granted that we have a beam that then enters the rest of the system. Let us also assume that many beam-control components are involved to alleviate beam alignment and beam wander due to motion of the beam due to motion of individual components along the beam train. Again, we will assume that these tasks are taken care of and finally the beam enters what we refer to as the target loop adaptive optics system that shapes the beam's wavefront, performs tracking/jitter management and delivers the beam to the exit pupil. Figure 8.1 shows a schematic of such a closed-loop adaptive optics system. The systems need a light from a guide star to correct for optical

Aero-Optical Effects: Physics, Analysis and Mitigation, First Edition. Stanislav Gordeyev, Eric J. Jumper, and Matthew R. Whiteley.

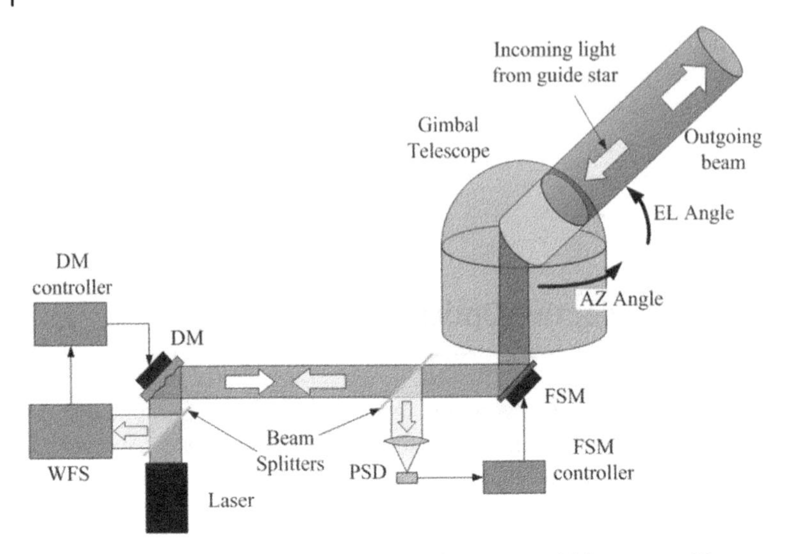

Figure 8.1 Target closed-loop adaptive optics system, which uses a guide star to compensate for both jitter and higher-order optical distortions. Gimbal Telescope showing outgoing High Energy Beam, HEL, and incoming light from guide star.

distortions. It can be a reflected glint from a target, for instance. The incoming light from the guide star passes through the telescope, bounces off a fast-steering mirror (FSM) and split by two beam splitters. One split beam is focused onto a Position Sensing Device (PSD), which measures the global unsteady tip/tilt (global jitter), imposed on the incoming beam. The information from PSD is sent to a FSM controller, which programs the FSM to add the opposite value of jitter on the beam. The second split beam is forwarded to a wavefront sensor (WFS), which measures the higher-order wavefront component of the incoming beam. A corresponding deformable mirror (DM) controller analyzes these wavefronts and sends commands to the DM surface to introduce the corrective (conjugate) optical distortions. Thus, assuming no delays, both systems actively compensate for overall optical distortions imposed on the incoming beam. The outgoing laser beam is reflected off the DM, where corrective anti-distortions are imprinted on the beam. The beam then travels to the FSM where the jitter is removed and then through and out the telescope where it finally arrives at the target undistorted. More details on closed-loop adaptive optics systems can be found in Tyson (1997) or Merritt (2012).

It should be pointed out that this schematic is a simplified version of an actual system. For one, there are many more fast-steering mirrors then indicated to compensate for beam-train movement, beam walk and such, and there may be multiple deformable mirrors that act as a "Woofer" and a "Tweeter," and many other possible components, but the schematic represents the group of components for each function in an actual system. From an aero-optic point of view, it is most important to establish the requirements for each group component of the overall system. The first of these requirements is tip/tilt removal. There are really two components that handle tip/tilt and make up what is often referred to as the target loop. These are the gimbal itself and the FSM; the gimbal typically offloads the FSM. The purpose of these components is to keep the beam centroid on the aim point to within some specified angular radius, whose angular error is usually specified in micro radians. The amplitude and frequency of this angular requirement is obtained from measurements and the stitching method applied to the measured individual-frame tip/tilt time series as described in Chapter 7. This amplitude/bandwidth requirement usually presents a stressing requirement on the adaptive-optic system with bandwidths typically in excess of 1 kHz.

The next component group is the beam splitter that splits the incoming beam from the guide star into the wavefront sensor. The wavefront sensor measures the beam, absent of tip/tilt, although some residual tip/tilt may remain on the beam, hopefully within the dynamic range that the DM group is able to remove. The remaining aberration on the beam after tip/tilt is removed is referred to as higher-order aberrations. In general, these high-order aberrations can be divided into pseudo-stationary “mean” aberrations, also often referred to as pseudo-steady lensing. Depending on the dynamic range of the DM, this could be removed with a single DM that also removes the highly unsteady higher-order disturbances; however, this lensing can also be removed by a relatively low-mirror-actuator count on a separate “Woofer” DM. The highest order aberrations still present are the most stressing requirements for an adaptive-optic system. Chapters 4, 6, and 7 have dealt with these higher-order aero-optic aberrations and the amplitude and bandwidth requirements can be determined from those chapters. Once the wavefront distortions have been measured, a conjugate wavefront figure is sent to the deformable mirror to be imposed on the outgoing HEL beam reflected off it. This wavefront figure on the outgoing beam is then “removed” by the aero-optic flow over the exit aperture. All these activities take time and typically the electro-mechanical, readout and concomitant computations impose a delay. Unfortunately, the aero-optic disturbances are changing so rapidly that once imposed on the beam the correction is late, which only makes the disturbance on the outgoing beam after passing through the aero-optic disturbance worse than if nothing had been done. Thus, the frontier areas of research and development lie in either mitigating the disturbances with flow control or developing robust feed-forward adaptive-optic control algorithms; fortunately, both areas have made excellent progress in recent years (Burns 2016; Burns et al. 2015).

8.2 How Much Correction Is Needed

In general the amount of correction really depends on how large of $OPD_{rms,\ uncorrected}$ is present to begin with. Starting with the Large Aperture Approximation Equation (2.32) in Chapter 2, recall that

$$SR = \frac{I}{I_0} = \exp\left[-\left(\frac{2\pi OPD_{rms}}{\lambda}\right)^2\right]$$

Thus, the ratio between the corrected $SR_{Corrected}$ to the uncorrected $SR_{Uncorrected}$ yields

$$SR_{Uncorrected} = \exp\left[-\left(\frac{2\pi OPD_{rms,\ uncorrected}}{\lambda}\right)^2\right]$$

$$\begin{aligned} SR_{Corrected} &= \exp\left[-\left(\frac{2\pi OPD_{rms,\ corrected}}{\lambda}\right)^2\right] \\ &= \exp\left[-\left(\frac{2\pi OPD_{rms,\ uncorrected}}{\lambda}\frac{OPD_{rms,\ corrected}}{OPD_{rms,\ uncorrected}}\right)^2\right] \\ &= \left(SR_{Uncorrected}\right)^{\left(\frac{OPD_{rms,\ corrected}}{OPD_{rms,\ uncorrected}}\right)^2} \end{aligned}$$

The amount of the ratio of corrected to uncorrected OPD_{rms} is usually described in dB, where

$$dB = 10\log_{10}\frac{OPD_{rms,corrected}}{OPD_{rms,uncorrected}}$$

Thus, for example, -3 dB correction corresponds to a 50% reduction in OPD_{rms}. If the $SR_{Uncorrected}$ is relatively small to begin with, even a small correction can make a large difference in actual system performance. For example, a -3 dB correction will take an uncorrected Strehl ratio from 0.2 to a corrected Strehl ratio of 0.67. Figure 8.2 shows the effect of correction.

8.3 Flow-Control Mitigation

As discussed in Chapter 6, even in subsonic free-stream flow of ~ Mach 0.6, flow over a generic hemispheric turret will become critical, that is to say reach Mach 1.0. This results in the development of shocks, which create large adverse pressure gradients and adverse pressure gradients separate boundary layers.

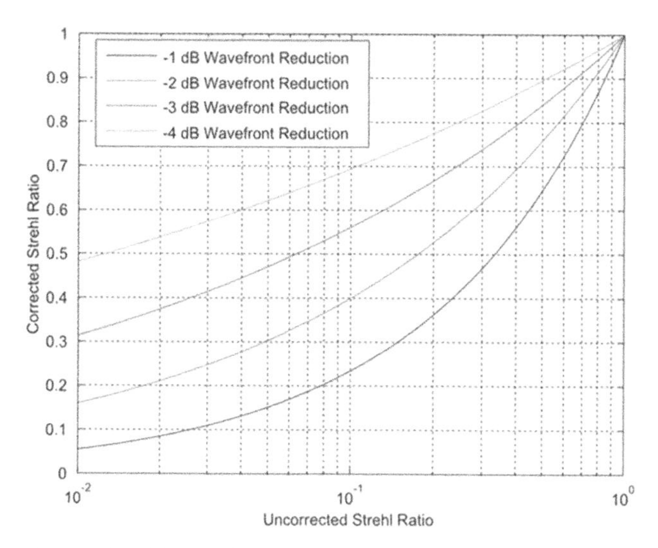

Figure 8.2 Effect on Strehl ratio by reducing OPD_{rms}. *Source:* Burns et. al. (2015), figure 15 / American Institute of Aeronautics and Astronautics.

Thus, the shock not only imposes a wavefront discontinuity, but also separates the flow, which becomes turbulent and badly aberrates a laser propagated through it. This, in turn, restricts the useful field of regard to only a portion of the forward quadrant. From a flow-control point of view, the most-pressing objectives are to prevent shocks from forming and keeping the flow attached over the aft portion of a turret. The latter objective is important even if the flow never becomes critical. There have been many attempts to control shocks and keep the flow attached through surface-flow disruptors, such as vortex generators and pins, and blowing and suction schemes, which are discussed in Crahan et al. (2012), Ponder et al. (2011), and Vukasinovic et al. (2013). At this writing new approaches are being tested. Fewer approaches have been investigated for shock control. In any event, if the shocks and separation can be suppressed, the remaining higher-order aero-optic aberrations are those imposed by an attached turbulent boundary layer and, as discussed in Chapter 6, these are usually relatively low. Aerodynamic lensing is still present and can be removed with presently available adaptive optics systems.

8.3.1 Non-Flow-Control Mitigation

If flow control has not been used or if it is used and leaves residual wavefront error, then electro-opto-mechanical solutions must be sought for mitigation. As mentioned above, if separated shear layers are present, these must be dealt with in order for the aft field of regard to be used. Fortunately, while appearing to be chaotic, the flow in the aft quadrant of the turret is much more organized than had been previously thought. Again, as discussed in Chapter 6, the breakdown into turbulence is initially driven by the Kelvin-Helmholtz instability. There are two important points about this breakdown. The first is that unperturbed, it has favored breakdown patterns. These patterns were also discussed in Chapter 6 and can be detected by using Proper Orthogonal Decomposition, POD. Further, most of the energy is contained in the first several pairs of modes so that only 8 or 12 modes need be compensated for (Burns et al. 2015). This fact has made feed-forward correction possible by using a few previous wavefront measurement frames; although a system has not yet been constructed, the requirements seem well within our present technical capability.

The second fact about the Kelvin-Helmholtz instability is that it can be robustly controlled using mechanical and acoustic actuators. This approach has not only been demonstrated on a transonic separated shear layer, but a feed-forward adaptive optics system has been shown to mitigate the shear layer for a fully subsonic shear layer (Nightingale et al. 2007; Rennie et al. 2007). Further, a control system for the approach has been developed. Figure 8.39(a), shows a schematic of a compressible shear-layer facility at the University of Notre Dame where a forced shear layer was robustly controlled using voice-coil actuators, which are shown in Figure 8.3(b). Once controlled, the wavefront pattern had a two-cycle repeat of the forcing frequency. A phase-averaged wavefront time series was created, and its conjugate patterns were imposed on to a DM, off which a laser beam was reflected and projected through the controlled shear layer. The results presented in Figures 8.4 and 8.5 show the uncorrected original wavefront at a single phase and the resulting far-field pattern and then the effect of correction and the average far-field pattern.

Because of this demonstrated feed-forward correction, many schemes for allowing the flow to separate off a turret but using flow control to "regularize" the resulting separated shear layer from a turret have been proposed and even attempted in wind tunnels but with only partial success.

8.3.2 Some Qualities of Separated Shear Layers

Once a shear layer separates and undergoes K-H breakdown, the turbulence in the shear layer evolves; however, it also convects so that over a short distance through the aperture, the convection is more rapid than the evolution. If the delay between measurement and application of the correction on the DM is relatively short, one feed-forward correction scheme is to simply shift the measurement in the flow direction by the convection velocity times the delay. This has been shown to work; however, the leading edge of the aperture is absent of information to convect from upstream of the aperture. On the other hand, if knowledge of the modes is known, as mentioned above for POD decomposition, then tracking only the

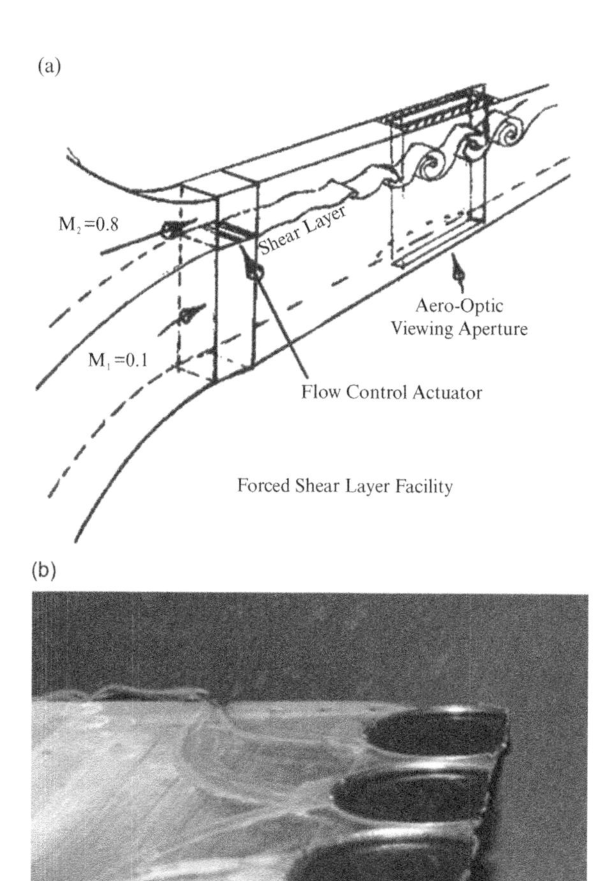

Figure 8.3 (a) Schematic of the Compressible Shear-Layer Facility. *Source:* Courtesy of Eric Jumper. (b) Photograph of the Voice-Coil Actuator. *Source:* Duffin (2009), Figure 3.7. Courtesy of Daniel Duffin.

evolution of the temporal eigenvalues can be used to predict not only the convection of the modes, their evolution, and the ability to predict the upstream edge of the aperture (Burns et al. 2015). Other modal decompositions are possible and have been used to feed forward the needed delayed correction on the DM (Faghihi et al. 2013; Goorskey et al. 2013b; Tesch et al. 2013).

Returning to POD decomposition, the way that POD decomposes is by developing the modes in descending order of the amount of information contained in that mode. The spatial content of the mode (two-dimensional spatial content in

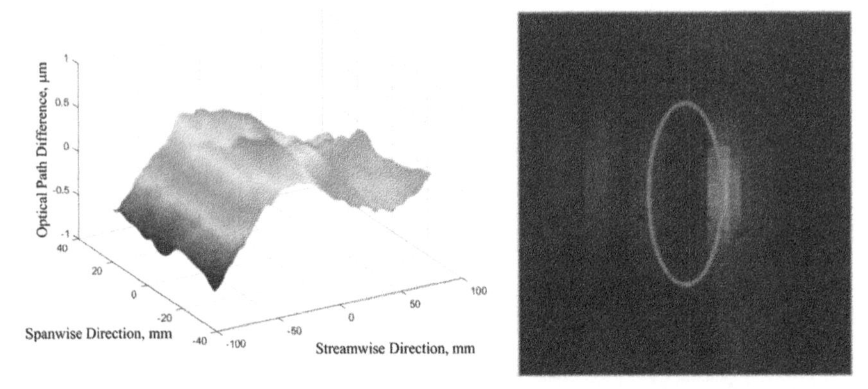

Figure 8.4 A phase-averaged wavefront at one phase angle and resulting far-field pattern. *Source:* Duffin (2009), Figures 7.19 and 7.20. Courtesy of Daniel Duffin.

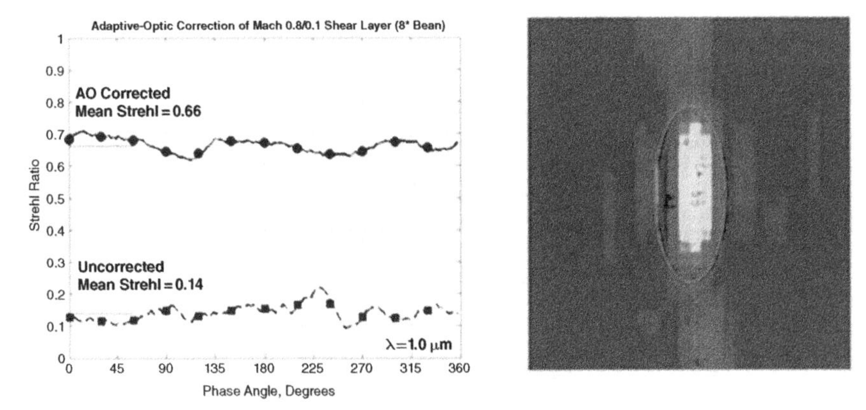

Figure 8.5 Effect of phase-locked correction and the resulting average far-field pattern. *Source:* Duffin (2009), Figures 7.21 and 7.22. Courtesy of Daniel Duffin.

the plane of the wavefront) can be used to determine the number of actuators in the DM that are required to correct that mode down to some determined reduction in amplitude (see section on p. 198). The temporal eigenvalues also turn out to be lowest frequency for the lowest order modes; this is because the lowest order modes not only contain the largest amplitudes but also the largest spatial extent (see below). Since all the modes convect through the aperture at the same speed, the largest spatial modes will also have the lowest frequency.

An example of actual wavefront modes at one instant in time from a time series of wavefronts from flight test on the Airborne Aero-Optics Laboratory at a viewing angle of $\alpha = 114$ degrees; Figure 8.6 shows the first 16 POD modes. Notice that the spatial patterns become smaller with increasing mode number; thus, for the full disturbance convecting through the aperture it is clear that as the mode

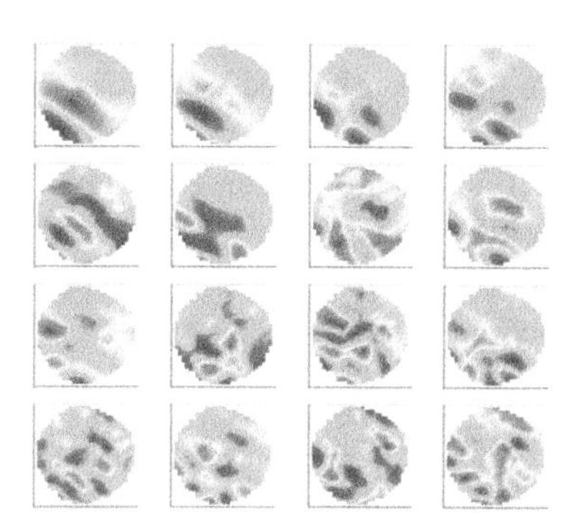

Figure 8.6 First 16 POD modes of a time series of wavefronts for a viewing angle of $\alpha = 114$ degrees. The mode numbers start at upper left (#1), left to right and down, to lower right (#16). *Source:* Robert et. al. (2015), American Institute of Aeronautics and Astronautics.

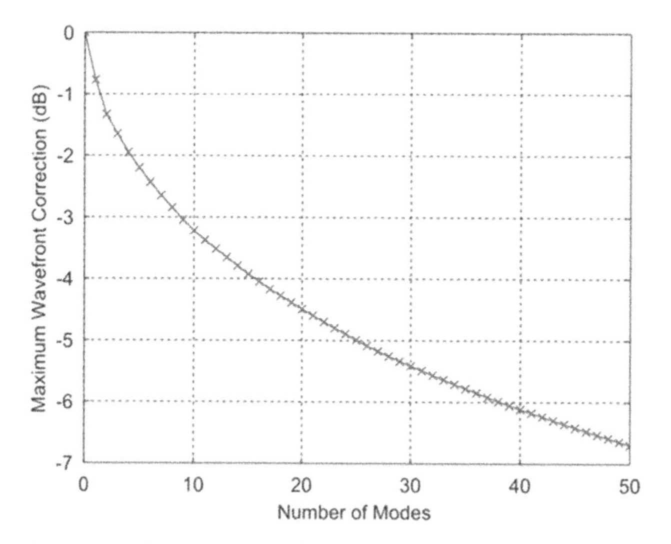

Figure 8.7 Demonstration of how correcting of only the First 8 to 12 Modes reduces the residual OPD_{rms}. *Source:* Burns et. al. (2015), figure 9 / American Institute of Aeronautics and Astronautics.

number increases the spatial patterns get smaller so that the frequencies needed to correct each of the higher order modes increase as do the required number of actuators.

Figure 8.7 shows how the correction of only a relatively small number of modes leads to a reduction in OPD_{rms} in dB, -3 dB achieved by correcting only about 8 modes.

A common assumption in the analysis of adaptive optics systems is that the response time of the deformable mirror and amplifiers is much smaller than the equivalent frequencies associated with disturbances. In reality, it is most often not the case, as many steps in the adaptive-optics system require a finite amount of time. Examples are the data offload from the wavefront sensor, the wavefront reconstruction algorithm and the conversion of the reconstructed wavefront to the commands to the DM. All of these and other effects will result in feedback latency, present in the adaptive-optics system. If not accounted for, the latency effect might severely affect the performance of the system (Burns et al. 2015). Using the POD decomposition of the wavefront data, it is possible to design a latency-tolerant architecture for adaptive-optics systems (Burns 2016; Burns et al. 2015, 2016).

8.3.3 Using the POD Analysis to Develop Requirements

The POD modal analysis is also very useful to help developing requirements for the number of actuators required for the DM and the concomitant number of subapertures needed for the wavefront sensor. POD analysis also is useful in developing the needed bandwidths. As discussed in the POD section of Chapter 4, POD analysis extracts stationary spatial modes of a long time series of unsteady wavefronts. These modes are ordered from most energetic to descending importance with increasing mode number; therefore, for example, in general, as few as 8 to 12 modes might account for 90% of the energy of the unsteady aberrations. The unsteadiness in the time series of wavefronts comes from the temporal eigenvalues, that is, the time-changing eigenvalues used as coefficients in the linear combination of the spatial POD modes. In other words, the POD analysis splits the spatiotemporal wavefront field, $W(x,y,t)$, into a series of statistically correlated, stationary, spatial patterns (spatial modes), $\phi_n(x,y)$, so that a wavefront can be reconstructed at any instant in time as a linear sum of the spatial modes times their temporal coefficients, $a_n(t)$, as shown in Equation (4.12). As a result, we have spatial information for all of the important modes, and thus a means of determining the number of actuators needed to produce a conjugate wavefront on the DM of the important modes (Abado et al. 2013).

The first four normalized POD modes of a 10 ms long time series of representative wavefronts taken at a back-looking angle on the AAOL are shown in Figure 8.8. The amount of "energy" contained in each specific mode (the first number in the parenthesis) and the cumulative "energy" up to that mode number (the second number in the parenthesis) are given for each spatial POD mode. It can be seen that the higher the spatial POD mode number, the higher the spatial frequencies it contains. It should be noted that the ordinate of the spatial-mode plots in Figure 7.7 are unitless; these modes do not actually represent a wavefront until they are multiplied by their associated temporal coefficients. The temporal coefficients of the first four POD modes for a time period of 10 ms are shown in Figure 8.9. It can be seen that the overall amplitude of the temporal coefficients, and hence their relative contribution to the total "energy," decreases as the mode number increases. It should be noted that the time-averaged component has been

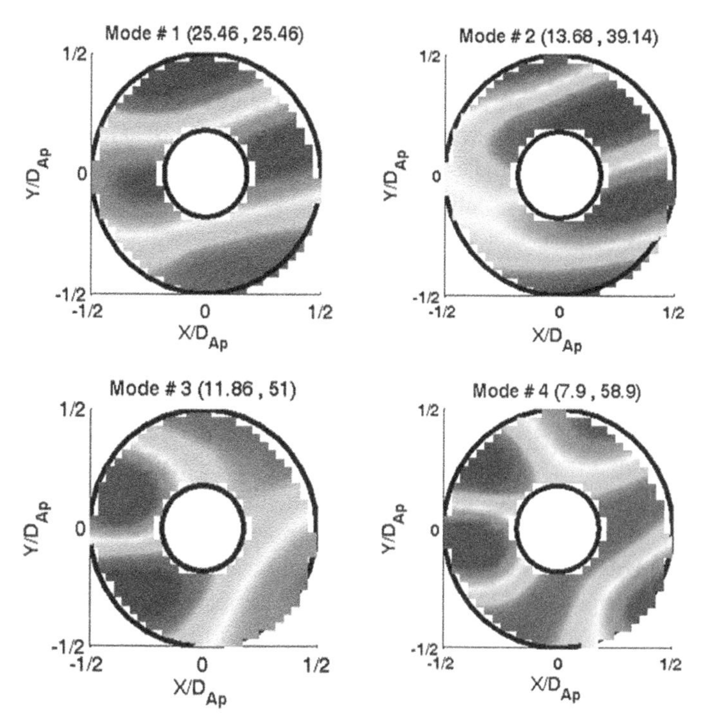

Figure 8.8 The first 4 normalized spatial POD modes. *Source:* Adapted from Abado et. al. (2013).

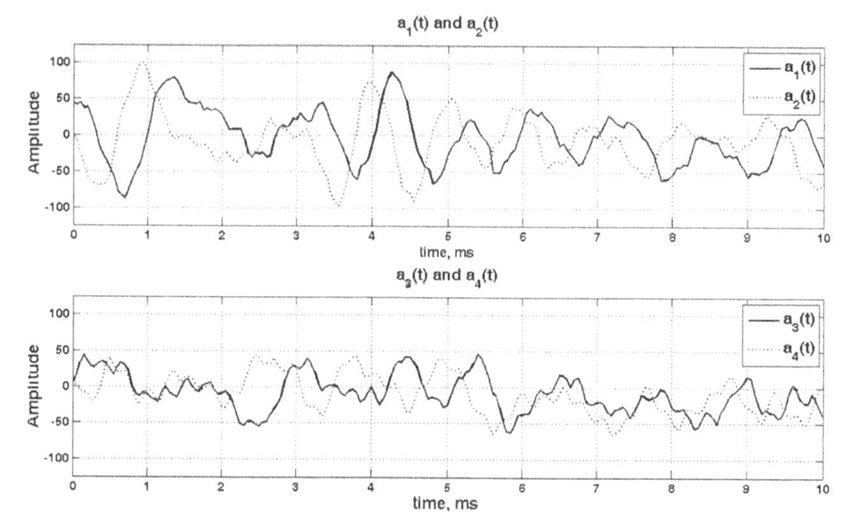

Figure 8.9 POD temporal coefficients of mode pairs for a time period of 10 ms. *Source:* Adapted from Abado et. al. (2013).

removed from the wavefront frames; this time-averaged component would represent the pseudo-steady lensing, for example.

As can also be noticed in Figure 8.8, the spatial POD modes operate in pairs. That is, the first and second modes have similar form, but appear shifted in phase, similar to sine and cosine waves, and are similar in shape. A similar trend can be seen for the third and fourth modes, and so on. As discussed in the POD section of Chapter 4, this clearly suggests that the phase-shifted pairs represent convecting structures. The interaction between these spatial modes and temporal coefficients allows for the "motion" of the structures in the POD reconstruction. This "motion" occurs as each given mode shifts its temporal amplitude to its partner with subsequent time steps. Also notable in Figure 8.8, is that third and fourth paired modes' temporal frequencies are higher than the first and second modes; this is a general trend, the higher the mode number the higher the temporal frequency.

It should be noted that the spatial patterns in Figure 8.8, are not aligned with x- and y-axis. Yet what we know about shear layer structures, the large coherent vortices that form in the separated shear layers are normal to the streamwise direction. Thus, the coherence lengths in the x-direction would be larger than if the wavefronts were aligned so that the x-direction were reoriented with the velocity convecting the structures. Using methods described in Chapter 4, it is possible to determine the velocity direction over the aperture. The velocity map over the aperture for this case was shown in Figure 4.20.

Now, having the orientation of a single, aperture-averaged velocity vector, the modes can be rotated so that the large coherent structures clearly visible in Figure 8.8, Modes 1 and 2 can be rotated so that they are normal to the aperture-averaged velocity vector. The same reoriented angle of rotation is then applied to all the modes. The first four reoriented modes are shown in Figure 8.10.

Once reoriented, the correlation function of each spatial POD mode can be computed. The two-dimensional, normalized, auto-correlation function in the x-direction is defined as,

$$R(\Delta x) = \frac{\langle W(x,y)W(x+\Delta x,y)\rangle_{x,y}}{\langle W(x,y)W(x,y)\rangle_{x,y}}$$

As a reminder, the angular brackets denote the spatially average operator and Δx is the distance between two points on the POD spatial mode. A similar correlation function in the y-direction can also be computed. To characterize the spatial extent of the correlation, and therefore characterize the coherent structures, a correlation length can be defined. The main advantage of defining a correlation length is that it provides a value for the size of the coherent structure. The correlation length is usually defined as the distance in space beyond which physical events are uncorrelated (Mela and Louie 2001). Here we define the correlation length as the location of the auto-correlation function's first minimum. Denoting

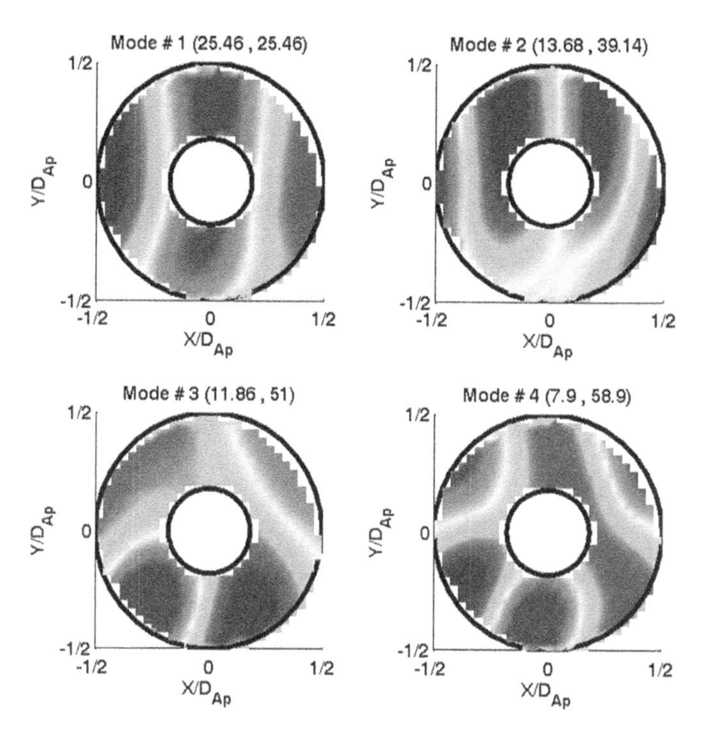

Figure 8.10 Reoriented POD modes from Figure 8.8 into the streamwise and spanwise directions given by the average convection velocity over the aperture. *Source:* Adapted from Abado et. al. (2013).

the correlation length of the spatial POD mode n as Λ_{CLn}, the number of structures per aperture size, D_{Ap}, can be computed as

$$\frac{1}{Aperture} = \frac{D_{Ap}}{\Lambda_{CLn}}$$

This can be done for the reoriented wavefront POD modes and for the original modes. For the 10 ms wavefront time series, the structures per aperture as a function of the mode number for rotated and original (unrotated) POD modes are shown in Figure 8.11.

The difference in the number of structures per aperture demonstrates the need to perform this reorientation. Also, it should be noted that Figure 8.11 shows only the first 200 POD modes. The spatial Nyquist frequency, which sets the spatial limit over the smallest structure size that can be resolved by the number of subapertures in the wavefront sensor used to collect the wavefronts, is plotted as a dashed line. The wavefront sensor used to collect these data has 29 × 29 subapertures. As in the example in Figure 8.11, the number of actuators for the number of

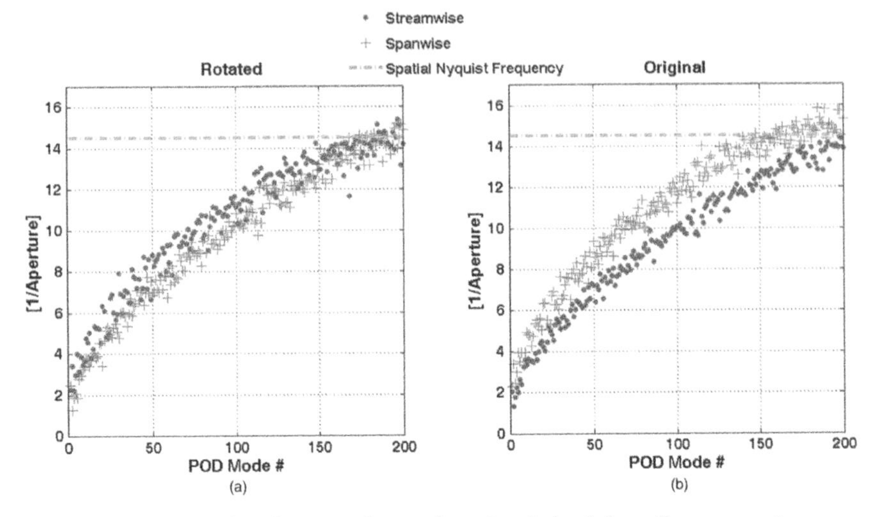

Figure 8.11 Comparison between the number of periods of aberration per aperture based on the definition of coherent length as the location of the first minimum of the auto-correlation function for the rotated and original spatial POD modes. *Source:* Adapted from Abado et. al. (2013).

modes required to reduce the amplitudes of the residual OPD_{rms} can be estimated. Suppose the number of modes needed to capture 90% of the wavefront energy is 12, then from Figure 8.11, the 12th mode would contain approximately 5 structures which would require at least a 10×10 actuator DM. The discussion of POD analysis and the number of modes required to mitigate the OPD_{rms} down to an acceptable level should be reviewed both to determine the number of modes and temporal bandwidth associated with the highest POD mode needed. As for the amount of energy needed to be removed, the earlier discussion in this chapter should be used to reduce the original OPD_{rms} and the ratio of that needed to raise the Strehl ratio to an acceptable mission requirement.

8.4 Proper Number of Wavefront Sensor Subapertures to Actuator Ratio

Once the number of actuators on the DM have been determined as discussed above, the arrangement of the DM's actuators and the wavefront sensor's subapertures (the registration) affects the accuracy with which an aberrated wavefront can be corrected. Some researchers in the aero-optics community have stated that this ratio of subaperture to actuator should be one (and registered) based on an analysis developed in the 1970s or 1980s (Weaver 1997); however, attempts to locate specific references for this guideline have been futile. The purpose of this section is to provide a new analysis that comes to the same conclusion using similar methods that we have used in developing other filters like that used for the aperture filter described in Chapter 7. The described approach makes use of a numerical simulation which was developed to analyze the accuracy of a Shack-Hartmann wavefront sensor-based adaptive-optics system as a function of the ratio between the number of deformable-mirror actuators and the number of wavefront sensor subapertures.

8.4.1 Numerical Simulation

In the simulation, the modeled adaptive-optics system is assumed to consist of a Shack-Hartmann wavefront sensor with $N_{WFS} \times N_{WFS}$ and a deformable mirror with $M_{DM} \times M_{DM}$ actuators. For $N_{WFS} = M_{DM}$, the spacing between the deformable mirror actuators is assumed to be equal/scaled to the spacing between the wavefront sensor's subapertures. For simplicity, the one-dimensional case is assumed here, that is, a one-dimensional array of wavefront sensor's subapertures and a one-dimensional array of deformable mirror's actuators; however, the approach can be easily extended to a two-dimensional wavefront sensor and a two-dimensional deformable mirror. A commonly used registration between the deformable mirror's actuators and the subapertures of an adaptive-optics system with a Shack-Hartmann wavefront sensor is the Southwell geometry (Southwell 1980). Therefore, a Southwell wavefront reconstruction algorithm was used to reconstruct the wavefront from the measured wavefront's gradients. As discussed throughout this book, the primary sources of aero-optical distortion (e.g., shear layers) are associated with a relatively narrow set of temporal and spatial frequencies. However, if additional sources of distortions exist, the distortion can be decomposed into a narrow set of spatial and temporal frequencies using Fourier transform. Therefore, it is possible to examine the temporal and spatial effects due to periodic correction and system latency by approximating the base wavefront disturbance as a sine function with a single disturbance temporal frequency, f_d, and a disturbance length scale, Λ_d, as

$$OPD^{Base}(x,t) = A \cdot \sin(2\pi x / \Lambda_d + 2\pi f_d t).$$

Assuming a time latency of τ_2 in applying the perfect conjugate correction to the deformable mirror, and a periodic correction update frequency, $f_U = 1/\tau_2$, of applying the correction, then the deformable mirror correction, $OPD^{DM}(x,t)$ can be expressed as,

$$OPD^{DM}(x,t) = A \cdot \sin\left[2\pi x / \Lambda_d + 2\pi f_d (t_0 - t)\right] \qquad (8.1)$$

Here t_0 is the starting point of the periodic correction. Therefore, the residual wavefront after correction, $OPD^{Corr}(x,t)$, will have the form

$$OPD^{Corr}(x,t) = OPD^{Base}(x,t) - OPD^{DM}(x,t)$$

For an aperture size D_{Ap}, the limit over the largest spatial frequency which can be resolved by the wavefront sensor is set by Nyquist criterion, which states that spatial frequency greater than half the sampling frequency cannot be observed. Defining the spatial Strouhal number, St, as the dimensionless ratio $St = D_{Ap} / \Lambda_d$, and recalling that the primary sources of aero-optical distortion are associated with a relatively narrow set of temporal and spatial frequencies, Nyquist criterion can be expressed as $St_{sub} \leq 1/2$. Here, $St_{Sub} = D_{sub} / \Lambda_d$, where Λ_d is the disturbance length scale, and D_{sub} is the subaperture size which is equal to $D_{sub} = D_{Ap} / N_{WFS}$. A small St_{sub} value corresponds with a low-frequency distortion; therefore, it is easier to mitigate. Whereas a large St_{sub} value corresponds to a high-frequency distortion; therefore, it is harder to mitigate. Figures 8.12 and 8.13 illustrate the difference between small and large St_{sub} values for constant Λ_d and N_{WFS} values, respectively.

In addition to Nyquist criterion, the system's accuracy depends on the ratio between the number of actuators and the number of subapertures, M_{DM} / N_{WFS}. Two methods were used to determine the deformable mirror conjugate correction surfaces based on the reconstructed wavefronts from the Southwell solver:

- For $M_{DM} / N_{WFS} < 1$: more wavefront sensor's subapertures than deformable mirror's actuators, a linear interpolation fit was used to infer the actuators' displacements from the reconstructed wavefront.
- For $M_{DM} / N_{WFS} > 1$: less wavefront sensor's measurements than deformable mirror's actuators a cubic spline interpolation was used to infer the actuators' displacements between the reconstructed wavefront subapertures.

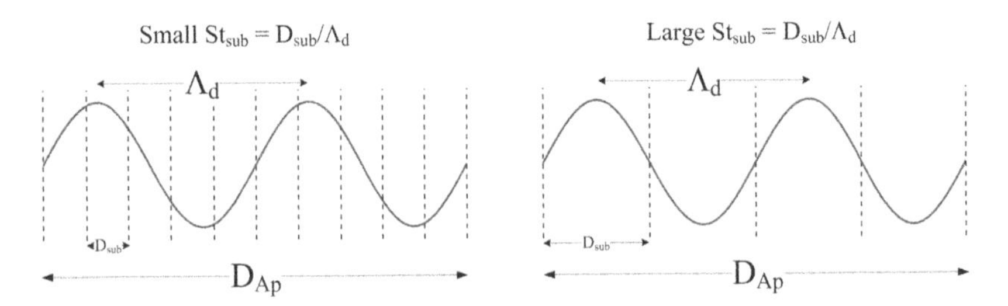

Figure 8.12 Illustration of small and large St_{sub} for constant Λ_d and variation in D_{sub} (different N_{WFS}).

8.4.2 Simulation Results

As in Equation (8.1), we represent the base wavefront disturbance, $OPD^{Base}(x,t)$, as a sine function with a single disturbance length scale, Λ_d, and a single disturbance temporal frequency, f_d. If we assume an infinite bandwidth system (no time latency in applying the conjugate correction to the deformable mirror), then the normalized residual wavefront variance is

$$\sigma_{fit}^2(t) = \frac{\int_{Aperture} \left[OPD^{Base}(x,t) - OPD^{DM}(x,t)\right]^2 dx}{D_{Ap}}$$

where $OPD^{DM}(x,t)$ is the conjugate correction surface introduced by the deformable mirror. The correction system's figure of merit is typically the mean residual variance, $\sigma_{Corr}^2 = \overline{\sigma_{fit}^2(t)}$.

To evaluate the system's correction accuracy, a gain function can be defined as the ratio between the residual wavefront, time averaged variance, σ_{Corr}^2, and the uncompensated wavefront, time-averaged variance, σ_{Base}^2, such that

$$G(St_{Sub}, M_{DM} / N_{WFS}) = \frac{\sigma_{Corr}^2}{\sigma_{Base}^2}$$

The defined gain function depends on the St_{sub} value, and the ratio between the number of actuators and the number of subapertures, M_{DM} / N_{WFS}. The results from a simulation for various St_{sub} values is shown in Figure 8.14. For presentation clarity, σ_{Corr}^2 was rescaled to logarithmic units, such that,

$$-dB = -10\log_{10} G(St_{sub}, M_{DM} / N_{WFS}).$$

As expected from Nyquist criterion, the simulation shows that for $St_{sub} \geq 0.5$ the use of an adaptive-optics system worsens the system's performance ($-dB < 0$). It

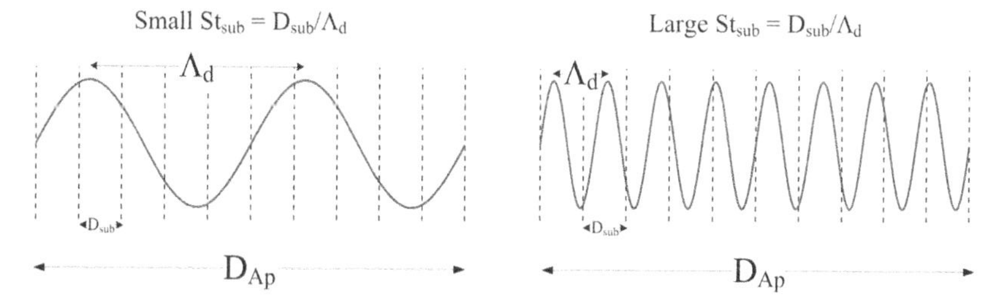

Figure 8.13 Illustration of small and large St_{sub} for constant D_{sub} (constant N_{WFS}) and variation in Λ_d.

is also shown that for a given constant M_{DM} / N_{WFS} value, the system's performance improves as the St_{sub} value decreases. This indicates that as the spatial frequency decreases, less sampling subapertures, and hence less correction elements, are required to achieve a given gain value.

In most adaptive-optic systems, the number of wavefront sensor's subapertures, N_{WFS}, is larger than the number of deformable mirror's actuators, M_{DM}, to be corrected ($M_{DM} / N_{WFS} < 1$). Such a system is said to be over-determined; in which case the compensation accuracy is limited by the deformable mirror. For an over-determined system, the deformable mirror cannot correct for all the high-frequency structures which are resolved by the wavefront sensor's subapertures. For a fixed number of subapertures, increasing the number of actuators will add additional degrees-of-freedom for the correction and increase the system's accuracy.

At the limit of $M_{DM} / N_{WFS} = 1$, the trend of the accuracy increase changes. The constrained optimal accuracy is achieved when the number of subapertures is equal to the number of actuators, $M_{DM} / N_{WFS} = 1$. A clear "knee" is seen in Figure 8.14 at this value. For this one-dimensional geometry, the actuators and subapertures are precisely matched such that the actuators are centered on the subapertures.

Beyond $M_{DM} / N_{WFS} = 1$, the improvement in the system's performance is negligible; therefore, there is no benefit from increasing the number of actuators beyond the number of subapertures. For $M_{DM} / N_{WFS} > 1$, the system is said to be underdetermined, and the compensation accuracy is limited by the wavefront

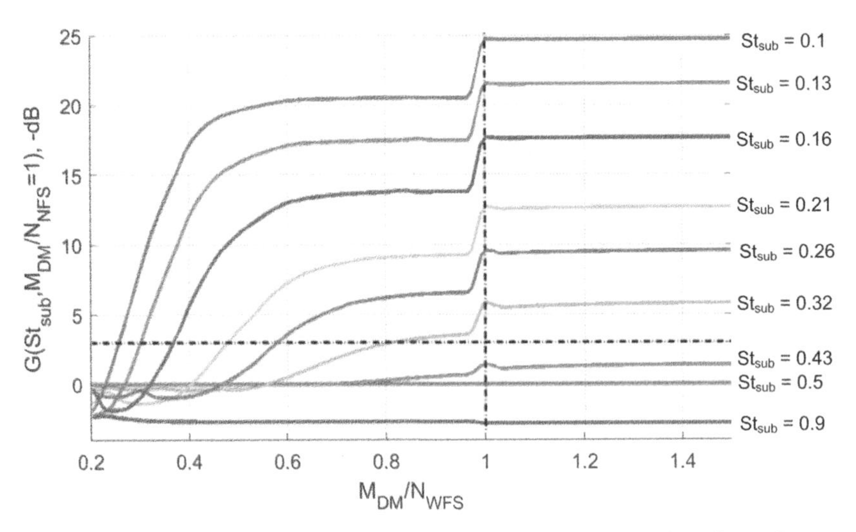

Figure 8.14 System gain as a function of the ratio between the number of wavefront sensor's subapertures and the number of deformable mirror's actuators. Courtesy of Shaddy Abado.

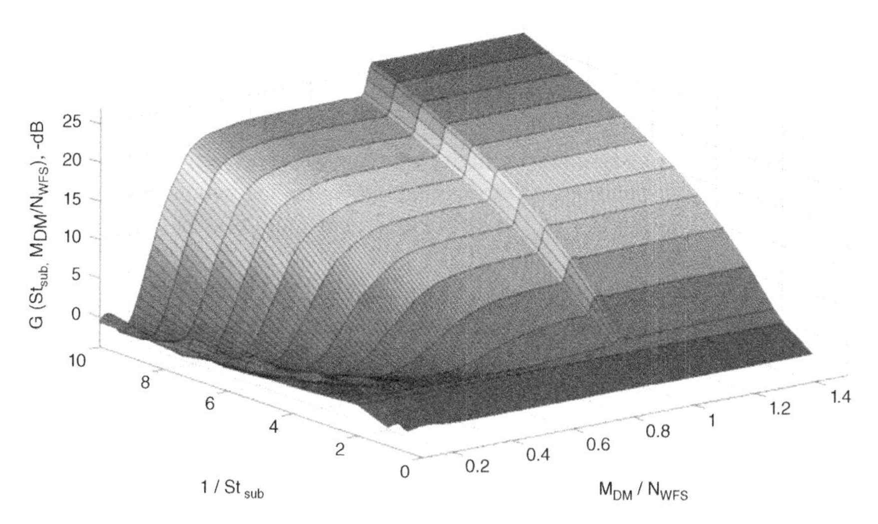

Figure 8.15 System gain as a function of the ratio between the number of wavefront sensor's sub apertures and the number of deformable mirror's actuators, and $1 / St_{sub}$. Courtesy of Shaddy Abado.

sensor. At this range, the gain function values become saturated since all the sampled information is used.

The curves in Figure 8.14 can be replotted as a three-dimensional surface, see Figure 8.15. Note that for presentation clarity, the y-ordinate is $1 / St_{sub}$. This three-dimensional graph provides a relationship between the system architecture (N_{WFS} and M_{DM}), the "aberration strength" (St_{sub}), and the correction accuracy (G).

The gain function, $G(St_{sub}, M_{DM} / N_{WFS})$, corresponds to the degree of aberration remaining after the wavefront sensor measurement and deformable mirror correction. Based on this, it is possible to use the graph in Figure 8.15 as a high-pass filter. The output of this filter for an incoming wavefront, which includes various disturbance length scales, will correspond to the aberrations remaining after the adaptive-optics correction. Such an analysis can assist in evaluating the designed system accuracy as a function of the ratio between the number of wavefront sensor's sub-aperture and deformable mirror's actuators.

8.4.3 Conclusion from the Simulation Results

Several conclusions can be drawn from Figures 8.14 and 8.15 for designing an adaptive-optics system with good accuracy, a minimum number of sub-apertures, and a minimum number of actuators. We can conclude that one needs, at least, as many wavefront sub-aperture measurements as actuators. In fact, the number of wavefront sensor's sub-apertures and the number of deformable mirror's actuators should be 1-to-1 matched as best as possible. This result is independent of the St_{sub} value; however, the system's gain depends on the St_{sub} value. A plot of the system's gain as a function of St_{sub} for $M_{DM}/N_{WFS} = 1$ is shown in Figure 8.16. The gain function plot resembles a high-pass filter where its output corresponds to the aberrations remaining after the adaptive-optics correction. As expected from Nyquist criterion, for $St_{sub} > 1/2$ the system amplifies the disturbances instead of suppressing them.

Although many analyses are available to account for both the temporal and spatial characteristics of the adaptive optics system exist (Fried 1967; Greenwood 1977; Karr 1991), most of the available analyses are mathematically complex and require detailed information regarding the wavefront sensor, the deformable mirror, and the control system characteristics. The presented analysis takes into account the spatial bandwidth limitations due to a finite number of subapertures for a wavefront sensor and deformable mirror. However, the unique aspect of this approach is the relative ease with which the gain function was derived.

The simple simulation derived in this section shows the very important criteria and guidelines necessary when designing an adaptive-optics system for aero-optics correction and can be generalized for additional adaptive-optic

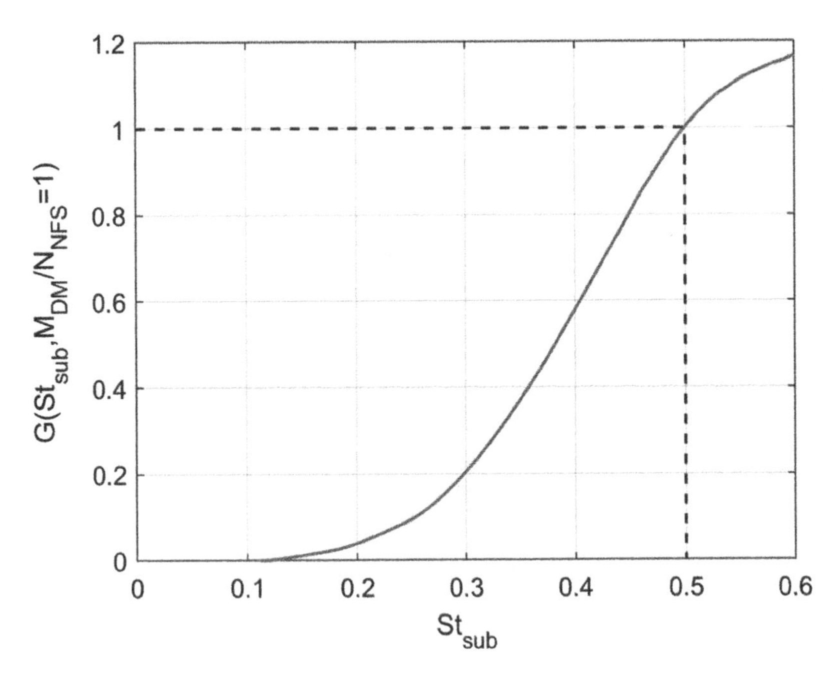

Figure 8.16 System gain as a function of St_{sub} for $M_{DM}/N_{WFS} = 1$. Courtesy of Shaddy Abado.

applications. In fact, the main results of the developed simulation agree with a simulation, which was presented in Band and Ben-Yosef (1994) for atmospheric turbulence. In their simulation, the authors used von Karman spectrum for the index-of-refraction spectral power density, and relied on Fried's coherence length, r_0, (Fried 1967), and the refractive index structure constant, C_n^2, to quantify the turbulence strength.

9

Adaptive Optics for Aero-Optical Compensation

We now transition to the subject of modeling the performance of an adaptive optics (AO) system for compensation of aero-optical wavefront disturbances. The application of adaptive optics to the aero-optical problem is a straightforward implementation of the hardware and control approaches typically employed for compensation of atmospheric free-stream turbulence, as illustrated earlier in Figure 8.1. Thus, the methods applied to quantify the performance of these systems proceed from analogy to free-stream compensation analysis, which has been worked out in detail within the literature (Roggemann and Welsh 1996; Sasiela 2007; Tyson 1997). Some key considerations for statistical optics related to aero-optics are addressed up front, leading to a thorough analysis of adaptive optics compensation of aero-optical disturbances in the sections that follow.

9.1 Analogies from Free-Stream Turbulence Compensation

9.1.1 Statistical Optics Theoretical Considerations

For this discussion, we rely heavily on the key conclusions of statistical optics (Goodman 2015), which have been generally articulated in the context of free-stream turbulence, but may be readily applied to the problem of aero-optics. The long exposure intensity, $I(x',y')$, on the focal (image) plane of an optical system at distance F is given by

$$I(x',y') = \frac{I_0}{(\lambda F)^2} \int_{Aperture} \mathrm{P}(\Delta x, \Delta y) \exp\left[-\frac{1}{2} D_{\varphi}(\Delta x, \Delta y)\right] \exp\left[-\frac{2\pi}{\lambda F}(x'\Delta x + y'\Delta y)\right] d(\Delta x) d(\Delta y). \tag{9.1}$$

Aero-Optical Effects: Physics, Analysis and Mitigation, First Edition. Stanislav Gordeyev, Eric J. Jumper, and Matthew R. Whiteley.

Here we assume that the light intensity over the aperture is constant and equal to I_0. $D_\varphi(\Delta x, \Delta y)$ is the phase structure function,

$$D_\varphi(\Delta x, \Delta y) = \langle [\varphi(x+\Delta x, y+\Delta y) - \varphi(x,y)]^2 \rangle_{(x,y)}$$
$$= 2\langle [\varphi(x,y)]^2 \rangle_{(x,y)} - 2\langle [\varphi(x+\Delta x, y+\Delta y)\varphi(x,y)]^2 \rangle_{(x,y)}, \tag{9.2}$$

where $\varphi(x,y) = \dfrac{2\pi W(x,y)}{\lambda}$ is the aero-optical phase. $\mathrm{P}(\Delta x, \Delta y)$ is the autocorrelation of the aperture pupil function, defined as a convolution of the uniform weighting function, $w(x,y)$,

$$\mathrm{P}(\Delta x, \Delta y) = \int_{Aperture} w(x,y) w(\Delta x - x, \Delta y - y) dxdy, \tag{9.3}$$

where

$$w(x,y) = \begin{cases} = C_{Ap}, & \text{inside aperture} \\ = 0, & \text{outside aperture} \end{cases}.$$

For instance, for a circular aperture of diameter, D, $C_{Ap} = 1/\pi$.

As discussed in Chapter 2, optical performance is quantified by the Strehl Ratio, SR, defined as the ratio of the on-axis far-field intensity to the diffraction-limited intensity, $SR = I(0,0)/I_0(0,0)$, where $I_0(0,0)$ can be obtained from Equation (9.1) by setting $D_\varphi(\Delta x, \Delta y) = 0$ (no aberrations). It provides an alternative way to compute SR as,

$$SR = \frac{\int_{Aperture} \mathrm{P}(\Delta x, \Delta y) \exp\left[-\frac{1}{2} D_\varphi(\Delta x, \Delta y)\right] d(\Delta x) d(\Delta y)}{\int_{Aperture} \mathrm{P}(\Delta x, \Delta y) d(\Delta x) d(\Delta y)}. \tag{9.4}$$

A key observation to make here is that the aero-optical structure function, $D_\varphi(\Delta x, \Delta y) = 0$, fully characterizes the aero-optical disturbance statistics needed for our calculations.

Another governing factor of the phase, φ, is the spatial covariance, $\mathrm{B}_\varphi(\Delta x, \Delta y)$, defined as

$$\mathrm{B}_\varphi(\Delta x, \Delta y) \equiv \langle [\varphi(x+\Delta x, y+\Delta y)\varphi(x,y)]^2 \rangle_{(x,y)}. \tag{9.5}$$

The spatial covariance can be used to compute the aero-optical structure function by plugging Equation (9.5) into (9.2) to get,

$$D_\varphi(\Delta x, \Delta y) = 2\left[\mathrm{B}_\varphi(0,0) - \mathrm{B}_\varphi(\Delta x, \Delta y)\right]. \tag{9.6}$$

Implicit in this discussion of the structure function and the covariance properties of φ is that the statistics only depend on the difference coordinate, $(\Delta x, \Delta y)$. Strictly speaking, aero-optical disturbances and the related variance and spatial statistics of the disturbance can vary within an aperture, and both the structure function and the spatial covariance will also be functions of the absolute position, (x, y), over the aperture. However, most of these aero-optical spatial variations are in a spatially varying mean, or aero-lensing – which we implicitly subtract off for these considerations. Further, we assume that the aero-optical disturbance is anisotropic and the spatial covariance depends only on the magnitude of the increment, $\Delta r = \sqrt{(\Delta x)^2 + (\Delta y)^2}$.

For convecting flow where there is very little change apart from motion through the aperture, a "frozen flow" hypothesis analogous to that of free-stream turbulence (Roggemann and Welsh, 1996) may be invoked to transform between spatial offsets, and temporal delays assuming $\Delta r = U_\infty \tau$, where U_∞ is the platform speed. Thus, we may transform the spatial covariance into a temporal autocorrelation as

$$R_\varphi(\tau) = \mathrm{B}_\varphi(U_\infty \tau). \tag{9.7}$$

Given the temporal autocorrelation of the aero-optical phase, the aperture-averaged temporal power spectral density (PSD), $\Phi_a(f)$, can be computed from the Fourier transform of the temporal autocorrelation, $R(\tau)$ (Whiteley et al. 1998). This PSD is proportional to the aperture-averaged wavefront spectrum, $S(f)$, introduced in Equation (4.2), such as $\Phi_a(f) = (2\pi / \lambda)^2 S(f)$. PSD becomes the driving factor in our analysis of AO compensation for aero-optical disturbances. As mentioned before, to the extent to which the aero-optical aberration statistics vary over the aperture, these variations have been averaged out over the aperture to quantify the net power spectrum of relevance to AO compensation. While this is a simplification of the full spatial-temporal treatment of the problem, studies have shown the sufficiency of this characterization for quantifying the performance of AO compensation of aero-optical disturbances (Brennan and Wittich 2013).

9.1.2 Power-Law Observations from Aero-Optical Wavefront Data

The aperture-averaged temporal PSD plays a key role in the performance limitations for AO compensation of aero-optical disturbances, as it does for free stream turbulence. With proper knowledge or empirical observations of the temporal PSD, much can be determined about the requirements for adaptive optics compensation of aero-optical disturbances. For free-stream turbulence, the temporal PSD follows directly from the Kolmogorov power-law for index of refraction which goes as $f^{-11/3}$, but when integrated to wavefront, scales as $f^{-8/3}$. This simple theoretical model underlies the theory for free-stream turbulence. An analogy to this simple model for aero-optical wavefronts opens the door to making similar advancements here.

In his early work, Kyrazis developed what he considered to be a universal temporal PSD for shear-layer aero-optics (Kyrazis 1993). To arrive at this model, it was assumed that the coherent structures in the shear layer arrive as random, independent, non-overlapping events. The average number of these events per second, at a given location, is given by the mean velocity of the shear layer, U_C divided by the shear layer thickness, δ_ω. The probability density of the fluctuations is therefore given by a Poisson distribution. Correspondingly, the temporal autocorrelation for these shear layer disturbances is a decaying exponential function, that is,

$$R_{SL}(\tau) = \exp\left(-\frac{2U_C}{\delta_\omega}\tau\right). \tag{9.8}$$

The temporal PSD is then computed as the Fourier transform of the autocorrelation function. Given the form of Equation (9.8), the power spectrum would then be

$$\Phi_{SL}(f) \propto \frac{U_C / \delta_\omega}{(U_C / \delta_\omega)^2 + (\pi f)^2}. \tag{9.9}$$

Note that for $f \to 0$, then $\Phi_{SL} \to 1$. When f becomes large compared to U_C / δ_ω, the power spectrum goes as f^{-2}, according to the Kyrazis model.

Flight testing of the Airborne Aero-Optics Laboratory (AAOL) turret has provided a wealth of high-quality data to address the nature of the aero-optical temporal PSD. Figure 4.3 presented in an earlier chapter shows a collection of aperture-averaged PSDs measured during flight tests from the AAOL turret with a conformal aperture window measured at Mach 0.65 with the increasing viewing or line-of-sight (LOS) angle, α, defined in Equation (6.47) (Goorskey et al. 2013). In these PSDs, as the LOS angle increases, we see a characteristic peak in the PSD at frequencies in the range of $(1-2)\times U_\infty / D_t$, where D_t is the turret diameter. At frequencies above this peak, the PSD follows an approximately linear trend in this plot with logarithmic axes. Thus, it follows that the power spectrum $\Phi_\alpha \propto f^a$, where a is the observed power law.

As an example of fitting this power-law behavior, Figure 9.1 shows the aperture-averaged power spectrum for AAOL flight test wavefront data at LOS angle $\alpha = 114°$. If we consider the power spectrum for $f > U_\infty / D_t$, then a simple power law is fit to the high-frequency portion of the PSD. In this case, we find $a = -1.9$. When this type of power-law fitting analysis is extended to a larger data set of AAOL conformal window flight test data, we find the values of a shown in Figure 9.2 as a function of LOS angle. The power-law fitting parameter starts near -0.5 for LOS angles less than 80° and drops with increasing LOS angle up to about 130° where it appears to level off at just below -2.0. This means that when looking through a shear layer, the AAOL data indicate the higher-order power spectrum is dropping off at approximately f^{-2}, that is, two orders of magnitude per decade.

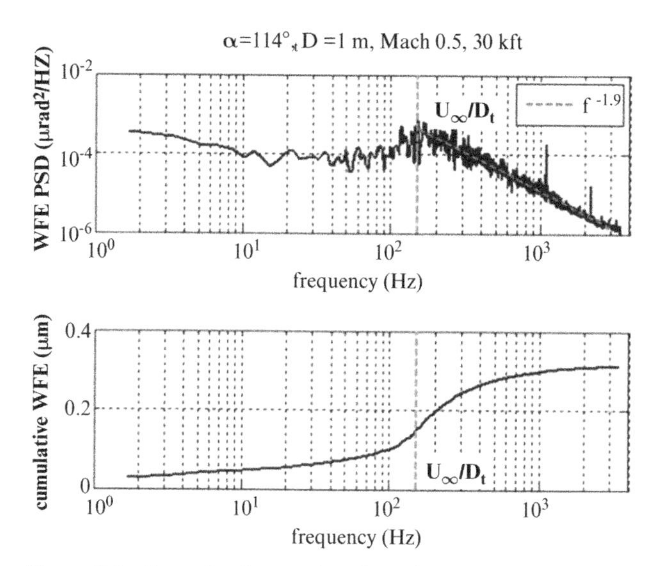

Figure 9.1 Example of power-law fit to the aperture-averaged temporal PSD for $f > U_\infty / D_t$. *Source:* MZA Associates Corporation, reproduced with permission of MZA Associates Corporation.

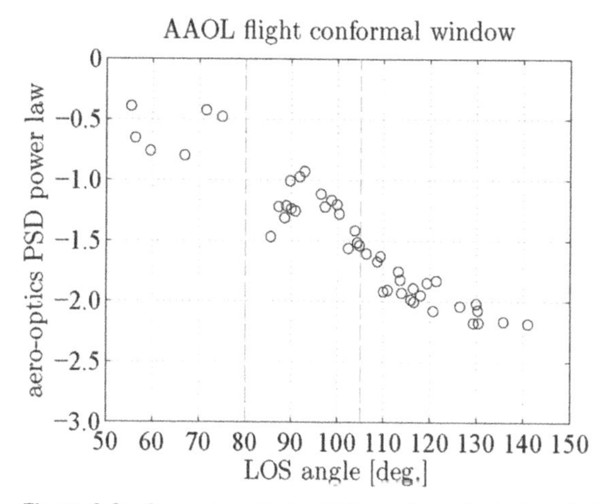

Figure 9.2 Power law fits to AAOL conformal window flight-measured higher-order wavefront temporal PSD as function of LOS angle. *Source:* MZA Associates Corporation, reproduced with permission of MZA Associates Corporation.

We note here that the AAOL wavefront data essentially confirms the fundamental argument of the Kyrazis heuristic model in Equation (9.9), which indicates a f^{-2} power law for shear layer disturbances. We note that for larger angles, a appears decrease further, suggesting a value in the range $-2.5 < a < -2.0$. We revisit this power-law dependence in the next section when considering the implications for AO compensation of such disturbances.

9.2 Compensation Scaling Laws for Aero-Optics

9.2.1 Adaptive Optics Control Law and Error Rejection Transfer Function

Typical AO systems implement a classical discrete integrator control law where the DM actuator commands c at time t_{k+1} are given by

$$c(t_{k+1}) = c(t_k) + \beta\varepsilon(t_k), \tag{9.10}$$

where β is the loop gain applied to the reconstructed wavefront error $\varepsilon(t_k)$ at each time step t_k as measured by the control wavefront sensor (WFS) shown in Figure 8.1. Note that the DM and WFS configuration shown in Figure 8.1 is a feedback control system, in that the WFS sees the difference between the incident wavefront and the current figure of the DM.

The temporal PSDs for each disturbance sequence are used with an error rejection model for a conventional controller to determine the ability for a standard AO loop to compensate each sequence. For a simple integrator of sample rate f_s with loop gain β and latency Δt, the error rejection for the controller is modeled theoretically as

$$ERR\left(f; f_s, \beta, \Delta t\right) = \left[1 + \left(\frac{\beta f_s}{2\pi f}\right)^2 - 2\left(\frac{\beta f_s}{2\pi f}\right)\sin(2\pi f \Delta t)\right]^{-1}, \tag{9.11}$$

The error rejection function quantifies the fraction of variance for a disturbance of frequency, f, that remains after application of the control. Given the error rejection function for the conventional AO controller in Equation (9.11), the residual phase variance for a particular disturbance can be computed as

$$\varepsilon_\phi^2(f_s, \beta, \Delta t) = \int_0^\infty ERR(f; f_s, \beta, \Delta t)\Phi_a(f)df, \tag{9.12}$$

where $\Phi_a(f)$ is the PSD of the aero-optical phase disturbance to be compensated.

If the control system has no latency and the response time of the mirror is instantaneous, then the error rejection function for the controller is modeled as

$$ERR(f) = \left[1 + \left(\frac{f_s}{f_{3dB}}\right)^{-2}\right]^{-1}, \tag{9.13}$$

where the error rejection bandwidth f_{3dB} is commonly (Ellerbroek 1994; Tyler 1994) defined as $f_{3dB} \equiv \beta f_s / (2\pi)$ and f_s designates the sampling frequency of the discrete-time system. By substituting Equation (9.13) into (9.12),

$$\varepsilon_\phi^2(f_s, f_{3dB}) = \int_0^\infty \left[1 + \left(\frac{f_s}{f_{3dB}}\right)^{-2}\right]^{-1} \Phi_a(f)df. \tag{9.14}$$

This form of the error rejection will be used in the following subsections where we consider the functional form of residual aero-optics disturbances with increasing error rejection bandwidth, f_{3dB}.

9.2.2 Asymptotic Results for Aero-Optics Compensation

Analysis of the residual phase variance may be conducted with aero-optical wavefront disturbances to assess the dependence of the residual phase variance on the bandwidth f_{3dB} of the AO system. Figure 9.3 shows an example of this analysis applied to AAOL flight test data for measurements made at Mach 0.8. To analyze AO compensation for the scaled flight test disturbances, and by analogy with the analysis of bandwidth specifications for compensation of free stream turbulence (Greenwood 1977), we tested the simulation results for an apparent scaling behavior of

$$\frac{\varepsilon_{\phi}^{2}}{\sigma_{A}^{2}} = K \cdot f_{3dB}^{-\gamma}, \tag{9.15}$$

where ε_{ϕ}^{2} is the residual phase variance with AO compensation, σ_{A}^{2} is the open-loop phase variance, γ is a power dependent upon the properties of the random disturbance and the assumed control law, and K is an arbitrary constant of proportionality. From the slope and intercept of a linear fit to the simulation data, the

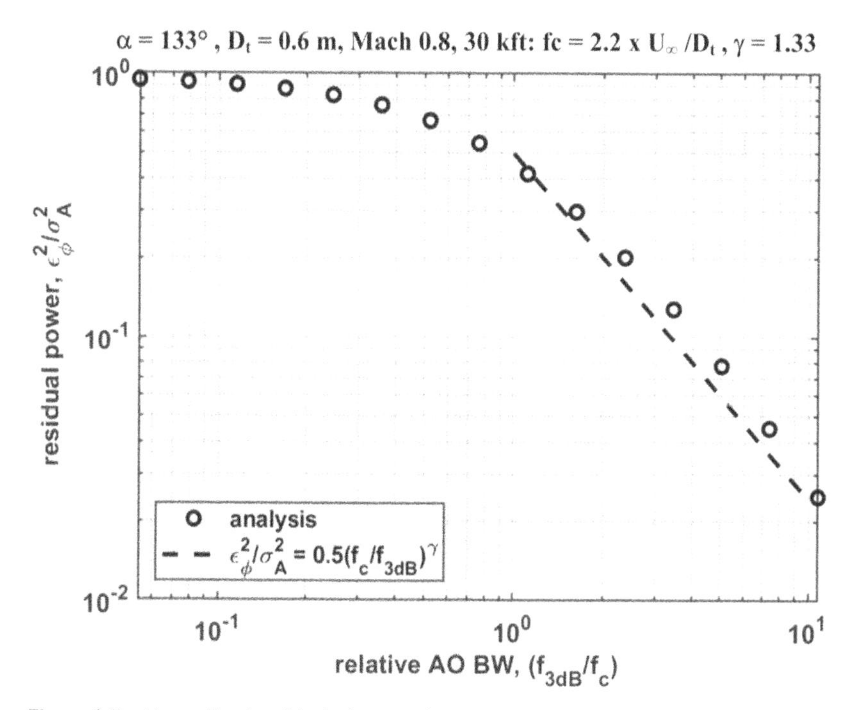

Figure 9.3 Normalized residual phase variance for AO compensation of AAOL Mach 0.8 wavefronts.

values of γ and K in Equation (9.15) may be determined. Substituting $K = f_s^{\gamma}/2$, it follows that

$$\varepsilon_{\phi}^2 = \frac{1}{2}\sigma_A^2\left(\frac{f_c}{f_{3dB}}\right)^{\gamma} \rightarrow SR \simeq \exp\left[-\frac{1}{2}\sigma_A^2\left(\frac{f_c}{f_{3dB}}\right)^{\gamma}\right], \tag{9.16}$$

where f_c may be interpreted as a characteristic scaling frequency for the aero-optical disturbance. By definition, when $f_{3dB} = f_c$, the aero-optics phase variance is reduced by a factor of 2.

Figure 9.4 shows this type of compensation analysis being applied to a number of AAOL flight test cases with varying LOS angle for the turret. From this analysis,

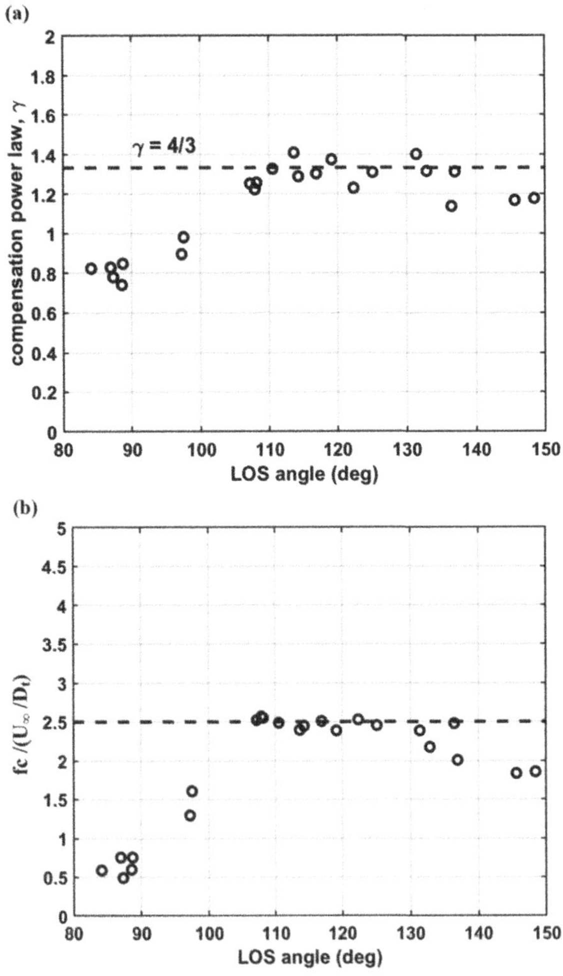

Figure 9.4 Power-law fitting of residual phase variance (a) fit power law as a function of γ and (b) characteristic scaling frequency, f_c, for AAOL Mach 0.8 flight test data with varying LOS angle.

we see that both the compensation power law and the characteristic scaling frequency increase with LOS angle up to approximately 110°. For larger angles in the aft field-of-regard of the turret, we see that $\gamma \simeq 4/3$ and $f_c \simeq 2.5U_\infty / D_t$. At the largest LOS angles, both quantities appear to drop somewhat, but these values are relatively constant throughout the aft field-of-regard.

9.2.3 Aero-Optics Compensation Frequency

We may bring our analysis of aero-optical compensation to a level of completion by moving toward a full analogy with similar scaling laws for free-stream turbulence compensation. To do this, we first make use of a functional form for the normalized strength of aero-optical wavefront disturbances which was suggested by Goorskey et al. (2013) as a fit to flight test data [9], as shown in Figure 9.5. The normalized wavefront error for aero-optical disturbances was found to be well captured by the following empirical dependence on the LOS angle, α, with $\alpha_0 = 85°$:

$$C_A(\alpha) = 3.55 \times 10^{-6} \frac{\tan(\alpha - \alpha_0)}{\cos(\alpha - \alpha_0)},$$

Using this approximation for the normalized strength of aero-optical disturbances, it follows from the aero-optical wavefront scaling relations for turrets stated earlier in Equation (6.47) that

$$\sigma_A^2 = \frac{2\pi^2}{\lambda^2} \left(\frac{\rho}{\rho_{SL}} \right)^2 M^4 D_t^2 C_A^2(\alpha),$$

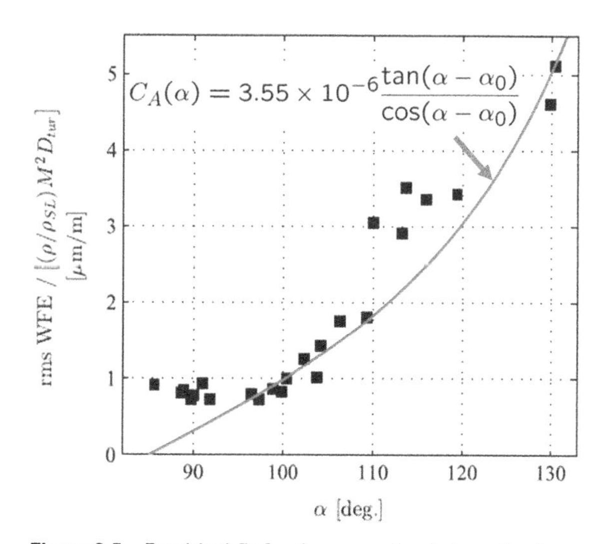

Figure 9.5 Empirical fit for the normalized strength of aero-optical wavefront disturbances. *Source:* Goorskey et al. (2013), figure 2. Reproduced with permission of SPIE.

where $\rho_{SL} = 1.225\ \text{kg/m}^3$ is the sea-level density. Now, the relationship established earlier in Equation (9.16) for the compensated Strehl ratio can be simplified as follows:

$$SR \simeq \exp\left[-\left(\frac{f_A}{f_{3dB}}\right)^{\gamma}\right], \tag{9.17}$$

where f_A, the aero-optics compensation frequency, is defined as:

$$f_A \equiv \left[\frac{2\pi^2}{\lambda^2}\left(\frac{\rho}{\rho_{SL}}\right)^2 M^4 D_t^2 C_A^2(\alpha)\right]^{1/\gamma} f_c. \tag{9.18}$$

Expressing f_c in units of U_∞ / D_t, we say $f_c = K_c / (U_\infty / D_t)$, Equation (9.18) can be rewritten in the following form to explicit bring in the platform speed U_∞:

$$f_A = \left[\frac{K_c^{\gamma}}{2}\left(\frac{2\pi}{\lambda}\right)^2\left(\frac{\rho}{\rho_{SL}}\right)^2 M^4 D_t^{2-\gamma} C_A^2(\alpha) U_\infty^{\gamma}\right]^{1/\gamma}. \tag{9.19}$$

With f_A written in this form, we can assess a wide range of adaptive optics compensation requirements over varying turret LOS angle and Mach number. Accepting the values of $\gamma = 4/3$ and $K_c = 2.5$ from the prior analysis, the f_A values shown in Figure 9.6 may be computed for varying Mach as a function of the

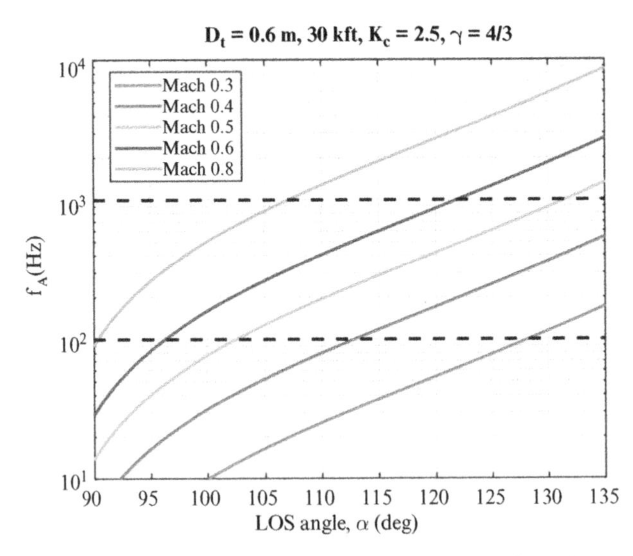

Figure 9.6 Aero-optics compensation frequency evaluated over a range of turret LOS angles for Mach 0.3–0.8 flight conditions with $D_t = 0.6$ m at 30 kft flight altitude.

turret LOS angle. For the low-speed platform at Mach 0.3, the required compensation bandwidth is less than 200 Hz across the range of LOS angles. For a platform at Mach 0.5, the required compensation bandwidth is less than 100 Hz at 100 deg LOS angle, and over 1 kHz at 135 deg LOS angle. For a platform at Mach 0.8, the minimum compensation bandwidth at 90 deg LOS angle is approximately 100 Hz, but reaches 6 kHz at 130 deg LOS angle.

9.2.4 Relation of Aero-Optics Scaling Laws to Free-Stream Turbulence

The scaling law for aero-optical Strehl ratio with adaptive optics compensation in Equation (9.17) was written in this form to call to mind the analogous scaling law for free-stream turbulence, expressed as:

$$SR \simeq \exp\left[-\left(\frac{f_G}{f_{3dB}}\right)^{5/3}\right],$$

where the compensation frequency for free-stream turbulence, known as the Greenwood frequency, f_G, for a path of length L with turbulence distribution $C_n^2(z)$ and wind speed profile $|V(z)|$ is given by:

$$f_G = \left[0.102\left(\frac{2\pi}{\lambda}\right)^2 \int_0^L C_n^2(z)|V(z)|^{5/3}\,dz\right]^{3/5}, \tag{9.20}$$

In the case of uniform turbulence C_n^2 and a constant wind speed V over the path, f_G is given by:

$$f_G = \left[0.102\left(\frac{2\pi}{\lambda}\right)^2 L\cdot C_n^2 V^{5/3}\right]^{3/5} \quad \text{(uniform path)}, \tag{9.21}$$

Comparing Equation (9.21) for free-stream turbulence to Equation (9.19) for aero-optics, we can conclude that the following quantities serve analogous roles in each of the regimes to quantify the severity of the turbulence propagation problem:

free-stream turbulence: $L\cdot C_n^2$

aero-optical turbulence: $\left(\frac{\rho}{\rho_{SL}}\right)^2 M^4 D_t^{2-\gamma} C_A^2(\alpha)$

To relate the aero-optical problem encountered here to free-stream turbulence, we have plotted the value of the quantity $\left(\frac{\rho}{\rho_{SL}}\right)^2 M^4 D_t^{2-\gamma} C_A^2(\alpha)$ in Figure 9.7 for Mach 0.3–0.8. For those familiar with freestream turbulence, these values are very

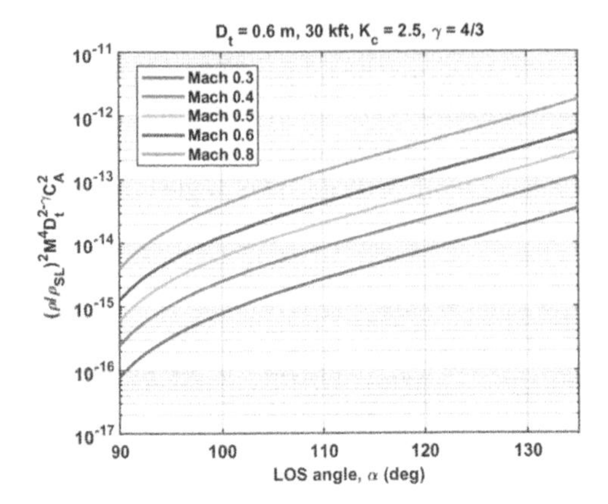

Figure 9.7 Aero-optical severity parameter equivalent to $L \cdot C_n^2$ for free-stream turbulence.

similar in quantity to typical values of C_n^2. For free-stream turbulence propagation, values of C_n^2 in the range $10^{-16} m^{-2/3}$ to $10^{-14} m^{-2/3}$ are relatively weak atmospheres. Values of C_n^2 in the range $10^{-14} m^{-2/3}$ to $10^{-12} m^{-2/3}$ are relatively strong atmospheres. Given the f_A values shown in Figure 9.6, a similar statement can be made for aero-optical disturbances.

9.3 Spatial and Temporal Limitations of Adaptive Optics

Having developed an appreciation for the challenges faced when compensating aero-optical disturbances with a basic adaptive optics controls model, we now turn to a more thorough treatment of the problem. Adaptive optics compensation is limited both by the temporal sampling as it affects the error rejection bandwidth of the AO control system, and also by the density of actuators for the deformable mirror (DM) corrector. We now bring in the full spatial and temporal limits to compensation of aero-optical wavefront disturbances, leading to methods for establishing AO requirements at a system performance level.

9.3.1 Framework for Analysis of Aero-Optical Compensation

In our treatment of this subject, we will use the empirical wavefront disturbance sequences applied in the previous section to establish scaling relations for AO compensation. Figure 9.8 shows the general framework in which our analysis of AO compensation proceeds. Moving from left-to-right in the diagram, we start with the desired engagement conditions and the properties of the AO control system, including sample frequency, f_s, control gains, and the latency Δt of the AO controller.

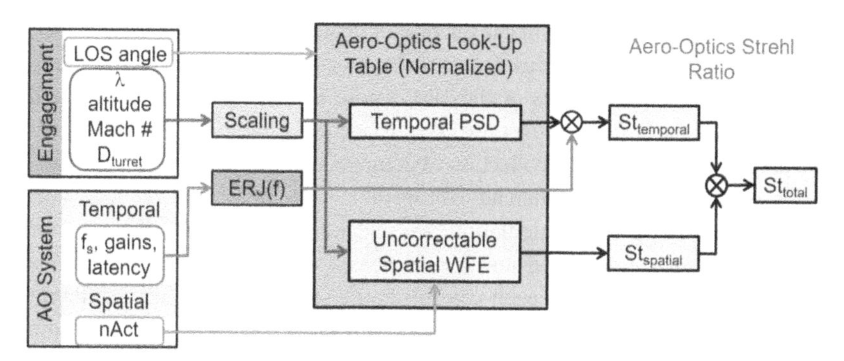

Figure 9.8 Analytic framework for aero-optics compensation modeling. *Source:* MZA Associates Corporation, reproduced with permission of MZA Associates Corporation.

The wavefront measurements were made during AAOL flight testing at Mach 0.8, 30 kft altitude with the aircraft test system described previously. However, we will want to apply them to arbitrary flight conditions. Thus, all the measurements made during flight testing have been cataloged for the test conditions, and each azimuth and elevation angle leading to a particular turret LOS angle. The wavefront statistics are normalized by the test conditions and stored in so-called look-up tables for scaling to new conditions. The scaling relations are applied to the desired flight conditions, and the wavefront statistics are scaled accordingly for analysis.

The wavefront statistics are then analyzed with the properties of the AO control system to determine the residual disturbances which are fit by the DM surface. This uncorrectable spatial wavefront error determines a spatially driven Strehl ratio components. The portion of the wavefront that can be affected by the DM then is subject to temporal rejection given the error rejection function of the AO controller. The residual wavefront error from the temporal analysis determines a temporally driven Strehl ratio component. The spatial and temporal Strehl ratio components are combined to present the full compensated wavefront error or combined aero-optics Strehl ratio including both spatial and temporal limits to the correction.

9.3.2 Deformable Mirror Fitting Error for Aero-Optical POD Modes

The first step in the analysis of compensation for a specified AO system is to determine the portion of the aero-optical wavefronts which can be corrected by the adjusting the deformable mirror actuators. With limited degrees of freedom, and a continuous reflecting surface, the DM cannot compensate the full disturbance. The portion of the wavefront error which is uncorrectable by the DM regardless of the temporal properties of the AO control is called the DM "fitting error." To quantify this fitting error, we will use the proper orthogonal decomposition (POD) modal decomposition of the wavefronts, introduced earlier in Chapter 4, along with the "influence" functions of the DM. These influence functions for each actuator specify the deflection over the DM surface as a function of the movement of the actuator. Thus, the full collection of actuator commands gives the full surface deformation of the DM when added together.

To begin the fitting error calculation, we select a DM for which the actuator influence functions are known and write each of its influence functions as a column in a matrix $\mathbf{P}$, where each column is the surface figure due a different actuator. Next, we compute the POD modes of the chosen wavefront data set and interpolate each mode onto the higher resolution grid of influence functions. These interpolated modes are arranged as columns of a matrix $\boldsymbol{\Psi}$ so that

$$\boldsymbol{\Phi} = \boldsymbol{\Psi}\mathbf{C}, \tag{9.22}$$

where each column of $\boldsymbol{\Psi}$ is a different mode, and each column of $\mathbf{C}$ is a set of modal coefficients at each time. We typically use 150–250 modes for aero-optic wavefront sequences in this process. The modes $\boldsymbol{\Psi}$ are projected onto the influence functions $\mathbf{P}$ with coefficients as columns in the matrix $\mathbf{D}$ according to

$$\Psi = \mathbf{PD} + \mathbf{E}. \tag{9.23}$$

The remainder is the modal fitting error as columns of E. Finally, we convert from modes back to spatial domain by substituting Equation (9.23) into (9.22) to produce

$$\begin{aligned} \phi &= \mathbf{PDC} + \mathbf{EC} \\ \phi_{OL} &= \phi_{FIT} + \phi_{ERR}, \end{aligned}$$

where $\phi_{FIT} = \mathbf{PDC}$ is the portion of the wavefront sequence that can be fit by the DM and $\phi_{ERR} = \mathbf{EC}$ is the portion of the wavefront sequence that cannot be fit by the DM due to spatial frequency limitations, that is, the DM fitting error. This process is done using DM models with increasing actuator density over the surface of the mirror. The actuator density is typically parameterized by the number of actuators, n_{act}, over the aperture diameter D, and assuming that the actuators are located at the corners of each AO wavefront sensor subaperture, as considered in Chapter 9 for optimal AO performance.

Applying the POD mode-fitting method to AAOL Mach 0.8 flight test data over a range of turret LOS angles, we varied n_{act}/D incrementally to quantify the residual wavefront error variance, ε_{DM}^2, which is the square of the aperture-averaged rms WFE for the fitting error component, ϕ_{ERR}. When normalized to the full open-loop aero-optical wavefront variance, σ_A^2, for each LOS angle, we find the fitting error plots shown in Figure 9.9. The data has been plotted in logarithmic axes to highlight an apparent power-law scaling to the DM fitting error, similar to what we saw earlier for the loop residual with increasing error rejection bandwidth. The apparent power-law scaling is seen at each LOS angle.

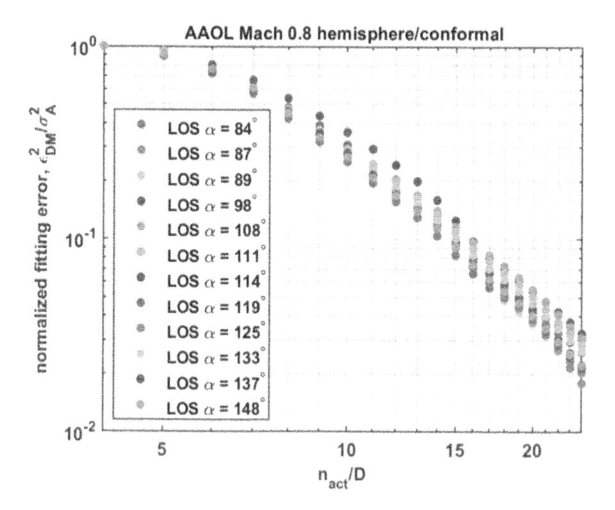

Figure 9.9 Residual wavefront variance from fitting a DM surface with increasing actuator count n_{act} / D to the POD modes of AAOL Mach 0.8 flight test wavefront disturbances.

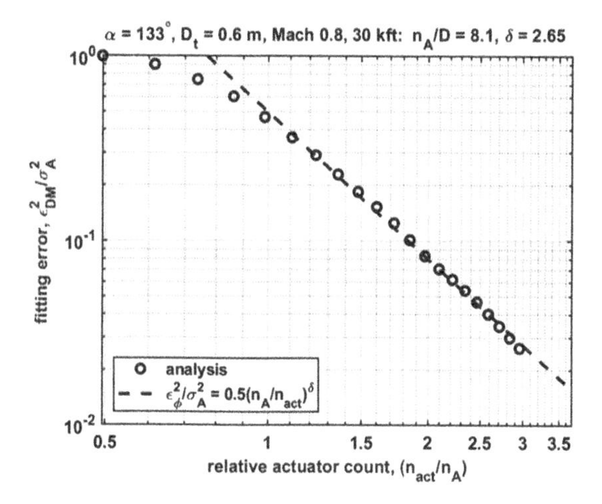

Figure 9.10 Normalized residual phase variance for POD-derived DM fitting error of AAOL Mach ~ 0.8 wavefronts.

Looking more closely at a particular LOS angle $\alpha = 133°$ in this data set as shown in Figure 9.10, we see more clearly the power-law reduction of the fitting error with increasing n_{act} / D. To test the power-law scaling hypothesis, we fit the DM fitting error variance in a form similar to the temporal residual in Equation (9.16), that is:

$$\varepsilon_{DM}^2 = \frac{1}{2}\sigma_A^2 \left(\frac{n_A}{n_{act}}\right)^\delta,$$

where δ is the apparent scaling power, and n_A is the characteristic actuator count for DM fitting error scaling. Analogously with the temporal scaling, n_A is defined as the number of actuators over the aperture at which the DM fitting error variance is 1/2 of the aero-optical wavefront variance for the given turret LOS angle. It is clear from Figure 9.10 that this DM fitting error power-law scaling applies consistently for values greater then n_A.

The power-law scaling was applied to all of the data from the AAOL Mach 0.8 flight testing. The fitted values of δ and n_A are shown in Figure 9.11. Here, we see that $\delta = 8/3$ and $n_A / D = 8$ fits the data consistently over the full range of LOS angles in the flight test data. Thus, it is clear that the DM fitting error follows regular scaling pattern regardless of the turret LOS angle. This scaling relation is

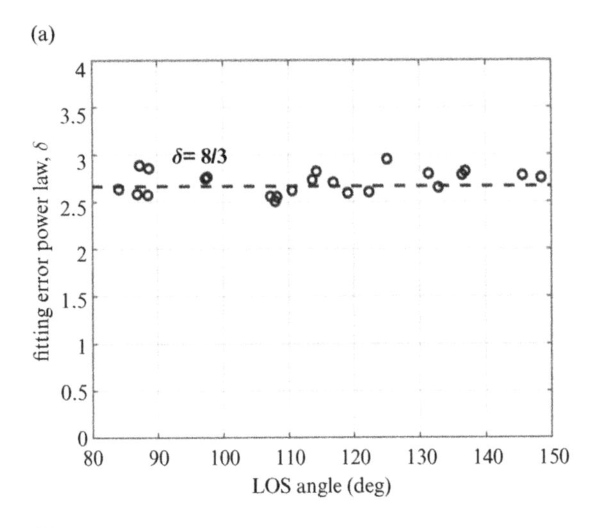

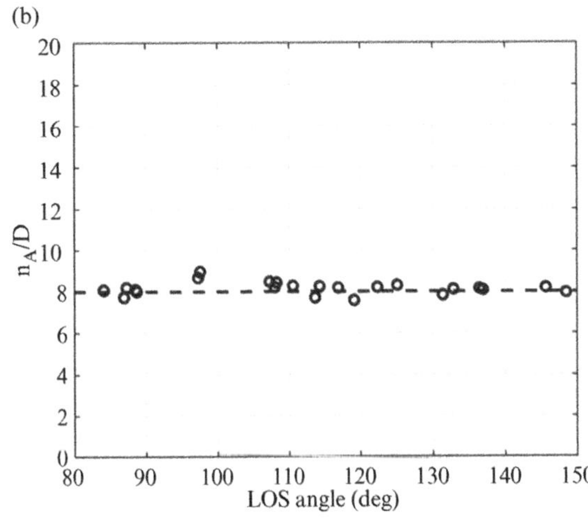

Figure 9.11 (a) Fit power law, δ, and (b) scaling actuator density, n_A / D, of DM fitting error residual phase variance for AAOL Mach 0.8 flight test data with varying LOS angle.

analogous to DM fitting error for free-stream turbulence which has been shown to scale as $(\Delta_{act}/r_0)^{5/3}$ for free-stream turbulence (Tyson 1997). Here r_0 is Fried's coherence length (Fried 1967), also known as the atmospheric coherence diameter, and Δ_{act} denotes the actuator spacing for the DM.

9.3.3 Decomposition of Correctable and Uncorrectable Power Spectrum

Having established the power-law scaling nature of DM fitting error, we turn our attention back to the subject of the properties of the fitted wavefront component ϕ_{FIT} and the error wavefront, ϕ_{ERR}. Since ϕ_{ERR} and ϕ_{ERR} are both individual wavefront sequences in time summing up to ϕ_{OL}, we can get the temporal power spectrum of each so that

$$\Phi_{OL}(f) = \Phi_{FIT}(f) + \Phi_{ERR}(f). \tag{9.24}$$

9.3.3.1 DM Sensitivity Transfer Function

To aid in our analysis of closed-loop adaptive optics performance, we define the DM sensitivity transfer function, $\eta(f)$ as

$$\eta(f) \equiv \frac{\Phi_{ERR}(f)}{\Phi_{OL}(f)}, \tag{9.25}$$

and it follows from Equation (9.24) that

$$\begin{aligned} \Phi_{FIT}(f) &= (1-\eta(f))\Phi_{OL}(f) \\ \Phi_{ERR}(f) &= \eta(f)\Phi_{OL}(f), \end{aligned}$$

which casts the fitted component and fitting error component in terms of the open-loop power spectrum $\Phi_{OL}(f)$ and the DM sensitivity transfer function $\eta(f)$. Figure 9.12(a) shows an example of decomposing the aperture-averaged aero-optical power spectrum into its correctable and uncorrectable components using the methodology described above. This analysis is shown for AAOL Mach 0.8 test data with LOS angle $\alpha = 131°$with $n_A / D = 9$. Figure 9.12(b) shows the derived $\eta(f)$. Here we see that the lower-frequency portion of the disturbance is fit nearly uniformly at $f < 2U_\infty / D_t$ and then climbs to near unity values for $f \sim 10U_\infty / D_t$. For $f > 10U_\infty / D_t$, there is little aero-optical disturbance remaining, and measurements at these frequencies are more affected by measurement noises. Thus, while $\eta(f)$ appears to decrease at high frequency, the underlying noise in the wavefront data is artificially reducing these values.

When the analysis of the DM sensitivity transfer function is extended to DM designs with larger n_{act} / D, then the DM sensitivity transfer function $\eta(f)$ changes accordingly. Figure 9.13 shows the variation of $\eta(f)$ with increasing

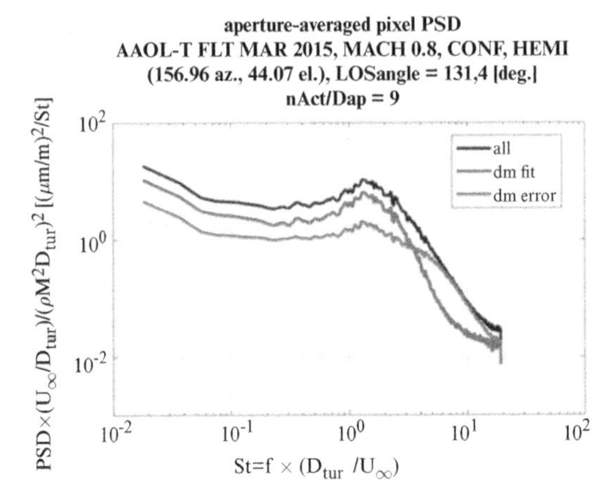

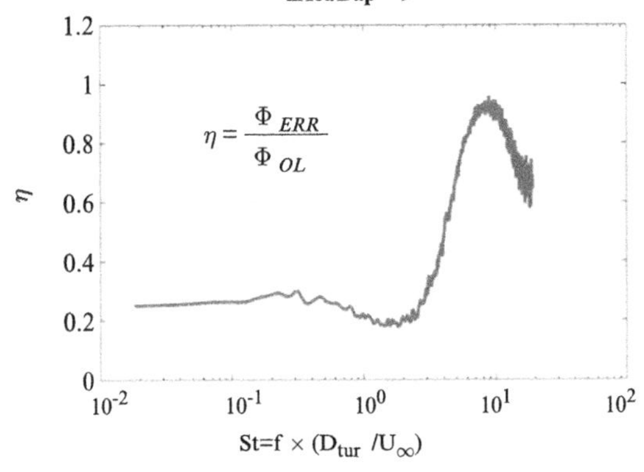

Figure 9.12 Decomposition of the aero-optical PSD into fitted and error components. (a) Components of PSD. (b) DM sensitivity transfer function, $\eta(f)$. *Source:* MZA Associates Corporation, reproduced with permission of MZA Associates Corporation.

n_{act}/D. As n_{act}/D increases, disturbances at lower frequency are reduced further, and the upper limit of this low-pass filter moves to higher frequencies relative to U_∞/D_t. In practice, the curves of $\eta(f)$ are regular enough to be interpolated in n_{act}/D for intermediate values not explicitly analyzed.

9.3.4 Closed-Loop Residual Wavefront Error

Having established the portion of the aero-optical disturbance which is correctable by the DM, we may then consider the residual power spectrum given AO correction. The AO controller can only correct temporally what the DM can fit

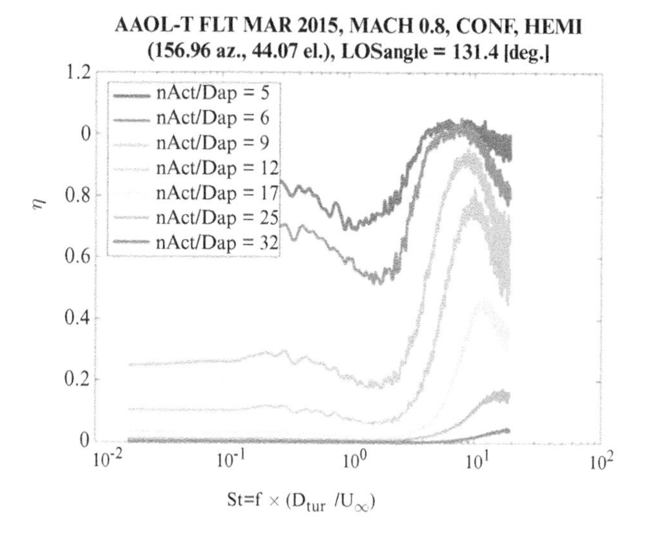

Figure 9.13 DM sensitivity transfer function with varying number of actuators per aperture n_{act} / *D*. *Source:* MZA Associates Corporation, reproduced with permission of MZA Associates Corporation.

spatially. Therefore, the temporal error rejection $ERR(f)$ for AO control can only act on the portion of the wavefront temporal power spectrum that can be fit by the DM, namely $\Phi_{FIT}(f)$. Therefore we model the closed-loop power spectrum as:

$$\Phi_{CL}(f) = ERJ(f)\Phi_{FIT}(f) + \Phi_{ERR}(f).$$

Using our definition for $\eta(f)$ in Equation (9.25), we can therefore write

$$\Phi_{CL}(f) = ERJ(f)[1-\eta(f)]\Phi_{OL}(f) + \eta(f)\Phi_{OL}(f)$$
$$\Phi_{CL}(f) = \Phi_{CL,TEMP}(f) + \Phi_{CL,SPAT}(f).$$

where $\Phi_{CL,TEMP}(f)$ is the portion of the residual closed-loop power spectrum that is associated with the temporal limitations of the AO system and $\Phi_{CL,SPAT}(f)$ is the portion of the residual closed-loop power spectrum that is associated with the spatial limitations of the AO system.

We explicitly write the closed-loop power spectrum in the following form to highlight the many factors affecting the loop residual:

$$\Phi_{CL}(f;\alpha, f_{BW}, \Delta t, n_{act}/D) = ERJ(f; f_{BW}, \tau_D)[1-\eta(f;\alpha, n_{act}/D)]\Phi_{OL}(f;\alpha) + \eta(f;\alpha, n_{act}/D)\Phi_{OL}(f;\alpha).$$

where α is the turret LOS angle, f_{BW} is the -3 dB error rejection AO bandwidth, and Δt is the AO loop latency. We note that as n_{act}/D increases, more of the total power spectrum is subject to the AO control loop properties, as characterized by the error rejection function, such as that given in Equation (9.11). Thus, the DM actuator properties affects how much of the disturbance is subjected to the temporal error rejection properties of the AO controller.

9.3.5 Effect of Latency in Aero-Optics Compensation

The previous analysis in the previous section considered only the case of zero latency in the AO control system, and hence was an idealized treatment of the error rejection characteristic of the AO system. In practice, no adaptive optics system operates without some degree of latency in the control loop. This latency may be the result of the time it tasks to read out data from the wavefront sensor, the time it takes to reconstruct the wavefront sensor data into an error wavefront signal, or the time it takes for the DM to assume the figure commanded by the controller. In most cases, these effects can be lumped together as a net latency for specifying a value of Δt in Equation (9.11).

Figure 9.14 shows the impact of varying latency, Δt with sample rate $f_s = 25$ kHz and control-loop gain $\beta = 0.5$ on the error rejection characteristic of the AO system. We note that for $\Delta t = 0$ μsec, we recover the simplified error rejection which is characterized only by the -3dB error rejection bandwidth. For the case shown in Figure 9.14 with zero latency, $f_{3dB} = 0.5 \times 25\ \text{kHz}/(2\pi) = 1.99$ kHz. Notice that the $\Delta t = 0$ μsec, latency error rejection crosses the -3dB line at

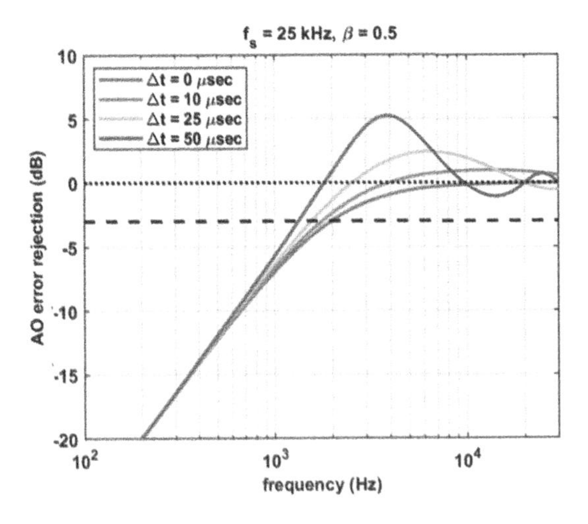

Figure 9.14 Example AO error rejection functions computed for varying latency Δt with sample $f_s = 25$ kHz and control-loop gain, $\beta = 0.5$.

approximately 2 kHz. As the latency of the AO system increases, the -3dB error rejection frequency is reduced. Further, we see that the high frequency error rejection no longer asymptotically approaches 0 dB as $f \rightarrow \infty$. Instead, the error rejection crosses 0 dB and peaks at intermediate frequencies before approaching 0 dB for large f. For the $\Delta t = 50$ µsec case shown in Figure 9.14, the error rejection peaks around 5 dB at $f \sim 4$ kHz.

The effect of this high-frequency amplification can be seen in Figure 9.15. Here, we have plotted the residual wavefront error as a function of varying sample rate fs for incrementally higher gains, $\beta = 0.1, 0.2, 0.3, 0.5, 0.7$. Figure 9.15(a) shows the

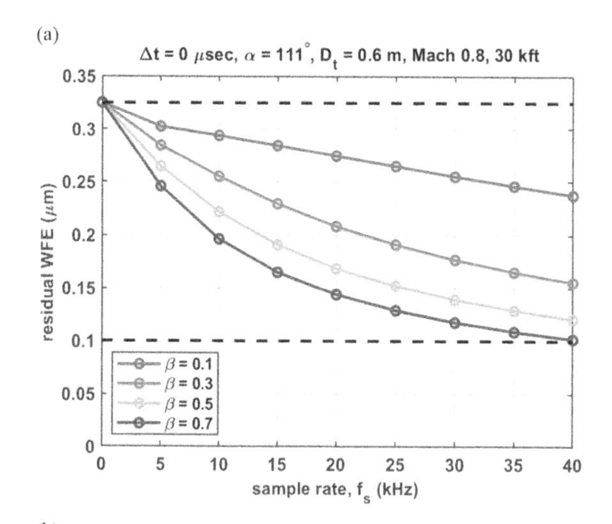

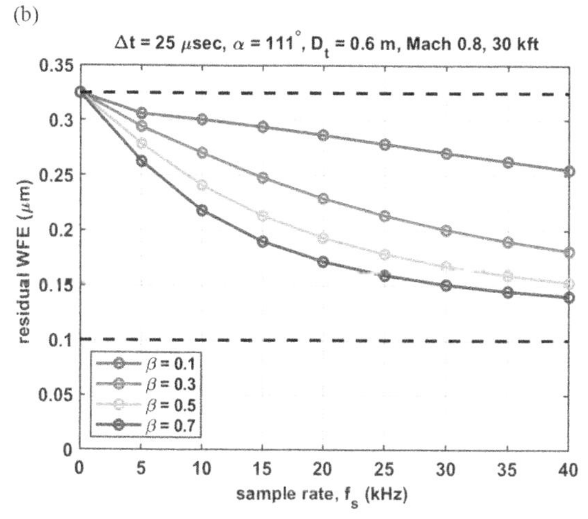

Figure 9.15 Effect of AO control latency on aero-optics compensation with (a) $\Delta t = 0$ µsec, (b) $\Delta t = 25$ µsec and (c) $\Delta t = 50$ µsec with $n_{Act} / D = 24$. Aero-optic scaling is for a Mach 0.8 platform at 30 kft altitude, with turret diameter $D_t = 0.6$ m.

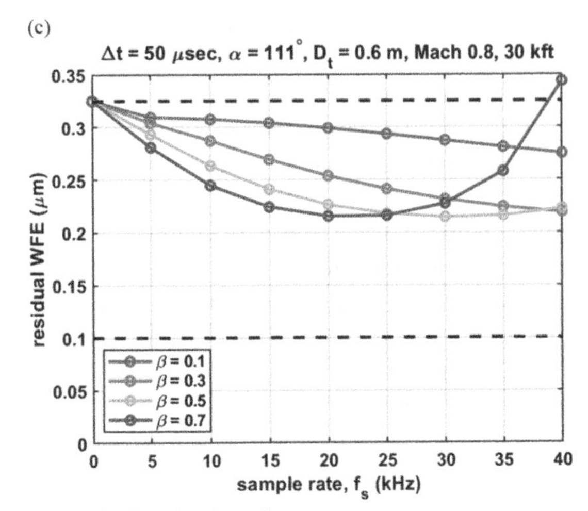

Figure 9.15 (Continued)

zero latency case where with $\beta = 0.7$ and increasing the AO sample rate up to 40 kHz, we can decrease the residual WFE to 0.1 μm. When latency is increased to $\Delta t = 25$ μsec in Figure 9.15(b), the reduction of residual WFE for the high-gain cases is diminished at high sample rates.

Figure 9.15(c) shows the case where the latency is increased to $\Delta t = 50$ μsec, the benefit of high sample rate is greatly reduced. In the high-gain $\beta = 0.7$ case, we see that the best performance of the AO loop is obtained for $f_s = 20$ kHz, and higher sample rates are deleterious for AO compensation. The disturbance is actually amplified at high sample rate when the latency is high, indicated by reduced stability in the control loop. In these cases, it is better to reduce the gain of the AO system to mitigate the degradation due to AO latency. While effects like these are observed in free-stream turbulence compensation, aero-optics compensation makes this degradation with latency more apparent since the disturbance frequencies are so high. Thus, latency-tolerant control methods such as predictive AO have been formulated to get the most benefit out of the available sample rate given practical levels of latency in AO systems (Goorskey et al. 2013b).

9.4 Application to System Performance Modeling

We now turn our attention toward applying the methods developed to assessment of AO compensation in the context of a practical laser system engagement. We first consider joint variations in actuator density and error rejection bandwidth to see the most suitable AO parameters for the chosen application. We then compare the residual aero-optical wavefront disturbances with AO compensation to other system degradation to put the aero-optics problem in better perspective.

For the results which follow, we have chosen to consider a hypothetical laser system operating on an aircraft platform with the following specifications:

- platform altitude: 30 kft ASL
- platform speed: Mach 0.8
- turret diameter: $D_t = 0.6$ m
- aperture diameter: $D = 0.2$ m
- laser wavelength: $\lambda = 1$ μm

9.4.1 Scaling of Aero-Optical Statistics to Flight Conditions

The engagement conditions outlined earlier differ in certain ways from the AAOL flight test data which supports our further analysis. The Mach 0.8, 30 kft flight condition is unchanged from the test condition, so scaling of the flight test data will not extend to a different Mach regime. Table 1 shows the full set of scaling relations for going from the AAOL flight test condition to the proposed laser engagement scenario.

The turret system indicated earlier represents a system which is a factor of 2 larger than the flight test system used for the AAOL aircraft. Thus, all the spatial scales of the data will increase by a factor of 2. The temporal frequencies will reduce by a factor of 2. In this case, $U_\infty / D_t = 404$ Hz. The scale factor for the magnitude of aero-optical disturbances in this case is $(\rho / \rho_{SL}) M^2 D_t = 0.144$ μm. Thus, a case with the nondimensional wavefront error of $C_A = 1$ μm/m as in Figure 9.5 would have an open-loop wavefront error of 0.144 μm and would increase steadily with increasing LOS angle. For LOS angles $\alpha > 120°$, the open-loop WFE would be greater than 0.5 μm. This level of aero-optical WFE represents an unacceptable degradation in the level of performance for such a laser system and must be reduced through AO compensation.

9.4.2 Joint Variations in Adaptive Optics Bandwidth and Actuator Density

Recognizing the severity of the aero-optics problem in this case, we will apply the AO compensation model developed to determine the requirements for an effective compensation system. Figure 9.16 shows the adaptive optics residual wavefront error for joint variations in DM actuator density and compensation bandwidth. This example is computed for the same LOS angle $\alpha = 133°$ as we have considered previously. The actuator density n_{act} / D has been varied between 4 and 30. The error rejection bandwidth of the AO system has been varied continuously from open-loop to 10 kHz. These calculations have been done with zero AO system latency, that is, $\Delta t = 0$ μsec.

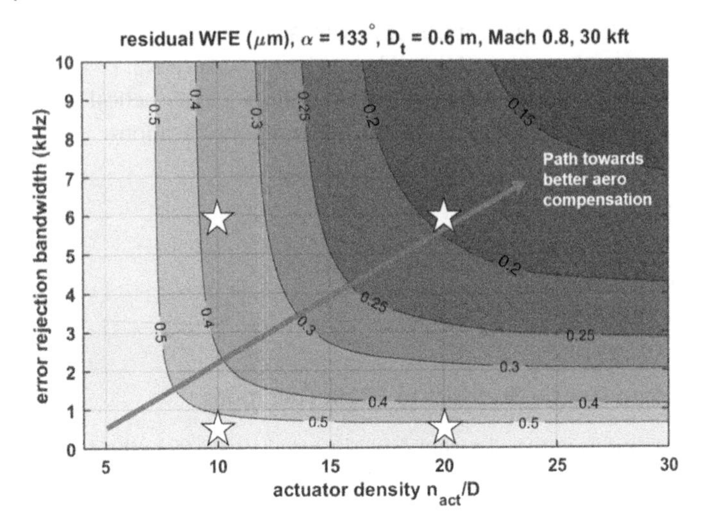

Figure 9.16 Adaptive optics residual wavefront error for joint variations in DM actuator density and compensation bandwidth.

As expected, we see in Figure 9.16 that the residual aero-optical wavefront error is reduced significantly while increasing error rejection bandwidth and actuator density. However, the shape of the contours indicates that *both* actuator density and bandwidth must increase consistently in order to achieve a reasonably reduced residual wavefront error. The annotation on the contours of Figure 9.16 indicate the approximate path toward better aero-optics compensation. We have chosen four locations in this contour plot to study more closely, as indicated by the four stars on the contour plot.

The first case to consider is that of low bandwidth and low actuator density of $f_{3dB} = 600$ Hz and $n_{act} / D = 10$ shown in Figure 9.17(a). The plot shows the limits imposed by the temporal residual and the spatial residual using the methods we have described earlier. The plot includes annotations at residual WFE values of 0.3 μm (poor), 0.2 μm (marginal), and 0.1 μm (ideal). For a laser with $\lambda = 1$ μm, these levels are 0.3λ, 0.2λ, and 0.1λ, respectively. Figure 9.18(a) indicates the Strehl ratio equivalents of these values. The "poor" WFE level corresponds to a Strehl ratio less than 5%. The "marginal" level gives a Strehl ratio of approximately 20%. The "ideal" level gives a Strehl ratio of nearly 70%. With $f_{3dB} = 600$ Hz and $n_{act} / D = 10$, Figures 9.17(a) and 9.18(a) show that performance to be limited both temporally and spatially, with temporal limitations being the largest factor.

Figures 9.17(b) and 9.18(b) show the case of keeping $n_{act} / D = 10$, but moving to higher bandwidth $f_{3dB} = 6$ kHz. While this reduces the temporal residual to the ideal level at large LOS angle, the spatial residual keeps the closed-loop performance near the "poor" level. Figures 9.17(c) and 9.18(c) show the opposite case of keeping the bandwidth at $f_{3dB} = 600$ Hz but increasing actuator density to $n_{act} / D = 20$. This case reduces the spatial residual to ideal levels, but it actually makes the temporal residual worse, as more wavefront disturbance at higher frequency is accessible to the temporal controls of the AO system.

To obtain reasonable compensation performance over the full ranges of LOS angles, both AO bandwidth and actuator density must be increased. Figures

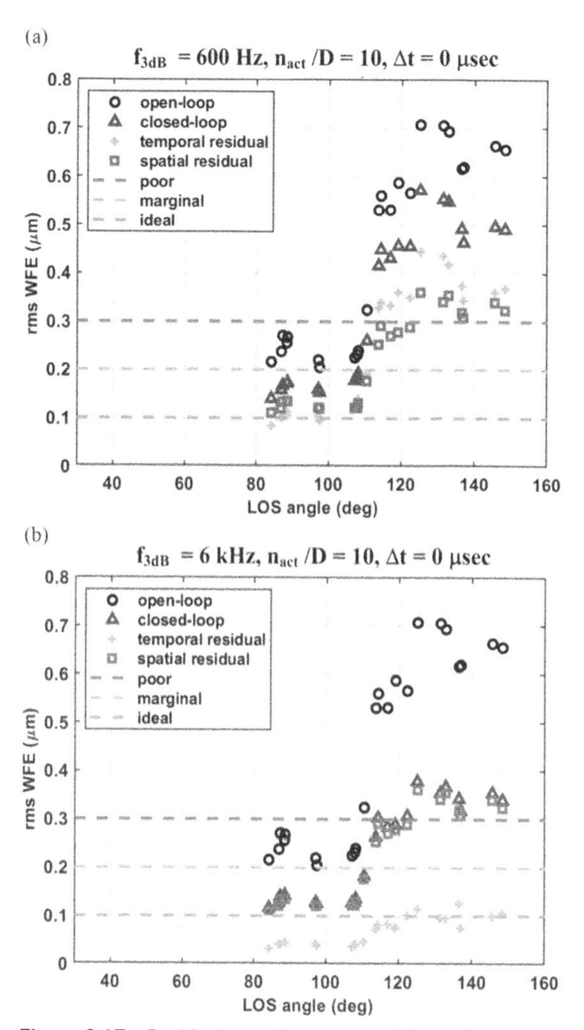

Figure 9.17 Residual wavefront error with varying compensation bandwidth and DM actuator densities: (a) $f_{3dB} = 600$ Hz, $n_{act} / D = 10$, (b) $f_{3dB} = 6$ kHz, $n_{act} / D = 10$.

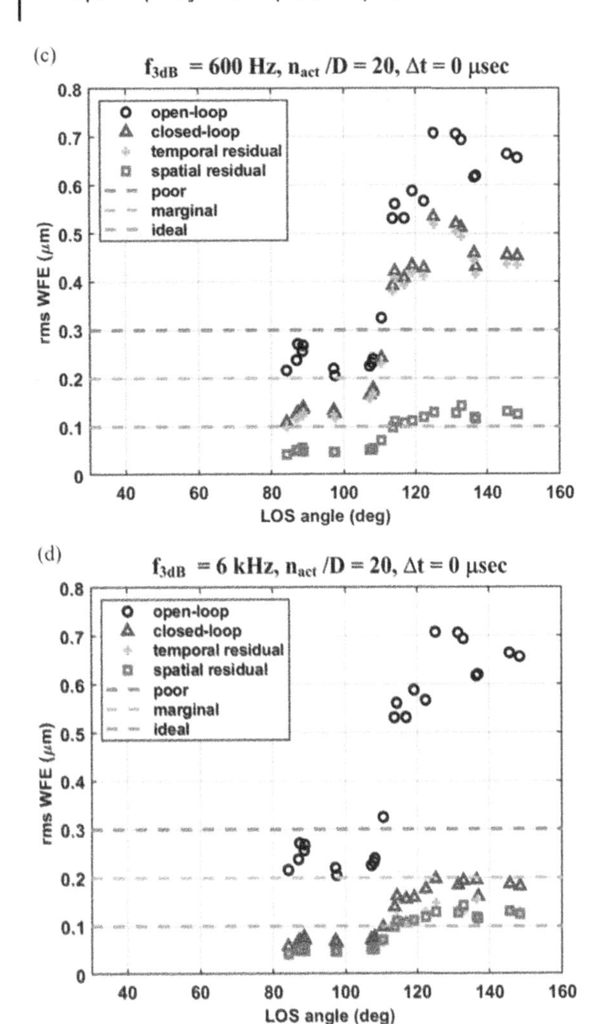

Figure 9.17 (Cont'd) (c) $f_{3dB} = 600$ Hz, $n_{act} / D = 20$ and (d) $f_{3dB} = 6$ kHz, $n_{act} / D = 20$.

9.17(d) and 9.18(d) show the case of $f_{3dB} = 6$ kHz and $n_{act} / D = 20$. In this case, both temporal and spatial residuals are brought down near the ideal levels. However, we note that when the temporal and spatial residuals are combined, the performance at large LOS angles is raised above the "marginal" level, but remains closer to "marginal" than "ideal." Thus, we see that improving the system-level performance with aero-optical disturbances in the path involves a balance between increased actuator density and increased bandwidth.

Considering the conclusions reached earlier for aero-optics compensation scaling, we recognize that with temporal limits going as $f_{3dB}^{-4/3}$ and spatial limits going as $n_{act} / D^{-8/3}$, a larger relative increase in temporal sampling will be required rather than DM actuators. When the operation of the AO system is supported by an illuminator laser and target beacon, this means that the required energy per illuminator pulse will not grow as quickly as the required average power level. The energy per pulse sets the number of photons available per subaperture, but the power level is driven by the rate at which the required photons must but propagated. Providing an illuminator laser and control wavefront sensor that meets these requirements may be challenging. Thus, advanced control methods for AO compensation combined mitigation of aero-optical effects is an approach that is

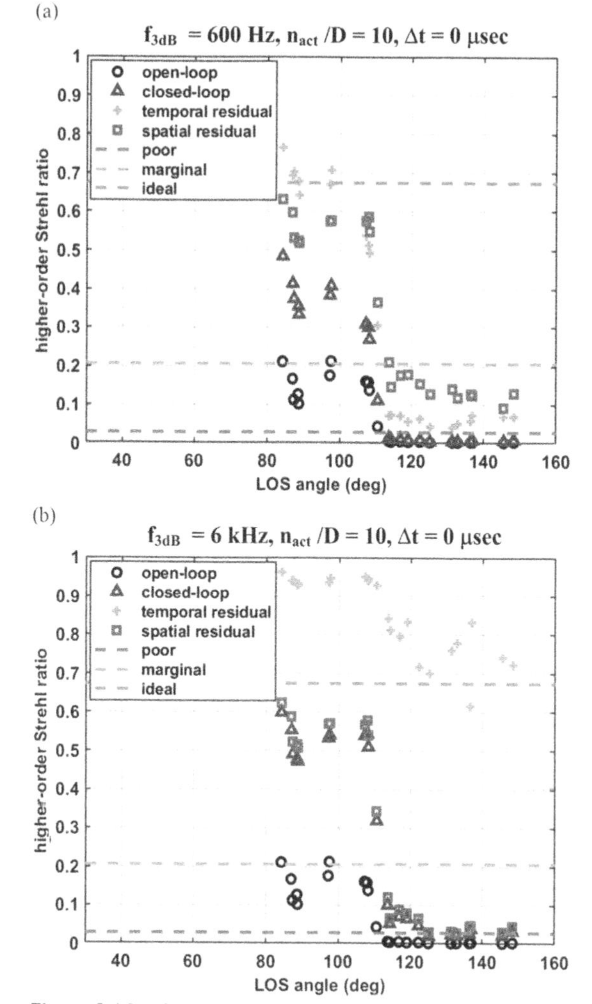

Figure 9.18 Compensated Strehl ratio with varying compensation bandwidth and DM actuator densities: (a) $f_{3dB} = 600$ Hz, $n_{act}/D = 10$, (b) $f_{3dB} = 6$ kHz, $n_{act}/D = 10$.

(c)

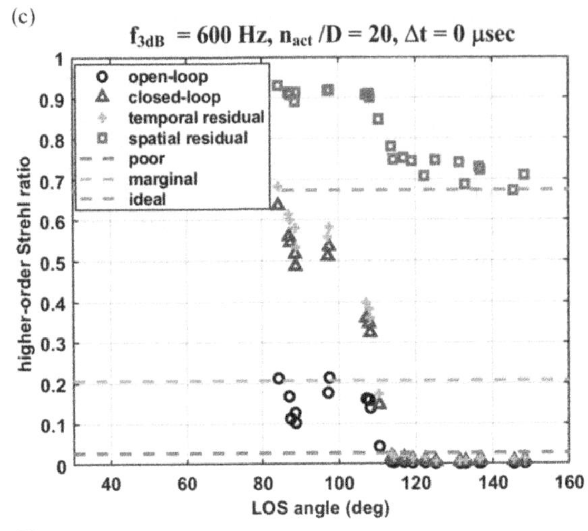

(d)

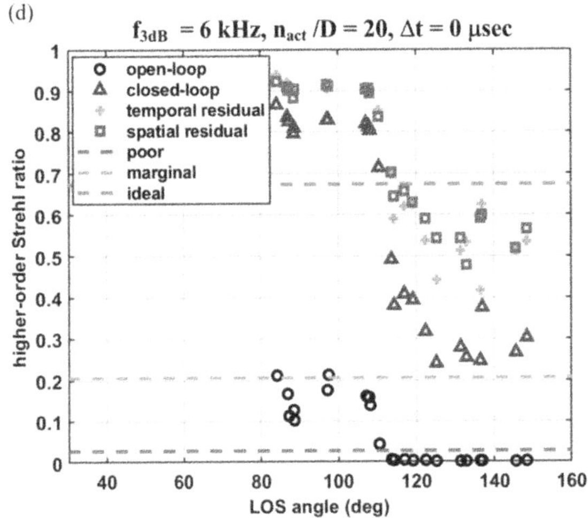

Figure 9.18 (Cont'd) (c) $f_{3dB} = 600$ Hz, $n_{act} / D = 20$ and (d) $f_{3dB} = 6$ kHz, $n_{act} / D = 20$.

required when working with flight regimes resulting in very high levels of aero-optical wavefront disturbance.

9.4.3 Relative Impact of Aero-Optics with Other Propagation and System Effects

We can place the degradations observed with aero-optical disturbances, and the achievable AO compensation results in better context when we consider effects other than aero-optics which also degrade system performance. In this section we will compare aero-optical degradation levels to both free-stream atmospheric turbulence, and system jitter. Since these effects must be addressed apart from the aero-optics problem, this analysis will highlight under what conditions aero-optical disturbances will dominate system performance, and where these other effects should be addressed as well.

9.4.3.1 Comparing Aero-Optics to Free-Stream Turbulence Propagation

By "free-stream" turbulence, we mean atmospheric turbulence that occurs naturally apart from the flow over the aircraft window or turret. Free-stream turbulence is present over all atmospheric paths at all altitudes (Roggemann and Welsh 1996). Near the surface, free-stream turbulence strength is typically driven by differential temperature gradients occurring when solar irradiance interacts with varying surface materials and features. At higher altitudes, free-stream turbulence is driven by wind-speed gradients, especially atmospheric shear layers which mimic the aero-optical shear layers we have discussed in earlier chapters.

Free-stream turbulence propagation effects are governed by the ratio of the aperture diameter D to the Fried's coherence length, r_0 (Fried 1966). For imaging or beam projection to a target at a finite range it is appropriate to calculate the spherical-wave coherence length (Sasiela 2007) as

$$r_0 = \left[0.423\left(\frac{2\pi}{\lambda}\right)^2 \int_0^L C_n^2(h(z))(1-z/L)^{5/3}dz\right]^{-3/5}, \tag{9.26}$$

where $C_n^2(h(z))$ is the refractive-index structure function coefficient at the beam altitude $h(z)$, which is a function of the position z along the path, L is the slant range, and the integral extends from the platform to the target. For targets at long range such as stellar objects, it is appropriate to calculate the plane-wave r_0 as

$$r_0 = \left[0.423\left(\frac{2\pi}{\lambda}\right)^2 \int_0^L C_n^2(h(z))dz\right]^{-3/5},$$

where the integral extends over the portion of the path of length L where turbulence is present.

The tilt-removed or "higher-order" Strehl ratio, SR for a given turbulence condition represents the effect of higher order (focus and above) aberrations on the on-axis intensity for an optical system. This free-stream turbulence propagation effect is the most analogous to the higher-order aero-optics disturbances we have considered here. The tilt-removed Strehl ratio is computed given D/r_0 by direct numerical integration of the turbulence-degraded optical transfer function relative to diffraction limit (Roggemann and Welsh 1996)

$$SR = \frac{\int H_0(\kappa)H_{SE}(\kappa)d\kappa}{\int H_0(\kappa)d\kappa}, \tag{9.27}$$

where the integral is over all two-dimensional spatial frequencies κ and the optical transfer function (OTF) of the diffraction-limited optical system is given by the auto-correlation of the pupil function (Roggemann and Welsh 1996),

$$H_0(\kappa) = \frac{\int w(\kappa'\lambda F) w((\kappa - \kappa')\lambda F) d\kappa'}{\int w(\kappa'\lambda F) w(-\kappa'\lambda F) d\kappa'}. \quad (9.28)$$

In Equation (9.28) $w(\kappa'\lambda F)$ is the pupil function of the aperture, defined in Equation (9.3) (ones inside pupil, zeros elsewhere,) and F is the distance between the pupil and image plane. The so-called short exposure atmospheric optical transfer function $H_{SE}(\kappa)$ in Equation (9.27) is given by

$$H_{SE}(\kappa) = \exp\left\{-\frac{1}{2}6.88\left(\frac{\kappa\lambda F}{r_0}\right)^{5/3}\left[1-\left(\frac{\kappa\lambda F}{D}\right)^{1/3}\right]\right\}.$$

Note that since the cut-off frequency of the diffraction-limited OTF is proportional to D, the governing parameter for turbulence degradation will be D/r_0.

Using the aero-optical characterization look-up tables from AAOL flight test data discussed earlier, we computed the aero-optical Strehl ratio over a range of LOS angles determined by the target azimuth angle for the hypothetical laser system reviewed earlier in the Mach 0.8, 30 kft flight condition. Figure 9.19 shows the aero-optical Strehl ratio for various AO configurations with increasing fidelity in terms of temporal bandwidth and DM actuators. The plot includes annotations for the Strehl degradation resulting from various levels of D/r_0. For LOS angle less than 110 degrees, the aero-optical effect is equivalent to $D/r_0 < 4$, even without compensation. The $D/r_0 < 4$ corresponds to lower turbulence levels where the effect on laser irradiance is above the "marginal" level indicated previously.

Figure 9.19 shows that the most challenging conditions occur at larger LOS angles where the effect is equivalent to free-stream turbulence with $D/r_0 > 10$, falling below the "poor" level indicated previously. As the AO compensation capability increases, the disturbance can be brought to levels similar to $D/r_0 < 6$, where both aero-optical and free-stream turbulence are contributing to system performance degradation. It should be noted that in a practical system, both aero-optical and free-stream turbulence disturbances will be seen in the AO control wavefront sensor and compensated through the same AO system. Thus, any advances made with AO compensation of aero-optical disturbances will also lead to better handling of free-stream turbulence in practical system applications.

9.4.3.2 Comparing Aero-Optics to System Optical Jitter

Optical jitter is a pervasive effect which cannot be ignored in any laser system (Merritt 2012). By "jitter," we mean any effect that results in a linear tilt mode being imparted to the wavefront. The most common sources of jitter in optical

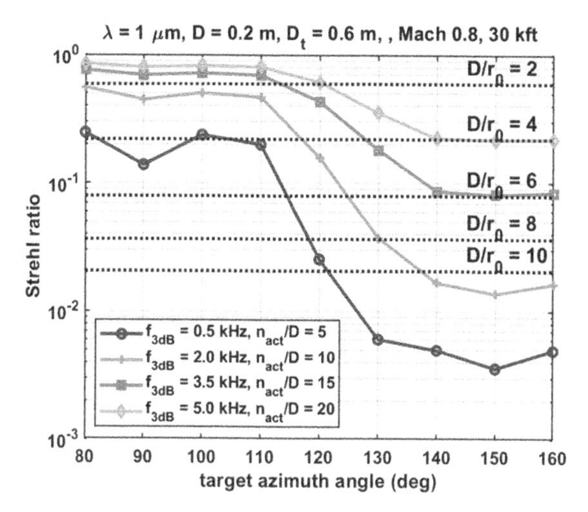

Figure 9.19 Comparison between compensated aero-optical disturbances with free-stream turbulence propagation. Aero-optic scaling is for a Mach 0.8 platform at 30 kft altitude, with turret $D_t = 0.6$ m and aperture diameter $D = 0.2$ m with $\lambda = 1$ μm.

systems are due to mechanical vibrations producing structural dynamic modes along the beam path. Unsteady loading of an aircraft turret also excites these structural modes and will generally increase in magnitude along with aero-optical disturbances. Further, aero-optical and free-stream turbulence also imparts jitter to a laser wavefront when propagating to the target (Whiteley et al. 2013). Optical tilt effects are usually observed to be Gaussian-distributed with zero mean with variance σ_j^2. The standard deviation of these effects is usually taken to be the "jitter" level of the system. The jitter σ_j in units of (λ / D) which results in the Strehl ratio degradation for a Gaussian fit with $\sigma_j = (\sqrt{2} / \pi)(\lambda / D) = 0.45(\lambda / D)$ to the diffraction-limited far-field irradiance for a uniform circular aperture (Merritt 2012). The Strehl ratio degradation due to jitter is computed as

$$SR_j = \frac{1}{1 + \frac{\pi^2}{2}\left[\sigma_j(\lambda / D)\right]^2}.$$

Figure 9.20 shows the same aero-optical Strehl values with AO compensation shown previously in Figure 9.19, but now with annotations for the Strehl ratio equivalents for various levels of system jitter, σ_j. Note that for the hypothetical laser system we are considering, $\lambda / D = 5.0$ μrad. Note that even with $1\lambda / D$ of jitter, system performance is already below the "marginal" value we established in our previous analysis. Even without AO compensation, aero-optical disturbances at LOS angles less than 110 degrees are less of an effect than $\lambda / D = 5.0$ μrad of jitter. It is not uncommon for optical systems to exhibit system jitter of $1 \times \lambda / D$ of

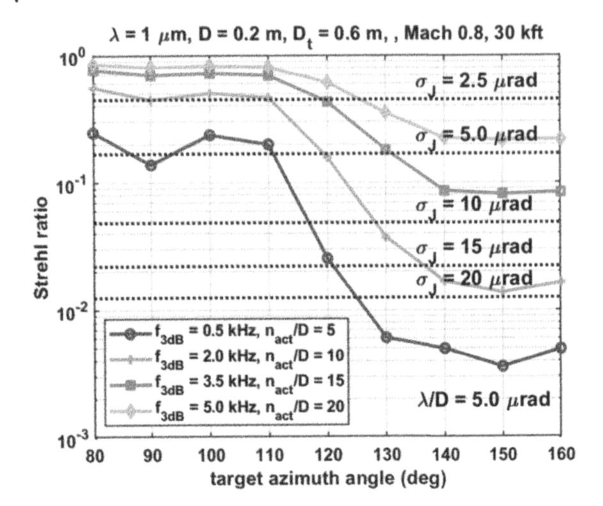

Figure 9.20 Comparison between compensated aero-optical disturbances with the Strehl impact of various levels of system jitter, σ_j. Aero-optic scaling is for a Mach 0.8 platform at 30 kft altitude, with turret diameter $D_t = 0.6$ m and aperture diameter $D = 0.2$ m with $\lambda = 1$ μm.

jitter or greater. Thus, jitter effects can easily dominate aero-optical degradations in limiting system performance. Once effective AO compensation is applied to the system, the aero-optical performance degradations are equivalent to levels of $(1-2)\times\lambda/D$ of jitter. Thus, effective aero-optical compensation is comparable to effective jitter control in its importance for preserving system-level optical performance for a laser system.

9.4.4 Tracker Performance Degradations Related to Aero-Optics

Our discussion of adaptive-optics compensation is not complete without mentioning the challenges of operating a "tracking" system in the presence of aero-optical disturbances in the line-of-sight. A tracker is an imaging system which controls the pointing of a laser to the target (Merritt 2012). In the context of our adaptive optics concept diagram in Figure 8.1, the tracker is equivalent to the position sensing device which controls a fast steering mirror (FSM) in the system. The tracker is an imaging camera which sees the correction of the FSM and any motion of the gimbal to reduce the pointing error relative to a selected aimpoint as much as possible. The tracker is to the FSM as the control wavefront sensor is to the DM in an AO system. Thus, aero-optical degradations in the track image will limit the ability for the tracker to operate effectively. Here, we consider two noteworthy effects on the tracker due to aero-optics: (1) loss of imaging resolution, and (2) breakup of a target illuminator as it propagates through the aero-optical volume and onto the target. The studies shown here have been conducted using AAOL flight test data at Mach 0.4 and using a 90 deg LOS pointing angle (Whiteley and Goorskey 2013). The wavefronts have been scaled to other Mach numbers using our conventional wavefront scaling techniques.

9.4.4.1 Track Sensor Aero-Optical Imaging Resolution Degradation

The contrast-reduction characteristic of aero-optical disturbances on imagery from the aircraft platform can be quantified by considering the modulation transfer function (MTF) of the optical system (Holst 1995) including aero-optical disturbances. The MTF is computed as the modulus of the optical transfer function (OTF) $H(\kappa_x,\kappa_y)$. The OTF is simply the Fourier transform of the point-spread function (PSF), $h(x,y)$ (Roggemann and Welsh 1996):

$$MTF(\kappa_x,\kappa_y) = \left|H(\kappa_x,\kappa_y)\right| = FT\{h(x,y)\}.$$

The MTF for each platform speed considered is shown in Figure 9.21. To present these data, we have radially averaged the MTF in the spatial-frequency plane, and normalized the scalar spatial frequency κ to the cut-off frequency of the optical system, $D/(\lambda F)$. Additionally, each MTF has been normalized to its value at

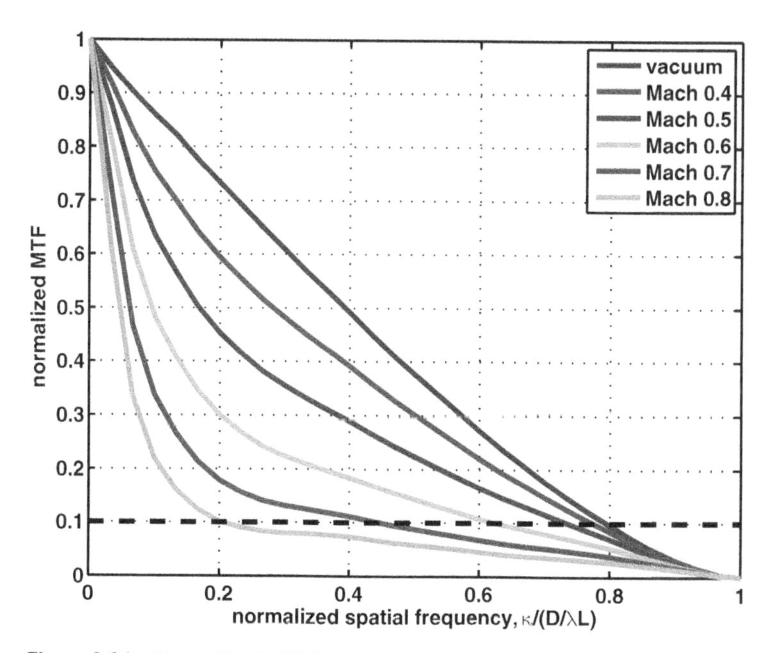

Figure 9.21 Normalized MTF for imaging with aero-optics. *Source:* Whiteley and Goorskey (2013), Figure 8. Reproduced with permission of SPIE.

$(\kappa_x, \kappa_y) = 0$ – which is proportional to the total energy in the PSF. When presented in this manner, we clearly see the influence of aero-optical disturbances with increasing Mach number. As U_∞ increases, we see that the lower spatial frequencies are severely attenuated in the imaging system. Put another way, since the MTF quantifies the contrast of sine waves at each spatial frequency, the reduction of the MTF with increasing Mach number shows that a major effect of aero-optical disturbances on imaging is to reduce the contrast of image details. When we consider an arbitrary limit where the MTF is 10% of its peak (indicated by the dashed line) the spatial frequency at which each MTF line falls below this value may be thought of as the effective reduction in resolution of the system. Given this interpretation, we see that for $M = 0.6$, imaging resolution is approximately 60% of diffraction-limit. The resolution falls below 50% at $M = 0.7$. For $M = 0.8$, the resolution is 20% of diffraction-limited, that is, the smallest object which can be resolved by the target is 5x larger including the aero-optical disturbance than the diffraction-limit.

To illustrate the effect of loss of resolution for an acquisition, tracking, and pointing (ATP) system due to aero-optical disturbances, we used the image of a tank at 10 km range and $D = 30\,\text{cm}$ aperture at $\lambda = 1\,\mu\text{m}$ wavelength, as shown in Figure 9.22. Figure 9.22(a) shows the diffraction-limited (no aero-optics) image of this target. Note the substantial detail in the tank features which is feasible for this target, including individual treads of the tank tracks. Figure 9.22(b) shows the aero-optically degraded image with $M = 0.8$. Here, we note that all of the fine detail in the image has been lost as expected due to the effective resolution loss we noted earlier. However, we can still see some small-scale features in the target, albeit at very low contrast. Figure 9.22(c) shows the same diffraction-limited image as in Figure 9.22(a) but this time with noise added to the image. Noise was added to the noise-free image to give image signal-to-noise ratio, $SNR = 10$. While we see a slight reduction in the clarity/contrast of the image, the fine scale features are still discernible. Figure 9.22(d) shows the aero-optically degraded image with $M = 0.8$ as in Figure 9.22(b), but with $SNR = 10$. Note that in this image, nearly all contrast features are effectively gone from the image. This example illustrates the substantial effect aero-optical disturbances with limited image SNR have on an ATP system using image-based tracking which relies on contrast features in the image. Given the reduction of contrast already imposed by the aero-optical disturbances, high SNRs are required to make any practical use of such imagery.

9.4.4.2 Illuminator Propagation and Active Imaging through Aero-Optics

We have seen that aero-optical disturbances can substantially reduce the resolution capabilities enabling ATP functions for a laser system. We note that often an ATP system does not work passively, but rather relies on laser illumination to provide signal to an imaging sensor. In this case, the illuminator laser is typically sent out the same aperture as is used for imaging. In such an illuminator/imager

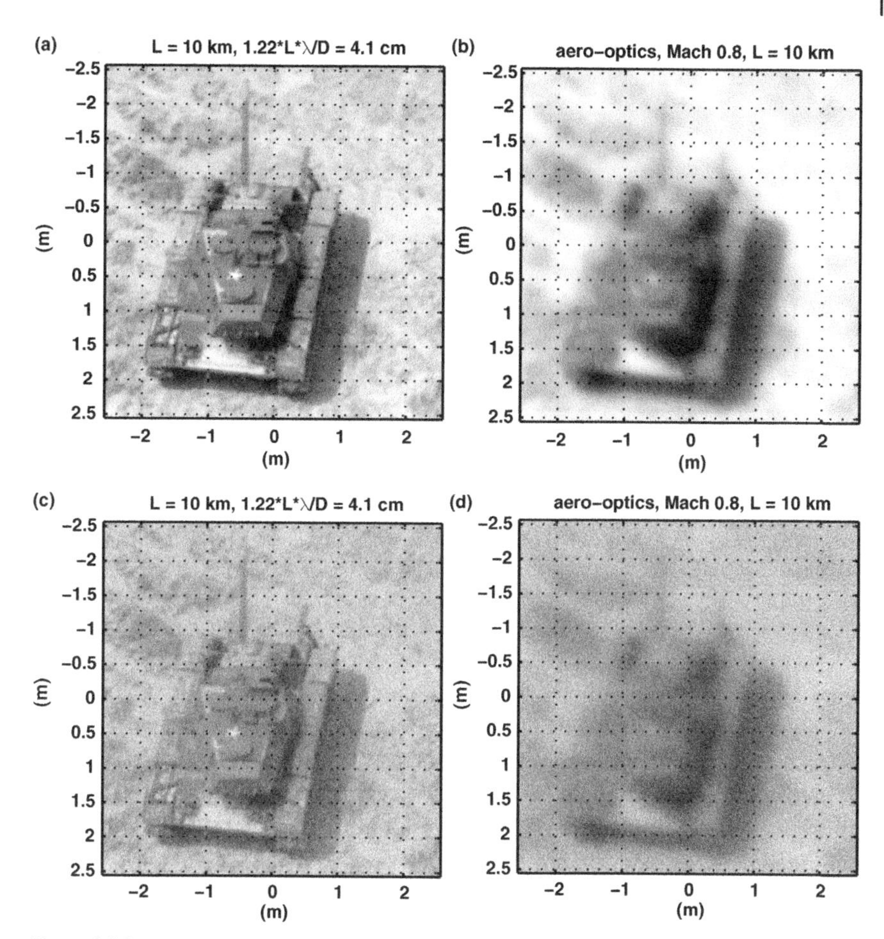

Figure 9.22 Image of a military vehicle at 10 km range. (a) Diffraction-limited image, (no noise). *Source:* Whiteley and Goorskey (2013), figure 9. Reproduced with permission of SPIE (b) Aero-optically degraded image with Mach ~ 0.8 platform speed (no noise). *Source:* Whiteley and Goorskey (2013), figure 9. Reproduced with permission of SPIE. (c) Diffraction-limited image, $SNR = 10$ *Source:* Whiteley and Goorskey (2013), figure 9. Reproduced with permission of SPIE (d) Aero-optically degraded image with Mach 0.8 platform speed, $SNR = 10$. *Source:* Whiteley and Goorskey (2013), Figure 9. Reproduced with permission of SPIE.

"shared aperture" configuration, the illuminator passes through the same aero-optical disturbance as is experienced by light entering the imaging aperture. To illustrate the effect of aero-optical disturbances on active imaging, we considered a single illuminator beam with Gaussian irradiance profile and an exp(−2) beam diameter of 28 cm and diverged to cover an area on the target approximately 1 m in diameter, as shown in Figure 9.23. We have centered this illuminator onto the star feature visible on the rear section of the tank turret, as shown in Figure 9.23(b). The star feature is approximately 25 cm (10 inches) in width.

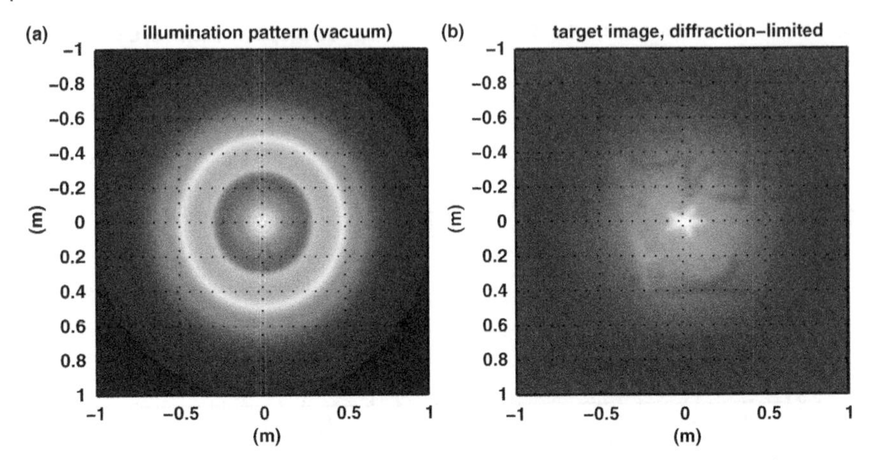

Figure 9.23 (a) Target illumination pattern. *Source:* Whiteley and Goorskey (2013), figure 10. Reproduced with permission of SPIE and (b) diffraction-limited imaging with vacuum propagation. *Source:* Whiteley and Goorskey (2013), Figure 10. Reproduced with permission of SPIE.

The same scaled aero-optical disturbances used to degrade passive imagery were applied to the illuminator laser for the active imaging case. Figure 9.24 (a)–(c) show an instantaneous illuminator pattern on target for $M = 0.4, 0.6, 0.8$, respectively. From these images we see that as the platform speed increases, the illuminator laser experiences increasingly severe beam breakup as it is incident on the target. Figure 9.24(d)–(f) show the illuminated target scene as it would appear back at the platform given illuminator breakup and the degraded resolution of the imaging system. Given the combination of these effects, we see that for $M = 0.6, 0.8$, the star feature is essentially unobservable in the imagery. In the image for $M = 0.6$ what appears as a bright feature similar in size to the star is actually just an intense illumination blob.

Figures 9.24(g)–(i) show the image that would be observed over a sensor integration time of $\tau_{\text{int}} = 1$ ms. The sensor integration time helps in discerning the star feature on the target as it smooths out the changing illuminator break-up pattern on target. When the 1 ms sensor integration time is used, the star feature is once again observable in the imagery with $M = 0.8$. However, we note that when sensor integration is applied, the illuminator breakup will cause a streaking/smearing pattern in the imagery because of the motion of the disturbances in the illuminator projection aperture. These elongated illuminator artifacts also flow within the imagery, and could easily be followed by a contrast-based image tracker instead of the desired target feature. When used to drive a feedback stabilization loop for laser pointing, this aero-optically induced "illumination clutter" may induce additional errors into the tracking process, thereby degrading overall ATP system performance.

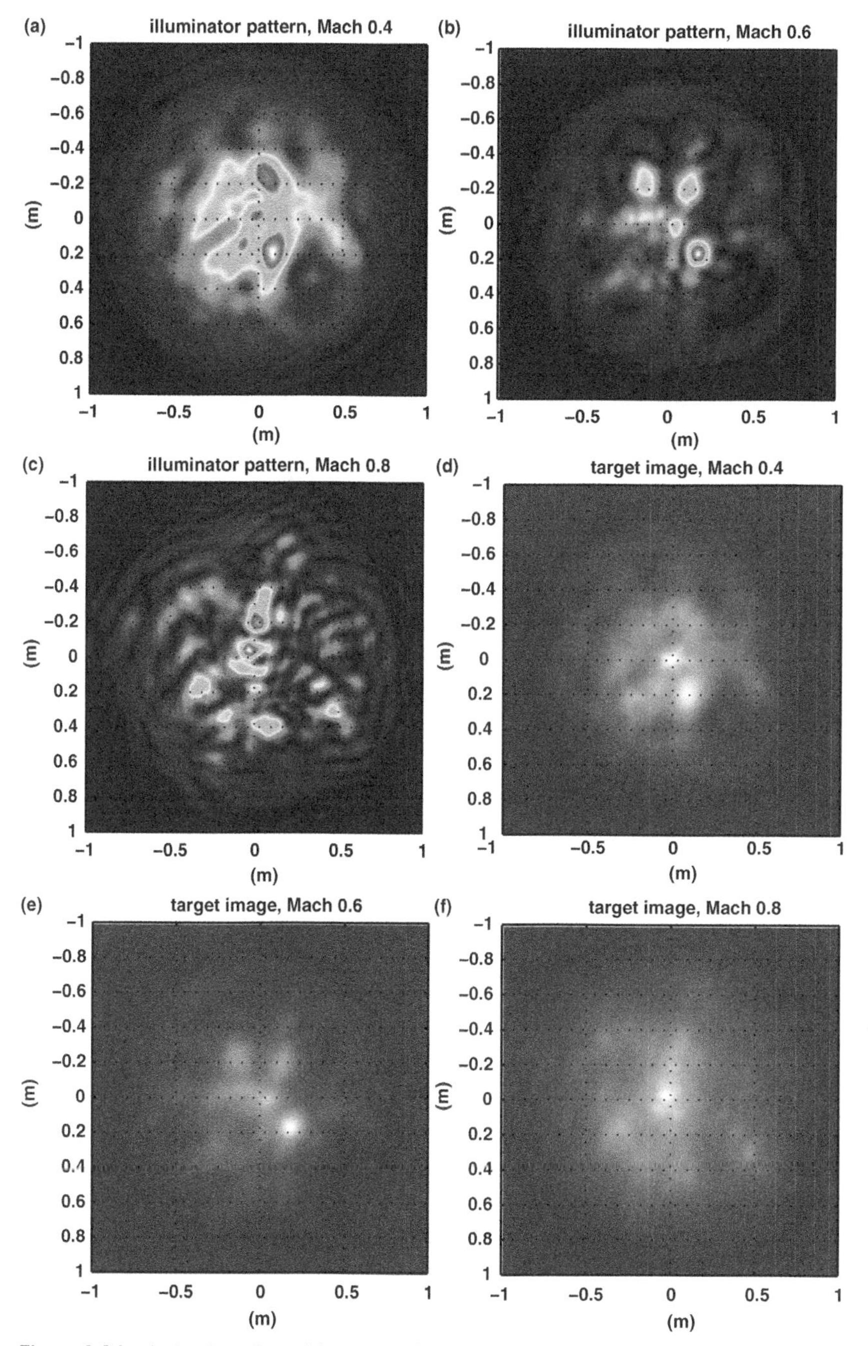

Figure 9.24 Active imaging with aero-optical disturbances for $M = 0.4, 0.6, 0.8$ (a)–(c) Instantaneous illuminator patterns on target. *Source:* Whiteley and Goorskey (2013), figure 11. Reproduced with permission of SPIE (d)–(f) Image at platform including illuminator propagation and image degradation. *Source:* Whiteley and Goorskey (2013), figure 11. Reproduced with permission of SPIE (g)–(i) Images with sensor integration, $\tau_{int} = 1$ ms. *Source:* Whiteley and Goorskey (2013), Figure 11. Reproduced with permission of SPIE.

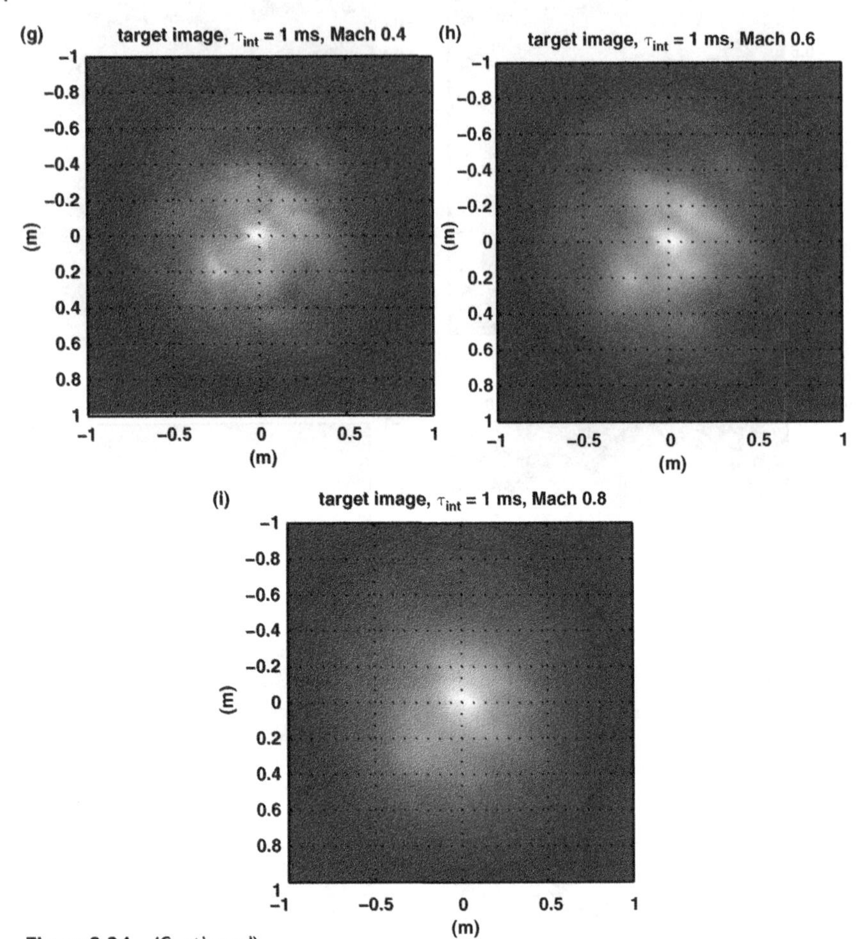

Figure 9.24 (Continued)

10

Concluding Remarks

This chapter really adds no technical content, but rather tries to highlight important developments and discoveries in the field of aero-optics in the last approximately 25 years. This field, while beginning to be considered mature, is still very active, from basic to applied research and development. We will mention some of the current avenues being pursued and then speculate about the possible future directions.

We think the most important achievement in aero-optical research in the last two decades is that aero-optical effects are no longer treated as a "minor penalty" on the performance of airborne systems. This paradigm switch from assuming that aero-optics presented only a small performance hit to the acceptance of their importance on airborne systems was neither fast nor easy. It should be noted that a luminary in the field, George Sutton, published a paper in 1985 (Sutton 1985), which suggested that the field of aero-optics was essentially mature, and that the opportunities for further research were to polish up the techniques already in existence and maybe apply them to a few more flows. Scientific and engineering communities across the globe now recognize that aero-optical effects can and do create significant optical distortions imposed on the outgoing laser beams from airborne optical platforms. For this reason, they are now well aware that aero-optical effects have to be taken into account, along with atmospheric optical effects, when designing and developing airborne laser systems.

Another significant advance in studying aero-optical effects is the instrumentation and the related data reduction techniques. While many wavefront sensors (interferometry and Shack-Hartmann sensors, for example) were available for decades, the advances in digital cameras and related software in the last decade opened up the possibility of collecting time-series of wavefront data with large, up to a few MHz, sampling rates. The availability of these time-resolved sequences led to many Fourier-based data reduction approaches and analysis techniques; these include spectral, cross-spectral and dispersion analyses. These methods were also

Aero-Optical Effects: Physics, Analysis and Mitigation, First Edition. Stanislav Gordeyev, Eric J. Jumper, and Matthew R. Whiteley.

particularly useful in identifying and removing a wide range of corrupting effects on collected data, from mechanical vibrations to acoustic contamination. Combined with better guidelines on how to properly set up optical benches, the advances in instrumentation and data analysis allow present and future researchers to collect useful spatiotemporally resolved wavefront datasets in many different tunnel and in-flight experiments. As a result, many characteristics of aero-optical effects around important applied geometries, like turrets, have been quantified over a range of subsonic and transonic speeds in realistic flight environments. The multitude of available experimental data helped in validating numerical simulations in attempts to predict aero-optical distortions. The data also provided crucial information about spatial and temporal characteristics of flight-relevant aero-optical distortions, leading to the development of several new approaches and algorithms for adaptive optics systems to improve beam optical quality.

Interestingly enough, it was not the availability of the high-speed digital cameras, which led to many major discoveries in the physics of the aero-optical distortions, but the development of simple, yet very effective analog-based wavefront sensors, like the Malley probe and the SABT sensor, at the turn of the century. As we discussed in Chapter 6, the collection of time-resolved aero-optical distortions in a high-Reynolds-number transonic shear layer using the SABT sensor in the mid-1990s was a pivotal moment in aero-optical research. The experimental results clearly contradicted the accepted theories, which were significantly underpredicting the shear-layer optical distortions. Just to emphasize the misconceptions that formed the basis of the earlier theories, aero-optical aberrations were thought to be due to relatively small spatial scale velocity fluctuations in the shear layer; this misconception also led to the idea that hot wires could be used to estimate the optical aberrations. The SABT collected wavefront data ultimately led to the identification of the large-scale structures and the associated pressure wells as the dominant physical mechanism of the aero-optical distortions in the shear-layer dominated flows, aided by the development of the Weakly Compressible Model. It took years for the AEDC results and conclusions to be fully accepted by the airborne-laser community. An even simpler wavefront sensor, the Malley probe, was used to collect the first time-resolved wavefronts in subsonic turbulent boundary layers in 2002. These data were used to develop a physics-based, semi-empirical model for aero-optical distortions, which has now been demonstrated to correctly predict aero-optical distortions by turbulent boundary layers over a wide range of Mach numbers from subsonic to hypersonic Mach numbers as high as six. The Malley-probe was also used to investigate non-adiabatic boundary layers and identified several issues in the underlying Extended Strong Reynolds Analogy, widely used to model non-adiabatic boundary layer flows. Currently, the analog position-sensing devices used in Malley probes have been replaced with digital cameras, but the data reduction techniques, developed to analyze the time-resolved wavefront data from earlier analog sensors, are still widely employed to post-process wavefronts collected by digital cameras.

As we demonstrated throughout the book, aero-optical effects also provide valuable information about the underlying physics of turbulent flows. Combined with the non-intrusive nature of optical measurements, wavefront sensors are becoming a standard tool of the instrumentation suite used to conduct fundamental research of the turbulent flows. It is the study of the fundamental flows, which led to a better understanding of the physical mechanisms behind aero-optical effects, helped in developing physics-based models and many useful scaling laws of aero-optical effects for different speed regimes and flow types. For several fundamental flows, like turbulent shear layers and boundary layers, the scaling laws have been demonstrated to collapse essentially all the available experimental data. More importantly, these scaling laws are essential to properly rescale aero-optical effects from subscale models tested in wind tunnels, to estimate the performance of larger airborne systems.

Even with the tremendous progress in aero-optical research in the last two decades, we still have many unsolved problems. With the exception of turbulent boundary layers, all other developed models of aero-optical distortions are limited to subsonic and transonic speeds. Considering the current push into supersonic and especially hypersonic regimes, with strong compressibility and heat transfer effects, these models should be either significantly adjusted or new models should be developed. In order to do so, wavefronts should be properly measured in these regimes. As advanced as the new instruments are, these higher-speed flows now stress the fastest digital cameras; thus, new developments in optical set-ups and related instrumentation are definitely needed. With the older cameras used for aero-optical research we were forced to use pulsed lasers and phase-locking techniques to freeze the moving aberrations until the new cameras were developed that not only had high frame rates, but had sub-microsecond exposure capabilities. With the now high supersonic and hypersonic speeds, presently available exposure times are too long. Again, we are having to resort to using pulsed lasers.

Although we are now armed with copious knowledge of aero-optical effects and their measurement, investigating and modeling aero-optical effects is only the first step in applying this knowledge to mitigating their deleterious effects. Furthermore, the field of aero-optics must now be applied to real systems. Since aero-optics is a multi-disciplinary field at the cross-section of fluid mechanics, controls, and optics, design of mitigating techniques in applied and exploratory research offers a variety of possible mitigation approaches. On the fluid mechanics side, many well-researched flow control approaches have and must continue to be explored to reduce the turbulence or break up the coherent structures, which are largely responsible for the significant aero-optical effects. From the optical side, adaptive optics systems provide a means of canceling the resulting optical effects by introducing conjugate or "anti"-distortions into the laser beam. However, this intersection also provides a third option of mitigating aero-optical effects: a hybrid approach. Here both flow control and adaptive control are implemented together to achieve a larger mitigation effect. One example of such an approach was provided in Chapter 8, where flow control was used not to break up the large-scale structures in a shear layer, but rather to enhance them and make them more regular. This flow regularity provides predictability of the underlying aero-optical effects, to remove these predictable aberrations by a simple fast feed-forward adaptive optics system. The important point of the demonstrated hybrid approach is the statement "feed-forward." In the Chapter 8, several approaches to feeding forward the correction so that the correction coincided with the aberration at the instant it is applied were mentioned. To our knowledge, on the hybrid approach just mentioned and the simple forward shift of the measured and constructed conjugate by the convection velocity times the time delay has been experimentally demonstrated. This leaves the others mentioned in Chapter 8 and new approaches to be developed and experimentally demonstrated.

In the end, let us reemphasize some comments made on the first sentence of Chapter 1: "A title that we considered for this book was Modern Aero-Optics, but because we believe that the book will become a quintessential reference far into the future, 'modern' will quickly lose its significance." Indeed, some progress and discoveries made in the last few years that it took to assemble this book are not contained in it. Thus, there is much that is now known, discovered in wind tunnels and flight tests, and the analysis and theory developed to understand these, are not contained. In particular, we have moved into fully supersonic and even hypersonic regimes and adapted and developed new measurement techniques suited for many areas of fluid mechanics that we had not anticipated when we started this book. While we believe this book is a first essential "modern" aero-optics collection of information that should be on the shelf of anyone working in this area now and in the future, it is not the only book that should be written and added to the shelf as time marches on.

The authors thank several people. First, in general, we thank all the many graduate students who helped to discover the material in this book; most of these former graduate students are now practicing PhDs in the laser field in academia, government laboratories, and industry. We would especially thank Dr. Shaddy Abado for direct help in performing simulations and contributing to writing Chapter 8. We also thank Dr. Matthew Kalensky for providing many useful comments and suggestions on how to improve the book, as well as for many fruitful discussions on aero-optics topics. Finally, we would also like to thank Timothy Bukowski and Matthew Orcutt, both graduate students in the Aero-Optics Group at the University of Notre Dame, for proofreading the manuscript and providing additional comments.

References

Abado, S., Gordeyev, S., and Jumper, E. (2013). Approach for two-dimensional velocity mapping. *Optical Engineering* 52 (7): 071402. doi: 10.1117/1.OE.52.7.071402.

Abramowitz, M. and Stegun, I.A. (eds.) (1972). *Handbook of Mathematical Functions with Formulas, Graphs, and Mathematical Tables*, 9th printing. New York: Dover. 928.

Achenbach, E. (1972). Experiments on the flow past spheres at very high Reynolds numbers. *Journal of Fluid Mechanics* 54 (3): 565–575.

Adrian, R.J. (1977). On the role of conditional averages in turbulent theory. In: *Turbulence in Liquids: Proceedings of the 4th Biennial Symposium on Turbulence in Liquids* (ed. G. Patteson and J. Zakin). Princeton: Science Press.

Adrian, R.J. (2007). Hairpin vortex organization in wall turbulence. *Physics of Fluids* 19 (4): 1–16.

Adrian, R.J., Meinhart, C.D., and Tomkins, C.D. (2000). Vortex organization in the outer region of the turbulent boundary layer. *Journal of Fluid Mechanics* 422: 1–54.

Andrews, L.C., Phillips, R.L., and Hopen, C.Y. (2001). *Laser Beam Scintillation with Applications*. SPIE Press.

Ashkenas, H.I. and Bryson, A.E. (1951). Design and performance of a simple interferometer for wind-tunnel measurements. *Journal of the Aeronautical Sciences* 18 (2): 82–90.

Baba, N., Tomita, H., and Miura, N. (1994). Iterative reconstruction in phase-diversity imaging. *Applied Optics* 33 (20): 4428–4433.

Baker, C.J. (1979). The Laminar Horseshoe Vortex. *Journal of Fluid Mechanics* 95: 347–367.

Band, O. and Ben-Yosef, N. (1994). Number of correcting mirrors versus the number of measured points in adaptive optics. *Optical Engineering* 33 (2): 466–472. doi: 10.1117/12.152199.

Aero-Optical Effects: Physics, Analysis and Mitigation, First Edition. Stanislav Gordeyev, Eric J. Jumper, and Matthew R. Whiteley.

Barre, S., Dupont, P., and Dussauge, J.P. (1997). Estimates of convection velocity of large turbulent structures in supersonic mixing layers. *Aerospace Science and Technology* 1 (5): 355–366.

Barrell, H. and Sears, J.E. (1939). The refraction and dispersion of air for the visible spectrum. *Philosophical Transactions of the Royal Society of London. Series A, Mathematical and Physical Sciences* 238 (786): 1–64.

Bell, J.H., Schairer, E.T., Hand, L.A., and Mehta, R.D. (2001). Surface pressure measurements using luminescent coatings. *Annual Review of Fluid Mechanics* 33: 155–206.

Bendat, J.S. and Piersol, A.G. (2010). *Random Data*. Wiley.

Berkooz, G., Holmes, P., and Lumley, J.L. (1993). The proper orthogonal decomposition in the analysis of turbulent flows. *Annual Review of Fluid Mechanics* 25: 539–575.

Blanchard, P., Fisher, D., Woods, S.C., and Greenaway, A.H. (2000). Phase-diversity wave-front sensing with a distorted diffraction grating. *Applied Optics* 39 (35): 6649–6655.

Bloembergen, N. et. al. (1987). Report to the American Physical Society of the study group on science and technology of directed energy weapons. *Reviews of Modern Physics* 59 (3): Part II, Chapter 3.

Born, M. and Wolf, E. (1999). *Principles of Optics: Electromagnetic Theory of Propagation, Interference, and Diffraction of Light*, 7e. Cambridge, England: Cambridge University Press.

Brennan, T.J., and Wittich, D.J. (2013). Statistical analysis of airborne aero-optical laboratory optical wavefront measurements. *Optical Engineering* 52 (7): 071416.

Brown, G.L. and Roshko, A. (1974). On density effects and large structure in turbulent mixing layers. *Journal of Fluid Mechanics* 64: 775–816.

Burns, W.R. (2016). Statistical learning methods for aero-optic wavefront prediction and adaptive-optic latency compensation. *PhD Thesis*, University of Notre Dame, Notre Dame, IN.

Burns, W.R., Jumper, E.J., and Gordeyev, S. (2015). A latency-tolerant architecture for airborne adaptive optic systems. *53rd AIAA Aerospace Sciences Meeting, AIAA Paper 2015-0679*.

Burns, W.R., Jumper, E., and Gordeyev, S. (2016). A robust modification of a predictive adaptive-optic control method for aero-optics. *AIAA Paper 2016-3529*.

Chen, K.K., Tu, J.H., and Rowley, C.W. (2012). Variants of dynamic mode decomposition: boundary condition, Koopman, and Fourier analyses. *Journal of Nonlinear Science* 22: 887–915.

Crafton, J., Forlines, A., Palluconi, S., Hsu, K.-Y., Carter, C., and Gruber, M. (2015). Investigation of transverse jet injections in a supersonic crossflow using fast-responding pressure-sensitive paint. *Experiments in Fluids* 56: 27.

Crahan, G., Rennie, M., Rapagnani, L., Jumper, E.J., and Gogineni, S. (2012). Aerodynamic shaping of spherical turrets to mitigate aero-optic effects. *AIAA-2012-0624*.

Dai, G. and Mahajan, V.N. (2007). Zernike annular polynomials and atmospheric turbulence. *Journal of the Optical Society of America A* 24 (1): 139–155.

De Lucca, N., Gordeyev, S., and Jumper, E. (2013). In-flight aero-optics of turrets. *Optical Engineering* 52: 071405.

De Lucca, N., Gordeyev, S., Jumper, E., and Wittich, D.J. (2018). Effects of engine acoustic waves on optical environment around turrets in-flight on AAOL-T. *Optical Engineering* 57 (6): 064107.

De Lucca, N., Gordeyev, S., and Jumper, E.J. (2012). The study of aero-optical and mechanical jitter for flat window turrets. *AIAA Paper 2012-0623*.

De Lucca, N., Gordeyev, S., Morrida, J., Jumper, E.J., and Wittich, D.J. (2018b). Modal analysis of the surface pressure field around a hemispherical turret using pressure sensitive paint. *AIAA Paper 2018-0932*.

De Lucca, N., Gordeyev, S., Smith, A.E., Jumper, E.J., Whiteley, M., and Neale, T. (2014). The removal of tunnel vibration induced corruption in aero-optical measurements. *AIAA Paper 2014-2494*.

Deron, R., Tromeur, E., Aupoix, B., and Desse, J.M. (2002). *Rapport d'Activités 2001 du Project de Recherche Fédérateur Effets Aéro-Optiques*. No. RT 4/06008 DOTA, ONERA, France.

Dimotakis, P.E. (1986). Two-dimensional shear-layer entrainment. *AIAA Journal* 24 (11): 1791–1796.

Drazin, P.G. and Reid, W.H. (1981). *Hydrodynamic Stability*. Cambridge, UK: Cambridge University Press.

Duan, L., Beekman, I., and Martin, M.P. (2010). Direct numerical simulation of hypersonic turbulent boundary layers. Part 2. Effect of wall temperature. *Journal of Fluid Mechanics* 655: 419–445.

Duffin, D. (2009). Feed-forward adaptive-optic correction of a weakly-compressible high-subsonic shear layer. *PhD Thesis*, University of Notre Dame, Notre Dame, IN.

Duffner, R. (1997). *Airborne LASER, Bullets of Light*. New York: Plenum Publishers.

Ellerbroek, B.L. (1994). First-order performance evaluation of adaptive-optics systems for atmospheric turbulence compensation in extended-field-of-view astronomical telescopes. *Journal of the Optical Society of America A* 11: 783–805.

Faghihi, A., Tesch, J., and Gibson, S. (2013). Identified state-space prediction model for aero-optical wavefronts. *Optical Engineering* 52 (7): 071419.

Fitzgerald, E.J. and Jumper, E.J. (2002). Aperture effects on the aerooptical distortions produced by a compressible shear layer. *AIAA Journal* 40 (2): 267–275.

Fitzgerald, E.J. and Jumper, E.J. (2002b). Scaling aerooptic aberrations produced by high-subsonic-Mach shear layers. *AIAA Journal* 40 (7): 1373–1381.

Fitzgerald, E.J. and Jumper, E.J. (2004). The optical distortion mechanism in a nearly incompressible free shear layer. *Journal of Fluid Mechanics* 512: 153–189.

Fried, D.L. (1967). Propagation of a spherical wave in a turbulent medium. *Journal of the Optical Society of America* 57 (2): 175–180.

Fried, D. L. (1966). Optical resolution through a randomly inhomogeneous medium for very long and very short exposures. *Journal of the Optical Society of America* 56 (10): 1372–1379.

Fried, D.L. (1977). Least-square fitting a wavefront distortion estimate to an array of phase-difference measurements. *Journal of the Optical Society of America* 67 (3): 370–375.

Fuhs, A.E. and Fuhs, S.E. (1982). Optical phase distortion due to compressible flow over laser turrets. In: *Aero-Optical Phenomena*, 80 (ed. K.G. Gilbert and L.J. Otten), 101–138. New York: Progress in Astronautics and Aeronautics, AIAA.

Gardiner, W.C., Jr., Hidaka, Y., and Tanzawa, T. (1980). Refractivity of combustion gases. *Combustion and Flame* 40: 213–219.

Gaviglio, J. (1987). Reynolds analogies and experimental study of heat transfer in the supersonic boundary layer. *International Journal of Heat and Mass Transfer* 30 (5): 911–926.

Geary, J.M. (1995). *Introduction to Wavefront Sensors*. SPIE Digital Library.

Gilbert, K.G. (1982). KC-135 aero-optical boundary-layer/shear-layer experiments. In: *Progress in Astronautics and Aeronautics: Aero-Optical Phenomena*, 80 (ed. K. Gilbert and L.J. Otten), 306–324. New York: AIAA.

Gilbert, K.G. (2013). The challenge of high brightness laser systems: a photon odyssey. *Optical Engineering* 52 (7): 071412.

Gilbert, K.G. and Otten, L.J. (eds.) (1982). *Aero-Optical Phenomena*. Prog. Astronaut. Aeronaut., 80, 412 pp. New York: AIAA.

Gladstone, J.H. and Dale, T.P. (1863). Researches on the refraction, dispersion, and sensitiveness of liquids. *Philosophical Transactions of the Royal Society of London* 153: 317–343.

Gonsalves, R.A. (1982). Phase retrieval and diversity in adaptive optics. *Optical Engineering* 21 (5): 215829.

Goodman, J.W. (1996). *Introduction to Fourier Optics*, 2e. The McGraw-Hill Companies, Inc.

Goodman, J.W. (2015). *Statistical Optics*, 2e. New York: Wiley.

Goorskey, D.J., Drye, R., and Whiteley, M.R. (2013). Dynamic modal analysis of transonic airborne aero-optics laboratory conformal window flight-test aero-optics. *Optical Engineering* 52 (7): 071414.

Goorskey, D.J., Schmidt, J., and Whiteley, M.R. (2013b). Efficacy of predictive wavefront control for compensating aero-optical aberrations. *Optical Engineering* 52 (7): 071418.

Gordeyev, S., Cress, J., and Jumper, E. (2013). Far-field laser intensity drop-outs caused by turbulent boundary layers. *Journal of Directed Energy* 5 (1): 58–75.

Gordeyev, S., Cress, J.A., Jumper, E., and Cain, A.B. (2011). Aero -optical environment around a cylindrical turret with a flat window. *AIAA Journal* 49 (2): 308–315.

Gordeyev, S., Cress, J.A., Smith, A.E., and Jumper, E.J. (2015). Aero-optical measurements in a subsonic, turbulent boundary layer with non-adiabatic walls. *Physics of Fluids* 27: 045110.

Gordeyev, S., De Lucca, N., Jumper, E., Hird, K., Juliano, T.J., Gregory, J.W., Thordahl, J., and Wittich, D.J. (2014b). Comparison of unsteady pressure fields on turrets with different surface features using pressure sensitive paint. *Experiments in Fluids* 55: 1661.

Gordeyev, S., De Lucca, N., Morrida, J., Jumper, E.J., and Wittich, D.J. (2018). Conditional studies of the wake dynamics of Hemispherical Turret using PSP. *AIAA Paper 2018-2048.*

Gordeyev, S. and Juliano, T.J. (2016). Optical characterization of nozzle-wall Mach-6 boundary layers. *AIAA Paper 2016-1586.*

Gordeyev, S. and Jumper, E. (2010). Fluid dynamics and aero-optics of turrets. *Progress in Aerospace Sciences* 46: 388–400.

Jumper, E.J., and Gordeyev, S. (2017). Physics and measurement of aero-optical effects: past and present. *Annual Review of Fluid Mechanics* 49: 419–441.

Gordeyev, S., Jumper, E., Ng, T., and Cain, A. (2003). Aero-optical characteristics of compressible, subsonic turbulent boundary layer. *AIAA Paper 2003-3606.*

Gordeyev, S., Jumper, E.J., and Hayden, T. (2012). Aero-optical effects of supersonic boundary layers. *AIAA Journal* 50 (3): 682–690.

Gordeyev, S. and Kalensky, M. (2020). Effects of engine acoustic waves on aero-optical environment in subsonic flight. *AIAA Journal* 58 (12): 5306–5317.

Gordeyev, S., Post, M., MacLaughlin, T., Ceniceros, J., and Jumper, E. (2007). Aero-optical environment around a conformal-window turret. *AIAA Journal* 45 (7): 1514–1524.

Gordeyev, S., Rennie, R.M., Cain, A.B., and Hayden, T. (2015a). Aero-optical measurements of high-Mach supersonic boundary layers. *AIAA Paper 2015-3246.*

Gordeyev, S. and Smith, A.E. (2016). Studies of the large-scale structure in turbulent boundary layers using simultaneous velocity-wavefront measurements. *AIAA Paper 2016-3804.*

Gordeyev, S., Smith, A.E., Cress, J.A., and Jumper, E.J. (2014). Experimental studies of aero-optical properties of subsonic turbulent boundary layers. *Journal of Fluid Mechanics* 740: 214–253.

Gordeyev, S., Smith, A.E., Saxton-Fox, T., and McKeon, B. (2015b). *Studies of the large-scale structure in adiabatic and moderately-wall-heated subsonic boundary layers*. TSFP-9, Melbourne, Australia, Paper 7A-3.

Greenwood, D.P. (1977). Bandwidth specification for adaptive optics systems. *Journal of the Optical Society of America* 67 (3): 390–393.

Gregory, J.W., Asai, K., Kameda, M., Liu, T., and Sullivan, J.P. (2008). A review of pressure-sensitive paint for high-speed and unsteady aerodynamics. *Proceedings of the Institution of Mechanical Engineers, Part G: Journal of Aerospace Engineering* 222: 249–290.

Gregory, J.W., Sakaue, H., Liu, T., and Sullivan, J.P. (2014). Fast pressure-sensitive paints for flow and acoustic diagnostics. *Annual Review of Fluid Mechanics* 46: 303–330.

Guarini, S.E., Moser, R.D., Shariff, K., and Wray, A. (2000). Direct numerical simulations of a supersonic turbulent boundary layer at Mach 2.5. *Journal of Fluid Mechanics* 414: 1–33.

Gureyev, T.E., Roberts, A., and Nugent, K.A. (1995). Phase retrieval with the transport of intensity equation: matrix solution with use of Zernike polynomials. *Journal of the Optical Society of America A* 12 (9): 1932–1941.

Harvey, J.E. and Hooker, R.B. (2005). *Robert Shannon and Roland Shack: Legends in Applied Optics*. SPIE Press.

Havener, G. and Heltsley, F. (1994). Design aspects and preliminary holographic-PIV measurements for a subsonic free shear layer flow channel. *AIAA Paper 94-2550*.

Hayashi, T. and Sakaue, H. (2017). Dynamic and steady characteristics of polymer-ceramic pressure-sensitive paint with variation in layer thickness. *Sensors* 17: 1125.

Hayashi, T. and Sakaue, H. (2020). Temperature effects on polymer-ceramic pressure-sensitive paint as a luminescent pressure sensor. *Aerospace* 7 (6): 80.

Holmes, P., Lumley, J.L., and Berkooz, G. (1996). *Turbulence, Coherent Structures, Dynamical Systems and Symmetry*. Cambridge University Press.

Holst, G.C. (1995) *Electro-Optical Imaging System Performance*. Bellingham, WA: SPIE Optical Engineering Press.

Huang, S., Liu, F., and Jiang, Z. (2012). Phase retrieval on annular and annular sector pupils by using the eigen function method to solve the transport of intensity equation. *Journal of the Optical Society of America A* 29 (4): 513–520.

Hudgin, R.H. (1977). Wavefront reconstruction for compensated imaging. *Journal of the Optical Society of America* 67 (3): 375–378.

Huerre, P. and Rossi, M. (1998). Hydrodynamic instabilities in open flows. In: *Hydrodynamics and Nonlinear Instabilities* (ed. C. Goldr`eche and P. Manneville), 81–294. Cambridge University Press.

Hugo, R.J. and Jumper, E.J. (2000). Applicability of the aero-optic linking equation to a highly coherent, transitional shear layer. *Applied Optics* 39 (24): 4392–4401.

Hugo, R.L., Jumper, E.J., Havener, G., and Stepanek, C. (1997). Time-resolved wave front measurements through a compressible free shear layer. *AIAA Journal* 35 (4): 671–677.

Hutchins, N., Hambleton, W.T., and Marusic, I. (2005). Inclined cross-stream stereo particle image velocimetry measurements in turbulent boundary layers. *Journal of Fluid Mechanics* 541: 21–34.

Jumper, E. and Hugo, R. (1992). Optical phase distortion due to turbulent-fluid density fields – Quantification using the small-aperture beam technique. *AIAA Paper 92-3020*.

Jumper, E., Zenk, M., Gordeyev, S., Cavalieri, D., and Whiteley, M. (2013). Airborne aero-optics laboratory. *Optical Engineering* 52 (7): 071408.

Jumper, E.J. and Fitzgerald, E.J. (2001). Recent advances in aero-optics. *Progress in Aerospace Sciences* 37: 299–339.

Jumper, E.J. and Hugo, R.J. (1995). Quantification of aero-optical phase distortion using the small-aperture beam technique. *AIAA Journal* 33 (11): 2151–2157.

Kalensky, M., Catron, B., Gordeyev, S., Jumper, E.J., and Kemnetz, M. (2021). Investigation of aero-mechanical jitter on a hemispherical turret. Proc. SPIE 11836, *Unconventional Imaging and Adaptive Optics 2021*, 1183606.

Kalensky, M., Gordeyev, S., and Jumper, E.J. (2019). In-Flight studies of Aero-Optical Distortions Around AAOL-BC. *AIAA Paper 2019-3253.*

Kamel, M., Wang, K., and Wang, M. (2016). Predictions of aero-optical distortions using LES with wall modeling. *AIAA Pap. 2016-1462.*

Karr, T.J. (1991). Temporal response of atmospheric turbulence compensation. *Applied Optics* 30 (4): 363–364.

Kaushal, H., Jain, V.K., and Kar, S. (2017). *Free Space Optical Communication.* Springer (India) Pvt. Ltd.

Kemnetz, M.R. (2019). *Analysis of the aero-optical component of the jitter using the stitching method.* PhD Dissertation, University of Notre Dame.

Kemnetz, M.R. and Gordeyev, S. (2016). Optical investigation of large-scale boundary-layer structures. *AIAA Paper 2016-1460.*

Kemnetz, M.R. and Gordeyev, S. (2017). Multiple aperture approach to wavefront prediction for adaptive-optic applications. *AIAA Paper 2017-1344.*

Kemnetz, M.R. and Gordeyev, S. (2022). Analysis of aero-optical jitter in convective turbulent flows using stitching method. *AIAA Journal* 60 (1): 14–30. doi: 10.2514/1.J060756.

Klein, M.V. (1970). *Optics.* USA: John Wiley & Sons, Inc.

Kyrazis, D.T. (1993). Optical degradation by turbulent free-shear layers. In: in *Proc. SPIE: Optical Diagnostics in Fluid and Thermal flow* (ed. S. S. Cha and J. D. Trolinger), 2005: 170–181.

Liepmann, H.W. (1952). Deflection and diffusion of a light ray passing through a boundary layer. *Tech. Rep. SM-14397*, Douglas Aircr. Co., Santa Monica, CA.

Liu, T. and Sullivan, J.P. (2005). *Pressure and Temperature Sensitive Paints.* Berlin: Springer.

Lumley, J. (1970). *Stochastic Tools in Turbulence.* New York: Academic.

Lynch, K.P., Spillers, R., Miller, N.E., Guildenbecher, D., and Gordeyev, S. (2021). Aero-optical measurements of a Mach 8 boundary layer. *AIAA Paper 2021-2831.*

Maeder, T., Adams, N.A., and Kleiser, L. (2001). Direct simulation of turbulent supersonic boundary layers by an extended temporal approach. *Journal of Fluid Mechanics* 429: 187–216.

Mahajan, V.N. (1982). Strehl ratio for primary aberrations: some analytical results for circular and annular pupils. *Journal of the Optical Society of America* 72 (9): 1258–1266.

Mahajan, V.N. (1983). Strehl ratio for primary aberrations in terms of their aberration variance. *Journal of the Optical Society of America* 73 (6): 860–861.

Mahajan, V.N. (2011). *Optical Imaging and Aberrations, Part II: Wave Diffraction Optics*, 2e. SPIE Press.

Majumdar, A.K. and Ricklin, J.C. (2008). Free-space laser communications: principles and advances. *Optical and Fiber Communications Reports* 2: Springer, 417 pp.

Malacara, D. (1978). *Optical Shop Testing*. New York: Wiley.

Malley, M.M., Sutton, G.W., and Kincheloe, N. (1992). Beam-jitter measurements of turbulent aero-optical path differences. *Applied Optics* 31: 4440–4443.

Masson, B., Wissler, J., and McMackin, L. (1994). Aero-optical study of a NC-135 fuselage boundary layer. *AIAA Pap. 94-0277*.

Mathews, E.R., Wang, K., Wang, M., and Jumper, E.J. (2016). Les of an aero-optical turret flow at high Reynolds number. *AIAA Pap. 2016-1461*.

Mela, K. and Louie, J.N. (2001). Correlation length and fractal dimension interpretation from seismic data using variograms and power spectra. *Geophysics* 66: 1372–1378.

Merritt, P. (2012). *Beam Control for Laser Systems*. Albuquerque, NM: Directed Energy Professional Society.

Monin, A.S. and Yaglom, A.M. (1975). *Statistical Fluid Mechanics, Vol. I, Mechanics of Turbulence*. Cambridge, MA: MIT Press.

Morkovin, M.V. (1962). Effects of compressibility on turbulent flows. In: *Mechanique de la Turbulence* (ed. A. Favre), 367–380. Paris, France: CNRS.

Morrida, J., De Lucca, N., Gordeyev, S., and Jumper, E.J. (2017). Simultaneous pressure and optical measurements around hemispherical turret in subsonic and transonic flight. *AIAA Paper 2017-3654*.

Morrida, J., Gordeyev, S., De Lucca, N., and Jumper, E.J. (2017). Shock-related effects on aero-optical environment for hemisphere-on-cylinder turrets at transonic speeds. *Applied Optics* 56 (16): 4814–4824.

Nagib, H.M., Chauhan, K.A., and Monkewitz, P.A. (2007). Approach to an asymptotic state for zero pressure gradient turbulent boundary layers. *Philosophical Transactions of the Royal Society A* 365: 755–770.

Navarro, R., López, J.L., Díaz, J.A., and Sinusía, E.P. (2014). Generalization of Zernike polynomials for regular portions of circles and ellipses. *Optics Express* 22 (18): 21263–21279.

Nightingale, A., Goodwine, B., Lemmon, M., and Jumper, E.J. (2007). Feedforward' adaptive-optics system identification analysis for mitigating aero-optics disturbances. *AIAA Paper 2007-4013*.

Nightingale, A.M., Gordeyev, S. (2013). Shack-hartmann wavefront sensor image analysis: a comparison of centroiding methods and image processing techniques. *Journal of Optical Engineering* 52 (7): 071413.

Papamoschou, D. and Roshko, A. (1988). The compressible turbulent shear layer: an experimental study. *Journal of Fluid Mechanics* 197: 453–477.

Peng, D., Jensen, C.D., Juliano, T.J., Gregory, J.W., Crafton, J., Palluconi, S., and Liu, T. (2013). Temperature-compensated fast pressure-sensitive paint. *AIAA Journal* 51 (10): 2420–2431.

Pond, J.E. and Sutton, G.W. (2006). Aero-optic performance of an aircraft forward-facing optical turret. *Journal of Aircraft* 43 (3): 600–607.

Ponder, Z., Gordeyev, S., and Jumper, E. (2011). Passive mitigation of aero-induced mechanical jitter of flat-windowed turrets. *AIAA Paper 2011-3281.*

Poon, T.-C. and Liu, J.-P. (2014). *Introduction to Modern Digital Holography: With Matlab.* Cambridge University Press.

Porter, C., Gordeyev, S., and Jumper, E. (2013a). Large-aperture approximation for not-so-large apertures. *Journal of Optical Engineering* 52 (7): 071417.

Porter, C., Gordeyev, S., Zenk, M., and Jumper, E. (2011). Flight measurements of aero-optical distortions from a flat-windowed turret on the airborne aero-optics laboratory (AAOL). *AIAA Paper 2011-3280.*

Porter, C., Gordeyev, S., Zenk, M., and Jumper, E. (2013b). Flight measurements of the aero-optical environment around a flat-windowed turret. *AIAA Journal* 51 (6): 1394–1403.

Ranade, P., Duvvuri, S., McKeon, B., Gordeyev, S., Christensen, K., and Jumper, E.J. (2019). Turbulence amplitude amplification in an externally forced, subsonic turbulent boundary layer. *AIAA Journal*, 57 (9): 3838–3850.

Rennie, M., Duffin, D., and Jumper, E.J. (2007). Characterization of a forced two-dimensional, weakly-compressible shear layer. *AIAA Paper 2007-4007.*

Robinson, S.K. (1991). Coherent motions in the turbulent boundary layer. *Annual Review of Fluid Mechanics* 23: 601–639.

Roeder, A.L. and Gordeyev, S. (2022). Wake response downstream of a spanwise-oscillating hemispherical turret. *Journal of Fluids and Structures* 109: 103470.

Roggemann, M. C., and Welsh, B. (1990) *Imaging Through Turbulence.* Boca Raton: CRC Press.

Rose, W.C. (1979). Measurements of aerodynamic parameters affecting optical performance. *Air Force Weapons Laboratory Final Report*, AFWL-TR-78-191.

Rose, W.C. and Johnson, E.A. (1982). Unsteady density and velocity measurements in the 6 x 6 ft wind tunnel. In: *Progress in Astronautics and Aeronautics: Aero-Optical Phenomena*, 80 (ed. K. Gilbert and L.J. Otten), 218–232. New York: AIAA.

Ross, T.S. (2009). Limitations and applicability of the Maréchal approximation. *Applied Optics* 48 (10): 1812–1818.

Sasiela, R.J. (2007). *Electromagnetic Wave Propagation in Turbulence Evaluation and Application of Mellin Transforms*, 2e. SPIE Monograph PM171, Bellingham, WA: SPIE.

Saxton-Fox, T., McKeon, B., and Gordeyev, S. (2019). *Effect of coherent structures on aero-optic distortion in a heated turbulent boundary layer, to appear in AIAA J.*

Schatzman, D.M., and Thomas, F.O. (2017). An experimental investigation of an unstready adverse pressure gradient turbulent boundary layer: embedded shear layer scaling. *Journal of Fluid Mechanics*. 815: 592–642.

Schmid, P.J. (2010). Dynamic mode decomposition of numerical and experimental data. *Journal of Fluid Mechanics* 656: 5–28.

Schnars, U., Falldorf, C., Watson, J., and Jüptner, W. (2015). *Digital Holography and Wavefront Sensing Principles, Techniques and Applications*, 2e. Springer-Verlag Berlin Heidelberg.

Settles, G.S. (2001). *Schlieren and Shadowgraph Techniques: Visualizing Phenomena in Transparent Media*. Berlin: Springer.

Shapiro, A.H. (1953). *The Dynamics and Thermodynamics of Compressible Fluid Flow*, 1. Ronald Press.

Siegenthaler, J., Gordeyev, S., and Jumper, E. (2003). Mapping the optically-aberrating environment in a partially-quieted Mach 0.6 free shear layer. *AIAA Paper 2003-3607*.

Sirovich, L. (1987). Turbulence and the dynamics of coherent structures, Parts I–III. *Quarterly of Applied Mathematics* 45 (3): 561–571.

Smits, A.J. and Dussauge, J.P. (1996). *Turbulent Shear Layers in Supersonic Flow*. Woodbury, New York: American Institute of Physics.

Sontag, J. and Gordeyev, S. (2022). Optical diagnostics of spanwise-uniform flows. *AIAA Journal* 60 (9): 5031–5045.

Southwell, W.H. (1980). Wave-front estimation from wave-front slope measurements. *Journal of the Optical Society of America* 70 (8): 998–1006.

Speaker, W.V. and Ailman, C.M. (1966). *NASA Contract. Rep. CR-486*.

Spina, E.F., Donovan, J.F., and Smits, A.J. (1991). Convection velocity in supersonic turbulent boundary layers. *Physics of Fluids A* 3 (12): 3124–3126.

Spina, E.F., Smits, A.J., and Robinson, S.K. (1994). The physics of supersonic turbulent boundary layers. *Annual Review of Fluid Mechanics* 26: 287–319.

Steinmetz, W.J. (1982). second moments of optical degradation due to a thin turbulent layer. In: *Aero-Optical Phenomena, Progress in Astronautics and Aeronautics*, 80 (ed. K.G. Gilbert and L.J. Otten), 78–110. New-York: AIAA.

Stine, H.A. and Winovich, W. (1956). *Light diffusion through high-speed turbulent boundary layers. Research Memorandum A56B21, NACA*, Washington.

Stratford, B.S. and Beaver, G.S. (1961). *The calculation of the compressible turbulent boundary layer in an arbitrary pressure gradient – a correlation of certain previous methods*. Reports and Memoranda No. 3207, Ministry of Aviation, Aeronautical Research Council, London.

Sun, Y., Liu, Q., Cattafesta, L.N., Lawrence, S., Ukeiley, L.S., and Taira, K. (2019). Effects of sidewalls and leading-edge blowing on flows over long rectangular cavities. *AIAA Journal* 57 (1): 106–119.

Sutton, G.W. (1969). Effects of turbulent fluctuations in an optically active fluid medium. *AIAA Journal* 7 (9): 1737–1743.

Sutton, G.W. (1985). Aero-optical foundations and applications. *AIAA Journal* 23: 1525–1537.

Taira, K., Brunton, S.L., Dawson, S.T.M., Rowley, C.W., Colonius, T., McKeon, B.J., Schmidt, O.T., Gordeyev, S., Theofilis, V., and Ukeiley, L.S. (2017). Modal analysis of fluid flows: an overview. *AIAA Journal* 55 (12): 4013–4041.

Tatarski, V.I. (1961). *Wave Propagation in a Turbulent Medium*. New York: McGraw-Hill. 285 pp.

Tesch, J., Gibson, S., and Verhaegen, M. (2013). Receding-horizon adaptive control of aero-optical wavefronts. *Optical Engineering* 52 (7): 071406.

Tromeur, E., Garnier, E., and Sagaut, P. (2006). Large-eddy simulation of aero-optical effects in a spatially developing turbulent boundary layer. *Journal of Turbulence* 7 (1): 1–27.

Tromeur, E., Garnier, E., Sagaut, P., and Basdevant, C. (2002). LES of Aero-Optical effects in turbulent boundary layer. In: *Engineering Turbulence Modelling and Experiments 5: Proceedings of the 5th International Symposium on Engineering Turbulence Modelling and Measurements, Mallorca, Spain, 16–18 September 2002* (ed. W. Rodi and N. Fueyo), 327–336. Elsevier Science Ltd.

Tromeur, E., Garnier, E., Sagaut, P., and Basdevant, C. (2003). Large eddy simulations of aerooptical effects in a turbulent boundary layer. *Journal of Turbulence* 4: 005.

Truman, C.R. (1992). The influence of turbulent structure on optical phase distortion through turbulent shear layer. *AIAA Paper 92-2817*.

Truman, C.R. and Lee, M.J. (1990). Effects of organized turbulence structures on the phase distortions in a coherent optical beam propagating through a turbulent shear layer. *Physics of Fluids* 2 (5): 851–857.

Tu, J.H., Rowley, C.W., Luchtenburg, D.M., Brunton, S.L., and Kutz, J.N. (2014). On dynamic mode decomposition: theory and applications. *Journal of Computational Dynamics* 1 (2): 391–421.

Tyler, G.A. (1994). Bandwidth considerations for tracking through turbulence. *Journal of the Optical Society of America A* 11: 358–367.

Tyson, R.K. (1997). *Principles of Adaptive Optics*, 2e. New York: Academic. 345 pp.

Tyson, R.K. and Frazier, B.W. (2012). *Field Guide to Adaptive Optics*, 2e. Bellingham, WA: SPIE Press.

Van Dyke, M. (1982). *An Album of Fluid Motion*. Stanford, CA: The Parabolic Press. 92.

Vukasinovic, B., Glezer, A., Gordeyev, S., Jumper, E., and Bower, W.W. (2013). Flow control for aero-optics application. *Experiments in Fluids* 54: 1492.

Vukasinovic, B., Glezer, A., Gordeyev, S., Jumper, E., and Kibens, V. (2008). Active control and optical diagnostics of the flow over a hemispherical turret. *AIAA Paper 2008-0598*.

Wang, K. and Wang, M. (2012). Aero-optics of subsonic turbulent boundary layers. *Journal of Fluid Mechanics* 696: 122–151.

Wang, M., Mani A., and Gordeyev S. (2012). Physics and computation of aero-optics. *Annual Review of Fluid Mechanics*, 44: 299–321

Wang, K. and Wang, M. (2013). On the accuracy of Malley probe measurements of aero-optical effects: a numerical investigation. *Optical Engineering* 52 (7): 071407.

Weaver, L. (1997). *Scientist at the directed energy directorate*, Air Research Laboratory, Kirtland AFB, New Mexico, personal communication.

White, M.D. and Visbal, M.R. (2012). Aero-optics of compressible boundary layers in the transonic regime. *AIAA Pap. 2012-2984.*

White, M.D. and Visbal, M.R. (2013). Computational investigation of wall cooling and suction on the aberrating structures in a transonic boundary layer. *AIAA Pap. 2013-0720.*

Whiteley, M.R., Roggemann, M.C., and Welsh, B.M. (1998). Temporal properties of the Zernike expansion coefficients of turbulence-induced phase aberrations for aperture and source motion. *Journal of the Optical Society of America A* 15: 993–1005.

Whiteley, M.R., Goorskey, D.J., and Drye, R. (2013). Aero-optical jitter estimation using higher-order wavefronts. *Optical Engineering*, 52 (7): 071411–071411.

Whiteley, M.R., and Goorskey, D.J. (2013). Imaging performance with turret aero-optical wavefront disturbances. *Optical Engineering*, 52 (7): 071410–071410.

Wittich, D., Gordeyev, S., and Jumper, E. (2007). Revised scaling of optical distortions caused by compressible, subsonic turbulent boundary layers. *AIAA Pap. 2007-4009.*

Woods, C. and Greenaway, A.H. (2003). Wave-front sensing by use of a green's function solution to the intensity transport equation. *Journal of the Optical Society of America A* 20 (3): 508–512.

Wyckham, C.M. and Smits, A.J. (2009). Aero-optic distortion in transonic and hypersonic turbulent boundary layers. *AIAA Journal* 47: 2158–2168.

Zernike, F. (1934). Diffraction theory of the knife-edge test and its improved form, the phase contrast method. *Monthly Notices of the Royal Astronomical Society* 94: 377–384.

Index

Aero-Optical Effects: Physics, Analysis and Mitigation, First Edition. Stanislav Gordeyev, Eric J. Jumper, and Matthew R. Whiteley.

Printed and bound by CPI Group (UK) Ltd, Croydon, CR0 4YY

16/04/2025

14658586-0002